Recombinant DNA

A Short Course

Recombinant DNA

A Short Course

James D. Watson

COLD SPRING HARBOR LABORATORY

John Tooze

EUROPEAN MOLECULAR BIOLOGY ORGANIZATION

David T. Kurtz

COLD SPRING HARBOR LABORATORY

SCIENTIFIC
AMERICAN
BOOKS

Distributed by

W. H. Freeman and Company

New York

Cover illustration by Marvin Mattelson.

Library of Congress Cataloging in Publication Data
Watson, James D., 1928–
   Recombinant DNA.

   Includes bibliographies and index.
   1. Recombinant DNA.   I. Tooze, John.   II. Kurtz,
David T.   III. Title.   IV. Title: Recombinant D.N.A.
[DNLM: 1. DNA, Recombinant.   QU 58 W339r]
QH442.W373 1983   574.87'3282   83-9069
ISBN 0-7167-1483-3
ISBN 0-7167-1484-1 (pbk.)

Printed in the United States of America

Scientific American Books is a subsidiary of Scientific American, Inc.

Distributed by W. H. Freeman and Company, 41 Madison Avenue, New
York, New York 10010

1 2 3 4 5 6 7 8 9 0   KP   1 0 8 9 8 7 6 5 4 3

Contents

Preface   xv

1. Establishing the Role of Genes Within Cells   1

The Building Blocks of All Life Are Cells   1
Cells Are Tiny Expandable Factories That Simultaneously Synthesize
    Several Thousand Different Molecules   2
A Cell's Molecules Can Be Divided into Small Molecules and
    Macromolecules   2
Special Cellular Catalysts Called Enzymes Effectively Determine the
    Chemical Reactions That Occur in Cells   3
A Given Protein Possesses a Unique Sequence of Amino Acids Along
    Its Polypeptide Chain   3
The Functioning of an Enzyme Demands a Precise Folding of Its
    Polypeptide Chain   4
Activation of Molecules to High-Energy Forms Promotes Their
    Chemical Reactivity   4
Cellular Metabolism Can Be Visualized Through Metabolic Maps   5
Enzymes Cannot Determine the Order of the Amino Acids in
    Polypeptide Chains   5
Mendel's Breeding Experiments with the Pea Plant First Revealed the
    Discreteness of Genetic Determinants (Genes)   6
Chromosomes Are the Cellular Bearers of Heredity   6
The One-Gene-One-Protein Hypothesis   9

2. DNA Is the Primary Genetic Material   11

DNA Is Sited Exclusively on Chromosomes   11
Cells Contain RNA as well as DNA   11
A Biological Assay for Genetic Molecules Is Discovered   12
Viruses Are Packaged Genetic Elements That Move from Cell
    to Cell   14

Molecules with Complementary Sizes and Shapes Attract One
    Another   14
The Diameter of DNA Is Established   15
The Nucleotides of DNA and RNA Are Linked Together by Regular
    5'-3' Phosphodiester Bonds   16
The Composition of Bases of DNA from Different Organisms Varies
    Greatly   16
DNA Has a Highly Regular Shape   17
The Fundamental Unit of DNA Consists of Two Intertwined
    Polynucleotide Chains (The Double Helix)   17
The Double Helix Is Held Together by Hydrogen Bonds Between
    Base Pairs   18
The Complementary Nature of DNA Is at the Heart of Its Capacity for
    Self-Replication   19
Proof of Strand Separation During DNA Replication   20
DNA Molecules Can Be Renatured as well as Denatured   20
G-C Base Pairs Fall Apart Less Easily Than Their A-T
    Equivalents   21
Palindromes Promote Intrastrand Hydrogen Bonding   21
5-Methylcytosine Can Replace Cytosine in DNA   21
Chromosomes Contain Single DNA Molecules   22
Viruses Are Sources of Homogeneous DNA Molecules   23
Phage λ DNA Can Insert Itself into a Specific Site Along the *E. coli*
    Chromosome   23
Abnormal Transducing Phages Provide Unique Segments of Bacterial
    Chromosomes   24
Plasmids Are Autonomously Replicating Minichromosomes   24
Circular DNA Molecules May Be Supercoiled   25
Most Double Helices Are Right-Handed, but Under Special
    Conditions Certain DNA Nucleotide Sequences Lead to Left-Handed
    Helices   26

3.   Elucidation of the Genetic Code   31

Identifying the Amino Acid Replacement in a Mutant Hemoglobin
    Molecule   31
Development of Fine-Structure Genetics   31
The Gene and Its Polypeptide Products Are Collinear   32
RNA Carries Information from DNA to the Cytoplasmic Sites of
    Protein Synthesis   32
How Do Amino Acids Line Up on RNA Templates?   34
Roles of Enzymes and Templates in the Synthesis of Nucleic Acids and
    Proteins   34
Proteins Are Synthesized from the N-Terminus to the
    C-Terminus   35

Three Forms of RNA Are Involved in Protein Synthesis   35

Genetic Evidence That Codons Contain Three Bases   36

RNA Chains Are Both Synthesized and Translated in a 5'-to-3'
   Direction   36

Synthetic mRNA Is Used to Make the Codon Assignments   37

The Genetic Code Is Fully Deciphered by June 1966   38

"Wobble" Frequently Permits Single tRNA Species to Recognize
   Multiple Codons   38

How Universal Is the Genetic Code?   39

Average-Sized Genes Contain at Least 1200 Base Pairs   39

Suppressor tRNAs Cause Misreading of the Genetic Code   39

The Signals for Starting and Stopping the Synthesis of Specific RNA
   Molecules Are Encoded Within DNA Sequences   40

Increasingly Accurate Systems Are Developed for the *in Vitro*
   Translation of Exogenously Added mRNAs   42

4. The Genetic Elements That Control Gene Expression   45

Repressors Control Inducible Enzyme Synthesis   45

Bacterial Genes with Related Functions Are Organized into
   Operons   46

Promoters Are the Start Signals for RNA Synthesis   46

Constitutive Synthesis of Repressor Molecules   47

Repressors Are Isolated and Identified   48

Positive Regulation of Gene Transcription   48

Attenuation   48

Translational Control   49

Early Difficulties in Probing Gene Regulation in Higher Plants and
   Animals   50

Purification of *Xenopus* Ribosomal RNA Genes   51

Eukaryotic mRNAs Have Caps and Tails   52

Three Kinds of RNA Polymerases in Eukaryotes   53

Eukaryotic DNA Is Organized into Nucleosomes   53

Animal Viruses Are Model Systems for Gene Expression in Higher
   Cells   53

The RNA Tumor Viruses Replicate by Means of a Double-Stranded
   DNA Intermediate   54

5. Methods of Creating Recombinant DNA Molecules   58

Nucleic Acid Sequencing Methods Are Developed   58

Restriction Enzymes Make Sequence-Specific Cuts in DNA   58

Restriction Maps Are Highly Specific   60

Restriction Fragments Lead to Powerful New Methods for Sequencing
   DNA   61

Oligonucleotides Can Be Synthesized Chemically   63

*Eco*RI Produces Fragments Containing Sticky (Cohesive) Ends   64

Many Enzymes Are Involved in DNA Replication   65

Sticky Ends May Be Enzymatically Added to Blunt-Ended DNA
Molecules   66

Small Plasmids Are Vectors for the Cloning of Foreign Genes   67

DNA of Higher Organisms Becomes Open to Molecular
Analysis   67

Scientists Voice Concerns About the Dangers of Unrestricted Gene
Cloning   68

Guidelines for Recombinant DNA Research Are Proposed at the
Asilomar Conference   69

6. The Isolation of Cloned Genes   72

"Safe" Bacteria and Plasmid Vectors Are Developed   72

Why Use Drug-Resistance Plasmids?   72

Probes for Cloned Genes   74

Synthesis and Cloning of cDNA   75

Identifying Specific cDNA Clones   76

Genomic Fragments Are Cloned in Bacteriophage λ   78

Cosmids Allow the Cloning of Larger Segments of Foreign
DNA   79

Chromosome Walking Is Used to Analyze Long Stretches of Eukaryotic
DNA   80

Cloning DNA in Phage M13 Speeds Up Sanger Sequencing   81

Southern and Northern Blotting   83

Developing Procedures for Cloning Genes That Code for
Less-Abundant Proteins   84

Screening Gene Libraries with Oligonucleotide Probes   85

Computers Match Up cDNA Classes   86

Expression Vectors May Be Used to Isolate Specific Eukaryotic
cDNAs   86

Immunological Screening for the Products of Expression Vectors   87

7. The Unexpected Complexity of Eukaryotic Genes   91

Split Genes Are Discovered   91

Specific Base Sequences Are Found at Exon-Intron Boundaries   92

Discovery of Autonomous Splicing   94

Complete Sequencing of the First Mammalian Genes   95

Open Reading Frames in DNA Delineate Protein-Coding
Regions   96

Leader Sequences at the NH_2-Terminal Ends of Secretory
Proteins   96

Introns Sometimes Mark Functional Protein Domains   97

Alternative Splicing Pathways Generate Different mRNAs from a
   Single Gene   97
Control Regions Are Found at the 5' and 3' Ends of Genes   98
Clustered Gene Families May Include Vestigial Evolutionary
   Relics   99
Gene Families May Be Expanded by Reverse Transcription of mRNA
   Molecules   100
Eukaryotic DNA Contains Interspersed Repetitive Sequences   100
Polypeptide Precursors Give Rise to Protein Hormones   101

8. *In Vitro* Mutagenesis   106

Deletions   106
Insertions   108
Substitutions: Deamination of Cytosine   109
Substitutions: Incorporation of Nucleotide Analogs   110
Substitutions: Misincorporation of Nucleotides   111
Mutants May Be Constructed by Using Oligonucleotides with Defined
   Sequences   111

9. Rearranging Germ-Line DNA Segments to Form Antibody Genes   117

The Basic Structure of Antibody Molecules Is Established   117
Separate Genes for V and C Segments Are Proposed   118
Messenger RNA Probes Are Used to Obtain Support for the Joining of
   V and C Genes   118
Functional Antibody Genes Are Isolated from Myeloma Cells   119
Embryonic Cells Are Sources of Unjoined V and C Genes   119
Multiple J (Joining) Segments Are Attached to Genomic C (Constant)
   Segments   119
Three Discontinuous Regions of DNA Code for Heavy-Chain Amino
   Acids   120
A DNA Elimination Event Allows a V_H Gene to Be Attached to Two
   Different C_H Genes   120
Alternative Splicing Allows Single Cells to Make Both μ- and δ-Class
   Heavy Chains with Identical V_H Segments   121
Somatic Mutations Provide a Further Source of Immunoglobulin
   Diversity   122
Establishing the Genes of the Major Histocompatibility Complex
   (MHC) Proteins and Their Protein Antibodies Through Gene
   Cloning   123

10. Tumor Viruses   127

Cloning Integrated Forms of DNA Tumor Viruses   128
Tumor Proteins of SV40 and Polyoma   129

Overlapping Genes Code for the Structural Proteins That Surround the SV40 and Polyoma Chromatin   129

Different Control Signals for the Initiation of Early and Late SV40 and Polyoma mRNA Synthesis   130

A Region of Approximately 100 Base Pairs Encompasses the SV40 (and Polyoma) Origin of DNA Replication   131

The Complex Organization of RNA Tumor Viruses (Retroviruses)   131

Highly Oncogenic Retroviruses Contain Specific Oncogenic Sequences

Proviruses Maintain the Same Gene Order as RNA Genomes   132

The Promoter of RNA Synthesis Is Located Within an LTR   133

Retroviral Oncogenes Often Code for Protein Kinases   133

Normal Cellular Genes Are the Progenitors of Retroviral Oncogenes   134

Deciding Whether Retroviral Oncogenesis Results from the Overexpression or the Misexpression of Normal Cellular Genes   134

Cancer Induction by Weakly Oncogenic Retroviruses   135

A Generalized Theory for the Induction of Cancer   135

11.   Movable Genes   140

Transposon Movement Involves the Creation of a New Daughter Transposon   141

Movable Genetic Elements May Be Common Features of All Organisms   142

Using Mobile Elements to Genetically Engineer *Drosophila* Embryos   142

Isolation of the Ds Transposable Element from Maize   144

Have RNA Tumor Virus Genomes Evolved from Movable Genetic Elements?   145

Are There Functionally Two Different Classes of Transposons?   145

Sex Changes in Yeast by Gene Replacement: The Cassette Model   146

Antigenic Changes in Trypanosomes Through Gene Switching   147

Antigenic Changes in *Neisseria gonorrhoeae* Through Gene Rearrangement   149

12.   The Experimentally Controlled Introduction of DNA into Yeast Cells   152

Yeast Spheroplasts Take Up Externally Added DNA   152

Expression of Yeast Genes in *E. coli*   153

Shuttle Vectors   153

Yeast Also Contains a Plasmid   153

Increasing Transformation Efficiency by Addition of Replication
   Origins   154
Stabilizing Yeast Plasmids with Yeast Centromere DNA   155
Hairpin Loops at the Ends (Telomeres) of Yeast Chromosomes   156
Directed Integration of Cloned DNA into the Yeast
   Chromosome   158
Retriever Vectors   159
Gene Organization   160
Regulation of Gene Expression in Yeast   161

13. Genetic Engineering of Plants by Using Crown Gall Plasmids   164

Conventional Plant Breeding Methodologies   164
Plant Cells in Culture   164
Redifferentiation of Whole Plants from Plant Culture Cells   164
Plant Protoplasts Can Regenerate Whole Plants   165
Making Hybrid Plants by Protoplast Fusion   165
Genetically Engineering Plants   167
Crown Gall Tumors   168
Tumor-Inducing (Ti) Plasmids   168
Ti Plasmid Mutants   168
Integration of T DNA into the Plant Chromosome   169
Is T DNA a Transposon?   169
Mendelian Inheritance of T DNA   169
Ti Plasmid DNA Can Be Used as a Vector   170
Transformation of Plant Cells and Protoplasts   170
Mobilization of T DNA by the *vir* Segment of the Ti Plasmid   171
Attenuated T DNA Vectors Allow Regeneration of Whole Plants from
   Single Cells   171
T DNA Insertion Can Be Used to Isolate Plant Genes   173
Practical Applications of Plant Engineering Using Ti Plasmids   173

14. Transferring Genes into Mammalian Cells   176

Ca^{++} Stimulates the Uptake of DNA by Vertebrate Cells   176
Thymidine Kinase (Tk) Has Served as the Archetypal Selective Marker
   for Transfection Experiments   176
Dominant-Acting Markers for Transformation of Normal Cells   177
Cotransformation Following Intracellular Ligation   178
Microinjection of DNA into Mammalian Cells   178
Semistable Inheritance of Transfected Methylated DNA   179
Isolation of Transferred Genes   180
Regulation Following Gene Transfer   181
Establishing the Existence of Specific Human Cancer Genes by DNA
   Transfer Experiments   183
Cloning Human Oncogenes   185

15. Viral Vectors   189

SV40 Vectors   189
SV40 Virions as Vectors   189
SV40 Late-Region Replacement   189
SV40 Early-Region Replacement   190
Analysis of Cloned Surface Antigen Genes   192
Plasmidlike Replication of DNA in COS Cells   192
Rescue of Integrated SV40 DNA by COS Cell Fusion   192
Discovery of Enhancer Sequences Using SV40 Vectors   194
Papilloma Virus DNA Replicates like a Plasmid in Mouse Cells   195
RNA Tumor Viruses Can Be Used as Vectors   196

16. The Introduction of Foreign Genes into Fertilized Mouse Eggs   200

Chromosomal Integration of Foreign Genes   200
Integration of Foreign Genes Is Not Chromosome-Specific   200
Foreign DNA Is Stably Integrated into the Germ-Line Cells   201
Expression of Foreign DNA in Mice   201
Expression of the MK Fusion Gene Following Microinjection   202
The Tissue-Specific Expression of the MK Gene   203
Altered Expression of the MK Gene in Progeny   203
Functional Expression of the MGH Fusion Gene   203
Integration of MuLV DNA   204
Early Embryos Infected with MuLV   205
Microinjection of Proviral DNA   205
First Implications of Implanting Genes in Fertilized Eggs   206
"Cloning" Animals   206

17. Recombinant DNA and Genetic Diseases   211

Mendelian Inheritance   211
Inborn Errors of Metabolism   211
Treatment of Inborn Errors of Metabolism   212
Early Diagnosis and Abortion   213
DNA Analysis of Inherited Disorders   213
β-Thalassemias   214
Nonsense and Frame-Shift Mutations   214
Transcription Mutations   214
RNA-Processing Mutations   214
Sickle-Cell Anemia   215
Diagnosis of α_1-Antitrypsin Deficiency with a Synthetic
    Oligonucleotide   216
Citrullinemia Is Correlated with Aberrant Argininosuccinate
    Synthetase mRNA   217
Diagnosing Mutations by Linkage   217

The Search for Mutated Genes   219
Mapping Human Chromosomes   219
Somatic-Cell Genetics   220
Human–Mouse Hybrid Cells   220
Subchromosomal Localization of Genes   220
Chromosomal Translocations and Cancer   222
In Situ Hybridization   223
Cloning Individual Chromosomes   224
Bridging the Gap Between Cytogenetics and Molecular Genetics   224
Prospects for Gene Therapy   225
Induction of Fetal Hemoglobin in Thalassemias   226

18.   The Science Used in The Recombinant DNA Industry   231

Commercial Potential of Recombinant DNA   231
Methods of Commercial Gene Cloning   231
Human Insulin from Bacteria   232
The Structure of Human Insulin   232
Synthetic Insulin "Genes"   233
Proinsulin cDNA   234
Bacteria That Secrete Proinsulin   235
Cloning Human Growth Hormone   235
Making the Growth Hormone "Gene"   235
The Methionine Problem   236
Different Types of Interferons   236
High-Level Expression of Human α-Interferon in *E. coli*   238
Cloning of γ- (Immune) Interferon   238
Viral Proteins for Vaccines   238
Cloning Foot-and-Mouth-Disease Virus   239
Synthetic Peptide Vaccines   239
Vaccines for Human Hepatitis B Virus   239
Hepatitis B Antigen by Gene Cloning   239

Recombinant DNA Dateline   242

APPENDIX A   Restriction Enzymes   248

APPENDIX B   Other Enzymes Used in Recombinant DNA Research   254

Index   255

Preface

The early years following the production of the first recombinant DNA molecules were not marked by the frantic competitive happiness that usually accompanies the opening up of a new scientific era. Instead of dreaming largely about what marvelous new discoveries the next few days, months, or years might bring, scientists were for the most part concerned about whether these discoveries might be prevented from being made for an indefinite period, due to the existence of stringent regulations governing the use of recombinant DNA. Fortunately, our worst fears proved unfounded, and today most forms of recombinant DNA research are no longer subject to any effective form of regulation.

Those troubled years that were dominated first by the regulation and then by the deregulation of DNA research we now tend to associate with the name Asilomar. This conference center on the California coast was the site of the 1975 meeting where guidelines concerning the use of recombinant DNA were virtually unanimously recommended by leading members of the molecular biology community. The meeting was not the typical scientific conference, and a scrapbooklike compilation of the largely sociopolitical events that occurred there are recorded in *The DNA Story* (James D. Watson and John Tooze, W. H. Freeman and Company, 1981).

At the end of *The DNA Story,* there appeared an extended summary of the main scientific advances to which recombinant DNA procedures had already led. Upon reading this section, a number of university instructors told us that if it were modestly expanded, it would serve as a most needed supplement to texts written before the recombinant DNA revolution had begun to dominate biologists' research. We thus decided to prepare this "short course" on recombinant DNA. Our goal was to emphasize the types of experiments that recombinant DNA makes possible, and to explain some of the important new facts that such experiments have revealed. As so often happens, our little book has grown to be not so little, as we have tried to include accounts of many of the most recently developed experimental procedures. It thus stands the risk of being dominated by the excitement of today rather than displaying the balanced perspective that time and distance can give a text. And future developments may prove some of the ideas we report to be less solid than we now believe.

In developing this text we have sought the advice of many colleagues, especially those at the Cold Spring Harbor Laboratory. In particular we wish to thank Yasha Gluzman, Joe Sambrook, Mary-Jane Gething, Mitch Goldfarb, John Fiddes, Jim Hicks, Steve Delaporta, Fred Heffron, Elizabeth Lacy, Frank Constantini, and Lee Silver for reading various chapters. The final version of the book is necessarily our responsibility, and we take the blame for any mistakes or omissions.

Finally, we gratefully acknowledge the much needed editorial help given by Karen Herrmann, and the careful secretarial competence of Andrea Stephenson. All of the illustrations are the work of the noted illustrator George Kelvin, who also did the artwork for *The DNA Story.*

James D. Watson   John Tooze   David T. Kurtz
June 1, 1983

1

Establishing the Role of Genes Within Cells

There is no substance as important as DNA. Because it carries within its structure the hereditary information that determines the structures of proteins, it is the prime molecule of life. The instructions that direct cells to grow and divide are encoded by it; so are the messages that bring about the differentiation of fertilized eggs into the multitude of specialized cells that are necessary for the successful functioning of higher plants and animals. And because it has been present in virtually an infinite number of interchangeable chemical species, DNA has provided the basis for the evolutionary process that has generated the many millions of different life forms that have occupied the earth, since the first living organisms came into existence some 3 to 4 billion years ago.

This extraordinary capacity of altered DNA molecules to give rise to new life forms that are better adapted for survival than their immediate progenitors has made possible the emergence of our own species, with its ability to perceive the nature of its environment and to utilize this information to build the civilizations of modern man. As a result of our ability for rapid conceptual thought, we have for several centuries been asking ever deeper questions about the nature of inanimate objects like water, rocks, and air, as well as about the stars of surrounding space. And biology, the science of living objects, which only 30 years ago was generally perceived to be a much inferior science, has swiftly come of age. By now there exists an almost total consensus of informed minds that the essence of life can be explained by the same laws of physics and chemistry that have helped us understand, for example, why apples fall

to the ground and why the moon does not, or why water is transformed into gaseous vapor when its boiling point is exceeded.

The key to our optimism that all secrets of life are within the grasp of future generations of perceptive biologists is the ever accelerating speed at which we have been able to probe the secrets of DNA. Now we know so much that it is difficult to remember the intellectual chaos that still existed in 1944—when DNA was first reported to carry genetic information—and that was to disappear effectively only in 1953, when the structure of DNA was revealed to be a complementary double helix. Since then it has been clear to all that the "brain," so to speak, of all cells is DNA: From DNA issue the commands that regulate the nature and number of virtually every type of cellular molecule. So, increasingly, our attention has been devoted to unlocking the information within DNA. We know that if we can reveal the exact form of our genetic blueprints, we shall have taken a giant step toward eventually understanding the many complex sets of interconnected chemical reactions that cause fertilized eggs to develop into highly complex multicellular organisms.

Before we examine DNA itself in more detail, we shall first look briefly at the cells in which it resides, to determine the nature of the commands that DNA must generate.

The Building Blocks of All Life Are Cells

The smallest irreducible units of life are cells. They were first seen over 300 years ago, soon after the construction of the first microscopes. By the

middle of the 19th century, it had become clear that all living organisms are built from cells.

Generally, cells are very small with diameters much less than 1 mm, so they are invisible to the naked eye. In the simplest cells, bacteria, a cell wall surrounds a very thin fatty-acid-containing outer (plasma) membrane which in turn surrounds a superficially unstructured inner region. Within this inner region is located the bacterial DNA that carries the genetic information. The plasma membrane is effectively impermeable, except to selected food molecules and ions, so the inner cell contents are contained and are not lost to the outside. The intactness of this outer membrane is thus essential for the life of the cell, and it is constructed in such a way that minor accidental tears or openings are sealed automatically, like punctures in a self-sealing automobile tire.

In virtually all cells other than bacteria, the inner cellular mass is partitioned into a membrane-bounded, spherical body called the nucleus and an outer surrounding cytoplasm. In the nucleus is located the cellular DNA in the form of coiled rods known as chromosomes. Cells that contain a nucleus are referred to as eukaryotic cells, whereas the nuclei-free bacteria and their close relatives, the blue-green algae, are known as prokaryotic cells.

Cells Are Tiny Expandable Factories That Simultaneously Synthesize Several Thousand Different Molecules

The essence of a cell is its ability to grow and divide to produce progeny cells, which are likewise capable of generating new cellular molecules and replicating themselves. To perform these functions, cells must be chemically very sophisticated; indeed, even the very simplest cells contain nearly 1000 different molecules. Thus, cells are, in effect, tiny factories that grow by taking in simple molecular building blocks, like glucose and carbon dioxide, and somehow converting them into the many diverse carbon-containing molecules that are required for cellular functioning. In growing and dividing, cells also require an external source of energy to ensure that the cellular chemical reactions proceed in the desired direction of biosynthesis. Cells are therefore governed

by the same laws of thermodynamics that describe the energies of atoms to the physicist and the energies of molecules to the chemist. For most cells, the energy input necessary for their maintenance and growth is gained from breaking down food molecules, although cells that are capable of photosynthesis use the energy of sunlight directly.

A Cell's Molecules Can Be Divided into Small Molecules and Macromolecules

The molecules of a cell fall into two very different size classes. One class is the so-called "small molecules": the various sugars, amino acids, and fatty acids. At least 750 different types of small molecules are found in virtually every cell. The second class of cellular molecules is the "macromolecules," among which the proteins and nucleic acids are the most important. These are invariably polymeric molecules that are formed by joining together specific types of small molecules (such as amino acids) into long chains (such as proteins). Polymeric molecules contain a large number of these monomeric subunits, and are frequently hundreds to thousands of times larger than the small molecules, which typically contain somewhere between 10 to 50 atoms precisely linked together. The number of different macromolecules in most small cells is even larger than the number of different small molecules. Our best guess now is that there are in excess of 2000 distinct macromolecular species.

Thus, just the number and sizes of their molecules make even the simplest cells extraordinarily complex from a chemical viewpoint. The chemical uniqueness of cells lies less, however, in the inherent complexity of their individual molecules than in the nature of the chemical reactions that transform one cellular molecule into another. The most important of these reactions are those that (1) lead to the breakdown of food molecules into molecular units that can be reassembled into vital cellular components, (2) harness much of the energy released by the breakdown of food (or absorption of light) into energy-rich molecules that later pass on their excess energy to ensure that the chemical reactions of the cell proceed in the desired direction, (3) synthesize the small-molecule building

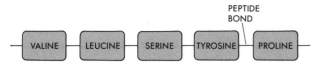

Figure 1-1
The amino acid building blocks of a protein. Chain lengths vary from 50 to more than 2000 amino acids.

blocks of the cellular macromolecules, and (4) assemble these monomeric building blocks into highly ordered macromolecules.

Special Cellular Catalysts Called Enzymes Effectively Determine the Chemical Reactions That Occur in Cells

Each of a cell's several thousand different molecules is potentially capable of a very large number of chemical reactions with other cellular molecules. Yet in any cell, only a very tiny fraction of these potential reactions take place at measurable rates, because the possibility of most chemical reactions occurring unaided at the temperatures at which life exists is very low. The cellular chemical reactions that do occur at rates compatible with life are greatly speeded up by enzymes, special catalytic molecules that are unique to cells. Like all other catalysts, enzymes are not used up in the course of chemical reactions, and a given enzyme molecule may function many thousands of times in a single second. Most enzymes are highly specific and only catalyze a single type of chemical reaction; conversely, most chemical reactions in cells are only catalyzed by one specific enzyme.

The chemical identity of enzymes was initially obscure, with many scientists believing that enzymes might represent a still undiscovered class of biological molecules. By 1935, however, it was clear that all enzymes were proteins, and today it is known that in fact the vast majority of proteins in all cells have enzymatic roles. Enzymes act by binding together on their surfaces the potential partners of a chemical reaction, thus greatly speeding up the rates at which these reactants can collide and undergo a chemical reaction. In enzyme-catalyzed reactions, certain key atoms of the enzyme often directly participate by temporarily forming

chemical bonds with the participating reagents (called enzyme substrates), thereby forming metastable chemical intermediates that have a heightened potential for chemical reactivity.

A Given Protein Possesses a Unique Sequence of Amino Acids Along Its Polypeptide Chain

The realization that all enzymes are proteins greatly heightened the interest of chemists in establishing the precise details of protein structure. Already by 1905, through the work of Emil Fischer in Germany, proteins were known to be polymeric molecules built up from amino acids linked to each other by peptide bonds to form linear polypeptide chains (Figure 1-1). The exact number of different amino acids remained in question until about 1940, when it was established that the vast majority of proteins were built up from a mixture of the same 20 amino acids, with the percentage of a given amino acid varying from one protein to another (Table 1-1). These findings made it likely that each polypeptide chain was characterized by the unique sequence of its amino acids, a conjecture that was first shown to be correct in 1951, when Frederick Sanger in Cambridge, England, reported the order of the amino acids along one of the two polypeptide chains that comprise the hormone insulin (Figure 1-2, page 4).

Most proteins contain only one polypeptide chain. However, there are many other proteins

Table 1-1

The 20 Amino Acids in Proteins

Glycine	GLY	Lysine	LYS
Alanine	ALA	Arginine	ARG
Valine	VAL	Asparagine	ASN
Isoleucine	ILE	Glutamine	GLN
Leucine	LEU	Cysteine	CYS
Serine	SER	Methionine	MET
Threonine	THR	Tryptophan	TRP
Proline	PRO	Phenylalanine	PHE
Aspartic acid	ASP	Tyrosine	TYR
Glutamic acid	GLU	Histidine	HIS

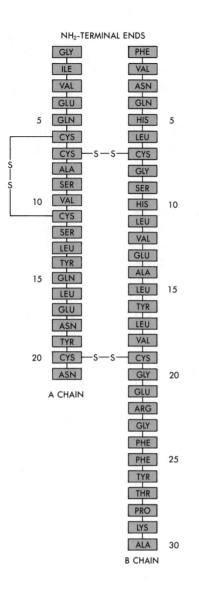

NH₂-TERMINAL ENDS

Figure 1-2
The amino acid composition of insulin. The insulin molecule has two amino acid chains linked by two chemical bonds between sulfur atoms.

ric of the outer plasma membrane and that of the nuclear membrane. Proteins such as the fibrous collagen aid in building the connective tissue between cells. Other proteins—for example, actin and tropomyosin—are constituents of muscle fibers. Still others, among them calmodulin, function to bind ions, like Ca^{++}, that regulate the activities of other enzymes. Finally, there are proteins such as insulin, which functions as a hormone.

The Functioning of an Enzyme Demands a Precise Folding of Its Polypeptide Chain

Once a polypeptide chain is put together, weak chemical interactions between the specific side groups of its amino acids cause it to fold up into a unique three-dimensional form whose exact shape is a function of its amino acid sequence. Whereas many proteins fold up into long, rodlike shapes, most enzymes are globular, with cavities into which their substrates can fit in the way that keys fit into locks. Such cavities, bounded by the appropriate amino acid side groups, enable enzymes to bring their substrates into the close proximity that will permit them to react chemically with one another. The specific amino acid sequence of a given enzyme is thus very important. If inappropriate amino acids are present, then the polypeptide chain cannot fold up to form the properly shaped catalytic cavity. This is why proteins need so many different amino acid building blocks for their construction. By having available 20 of these building blocks, each with a unique shape and with its own chemical properties, and by being able to form polypeptide chains of variable lengths, a cell has the potential to evolve new proteins possessing catalytic cavities with shapes complementary to virtually any potential substrate.

Activation of Molecules to High-Energy Forms Promotes Their Chemical Reactivity

Enzymes catalyze both the forward and backward directions of a given chemical reaction. Their presence alone does not cause the chemical reactions that they catalyze to move in directions that provide orderly cell growth and division. The favored direction of a chemical reaction is actually determined by the thermodynamic law that a reaction

that are formed through the aggregation of separately synthesized chains that have different sequences. The oxygen-carrying protein hemoglobin, for example, is formed by the aggregation of four polypeptide chains, two with a specific α sequence and two containing the β sequence. The sizes of different proteins vary greatly: Their component polypeptide chains contain between as few as 30 to 40 amino acids and as many as 3000.

Though the vast majority of proteins are enzymes, many proteins have structural roles as well. They help, for example, to form the essential fab-

moves in the direction that maximizes the generation of heat and molecular disorder. At first glance, many cellular chemical reactions appear to flaunt this rule. Within growing cells, for example, the polymerization of amino acids into polypeptides is greatly favored over the energetically more likely breakdown of polypeptide chains into their component amino acid monomers. In reality, however, the laws of thermodynamics are never even momentarily violated in cells. Instead cells have evolved specific enzymes that push energetically unfavorable reactions to completion, by coupling them to reactions that are much more energetically favorable. The net result of both reactions is the release of energy that usually becomes dissipated as heat. Coupling is brought about when an enzyme causes a substrate that is to be consumed (for example, one of the nucleotide monomers for DNA synthesis) to react with one of the cell's small-molecule energy donors (say, ATP) to give rise to an energized activated substrate molecule. The subsequent decomposition of this activated substrate in the synthetic reaction will release significant energy, which will result in the synthetic reaction (in our example, DNA synthesis) being favored over the reverse degradative reaction (DNA breakdown).

The need for cells to carry on simultaneously many different coupled reactions requires cells to devote a significant part of their total metabolic activity to the constant replenishment of their ATP supply. In animal cells and in most bacteria, the formation of ATP is coupled with the breakdown of food molecules that are even more rich in energy. In cells that carry out photosynthesis, the energy of sunlight is used directly to make ATP.

Cellular Metabolism Can Be Visualized Through Metabolic Maps

Certain key small molecules of the cell serve as substrates for many different enzymes and thus can be incorporated into many other molecules. However, there are a number of metabolites that function only as intermediates in the biosynthesis of a given complex amino acid, such as tryptophan, or as intermediary metabolites in a quite different metabolic pathway, such as the one through which tryptophan is broken down as a food source. Thus, when we diagram all the various known enzymatic pathways that link together the small molecules of

a cell, the pattern (metabolic map) we perceive is not one of hopeless complexity. Instead we find a most revealing picture of how cells use food molecules to provide the active building blocks, as well as the energy, for the biosynthetic reactions that build up the small-molecule monomeric precursors of the complex polysaccharides like glycogen, or of the even more complicated proteins and nucleic acids.

We also see that for every small molecule a cell possesses, there must exist at least one enzyme that directly catalyzes its synthesis, and quite often another enzyme that can degrade it when it is not needed. The number of different enzymes in a cell must thus greatly exceed the number of different small molecules. All the enzymes, like their respective small molecules, are not present in equal amounts; the enzymes whose substrates are present only in small amounts (for example, vitamins) are likewise present only in small amounts.

Enzymes Cannot Determine the Order of the Amino Acids in Polypeptide Chains

Although enzymes completely determine the specificity of the chemical reactions between small molecules, there is no way that they can be used to determine the order of amino acids in the thousands of different proteins that every cell possesses. To show this, we shall briefly consider first how cells put together polysaccharides, macromolecules that are made up of repeating sugar groups. Many polysaccharides are indeed very long, but they frequently contain only one type of sugar, and their assembly thus involves only one specific enzyme. In polysaccharides that contain more than one sugar, the sugar subunits are linked together in highly repeating patterns, so their synthesis would again only require a correspondingly small number of specific enzymes.

In contrast, an average-sized polypeptide chain contains several hundred amino acids arranged in a unique irregular sequence. If the ordering of the amino acids in such a chain were carried out by enzymes, there would have to be an enormous number of these enzymes, each capable of recognizing a large number of contiguous amino acids. In turn, each of these hypothetical "amino-acid-sequence-recognizing enzymes" would have to be put together by its own set of different "sequence-

recognizing" enzymes, and so forth. This type of scheme obviously cannot work, and we are led to the inescapable conclusion that cells must contain specific "information-bearing" molecules, analogous perhaps to the molds of the sculptor or to the master plates of the lithographer. Such molecules must encode the ordering information so that it can be used to select the correct amino acids in the course of polypeptide synthesis. These information-bearing molecules, moreover, must somehow also be able to synthesize new copies of themselves, so that when a growing cell splits into two daughter cells, each of the progeny cells possesses copies of the master molds (or templates).

As thus described, these putative information-bearing molecules are clearly very similar to the chromosomes that carry the genes that control our heredity. We shall in fact soon show that chromosomes contain the information that is used to determine the order of amino acids in cellular proteins. Here we simply want to emphasize that the cells' possession of a discrete hereditary system is not an accident of evolution. Rather, the existence of discrete genetic molecules that determine the specificity of proteins was an essential ingredient for the emergence of life itself.

Mendel's Breeding Experiments with the Pea Plant First Revealed the Discreteness of Genetic Determinants (Genes)

The knowledge that many human traits like the color of our eyes and the shape of our faces have been passed on to us from our parents must go back to the days of early man. Such inheritable characteristics are of course not limited to humans, and long before the current science of the study of heredity or genetics was born, plant and animal breeders sought to improve the quality of their domestic plants and animals. Hereditary traits were not effectively analyzed, however, until controlled matings were made between individuals with well-defined traits. This was first successfully accomplished by the Austrian monk Gregor Mendel, who in the early 1860s in Brno, Czechoslovakia, crossed pea plants of differing morphologies. Most importantly, he not only noted the appearance of specific traits in the first generation of progeny, but he went on to make and study crosses among these progeny as well as among the progeny and the original parent pea plants.

Mendel's experiments led him to conclude that many traits of the pea were under the control of two distinct factors (later named genes), one coming from the male parent and the other from the female parent. Mendel further noted that the various traits he studied did not assort together, and must thus have been borne on separate hereditary units. Of equal importance was Mendel's distinction between the physical appearance of an organism, which we now call the phenotype, and the exact genetic composition of an organism, its genotype. Mendel realized that some genes (dominant genes) express themselves when they are present in only one copy, whereas other genes (recessive genes) require two copies for expression.

Chromosomes Are the Cellular Bearers of Heredity

About the same time that Mendel made his observations, it became known that heredity is transmitted through the egg and sperm. And since sperm contain relatively little cytoplasm in comparison to their nuclei, the obvious conjecture was that the function of the nucleus was to carry the hereditary determinants of a cell. Soon afterwards, the chromosomes within the nucleus became visible through the means of special dyes (chromo- comes from the Greek for "color"), and the cells of members of a given species were found to contain a constant number of chromosomes. This number was seen to exactly double prior to the cell-division process (mitosis), which in turn reduced the chromosomes back to their exact original count. Within a given cell type, a variety of chromosomes of different sizes were observed, with each distinct type usually being present in two copies (for a total of $2N$ homologous chromosomes). A few years later, the number of chromosomes in the sex cells of sperm and eggs was shown to be exactly N (the haploid number), one-half the $2N$ (the diploid number) found in somatic (nonsex) cells. The partitioning of chromosomes during mitosis was found to be exact: Each daughter cell receives one copy of each chromosome present in the parental cell. In contrast, during the formation of sex cells (meiosis), the chromosome number is reduced to N (Figure 1-3). The fertilization process between sperm and egg thus restores the $2N$ chromosome number characteristic of somatic cells, with one

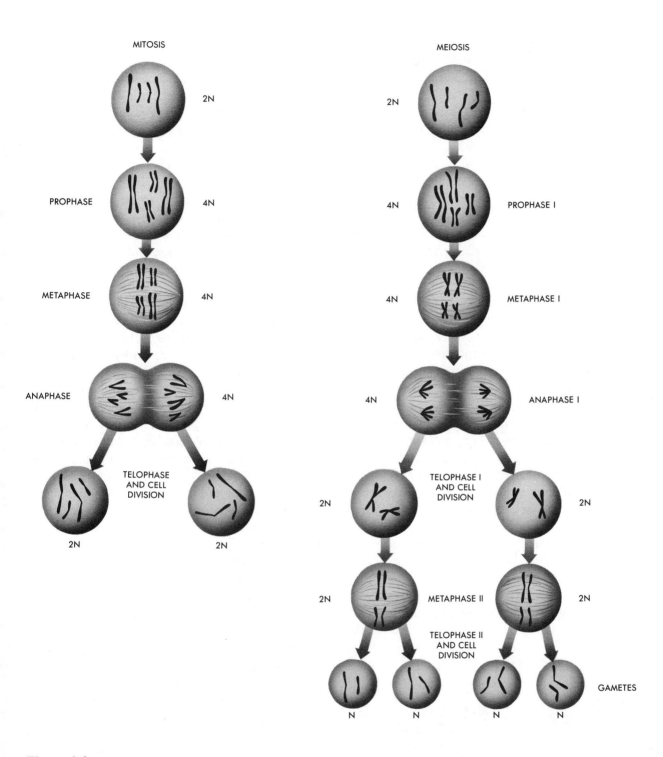

MITOSIS

2N

PROPHASE   4N

METAPHASE   4N

ANAPHASE   4N

TELOPHASE
AND CELL
DIVISION

2N   2N

MEIOSIS

2N

4N   PROPHASE I

4N   METAPHASE I

4N   ANAPHASE I

TELOPHASE I
AND CELL
DIVISION

2N   2N

2N   METAPHASE II   2N

TELOPHASE II
AND CELL
DIVISION

N   N   N   N   GAMETES

Figure 1-3
A comparison of cell division by mitosis and meiosis. Mitosis, somatic cell division, leads to two identical daughter cells that each have the same number of chromosomes as the parent cell. Meiosis, sex cell division, produces four daughter cells (gametes) that each have half the number of chromosomes as the parent cell.

chromosome in each pair coming from the male parent and the other from the female parent. Chromosomes thus behaved exactly as they should if they were the bearers of Mendel's genes. Surprisingly, however, the hypothesis that parents each contribute half of an offspring's genes was not proposed in this form until 1903.

The first trait to be assigned to a chromosome was sex itself. The work was done at Columbia University, where in 1905 Nettie Stevens and Edmund Wilson discovered the existence of the so-called sex chromosomes. Their research showed that one chromosome, the "X," is present in two copies in females but in only one copy in males, who also carry a morphologically distinct Y chromosome. During the halving of chromosome numbers as the sex cells are formed, all eggs necessarily receive single X chromosomes, while sperm receive either an X or a Y chromosome. Fertilization by a sperm containing an X chromosome generates female (XX) progeny, whereas fertilization by a sperm containing a Y chromosome yields male (XY) progeny. The proposal that sex traits are located on a single pair of chromosomes neatly explained the 1:1 ratio of males and females.

It was natural to speculate that all traits, not just those determining sex, might be chromosomally located. The initial confirmation came very soon, between 1910 and 1915, from genetic crosses with the red-eyed fruitfly *Drosophila* in T. H. Morgan's laboratory at Columbia University (Figure 1-4). In the first such experiments, a mutated gene leading to white eyes was located, or mapped, on the X chromosome. Over the next

several years, many more mutants were mapped to either the sex chromosome or to one of the nonsex chromosomes (autosomes). The coexistence of two mutants on the same chromosome was shown by their tendency to reassort together (to be linked) in genetic crosses. But this linkage was never complete; it was sometimes broken as a consequence of a physical exchange of chromosome parts (called crossing over) that occurred when homologous chromosomes came together in pairs before forming haploid sex cells. Most important, crossing over provided a way of ordering genes along chromosomes, because those genes that are far apart on a chromosome have a better chance of being separated due to crossing over than do closely spaced genes (Figure 1-5). The arrangements of genes along chromosomes were always found to be linear, and by 1912 the hypothesis that chromosomes are the bearers of heredity was virtually unassailable.

Early breeding experiments utilized genetically stable variants (mutants) that appeared spontaneously at low frequency. We now know that many spontaneous mutants result from changes in single genes. Such genes are mutant genes, as opposed to normal or "wild-type" genes. In 1926 Hermann Muller and Lewis Stadler independently discovered that x-rays induce mutations, and thereby provided geneticists with a much larger number of mutant genes than were available when the so-called spontaneous mutations were the only source of variability.

As genetic experiments grew in momentum, it became apparent that a very large number of different genes were located on each chromosome; estimates for the number of genes present on all the chromosomes in a given sperm or egg soon exceeded 500. By the late 1930s, it seemed probable that at least 1500 genes were contained within the four chromosomes of *Drosophila*.

At first it was taken for granted that crossing over occurred only between genes, because chromosomes were conceived as being constructed like children's beads; that is, genes were thought to be held together by connectors that easily broke and rejoined. By the late 1930s, though, when many thousands of progeny from single *Drosophila* crosses were examined, scientists discovered rare examples of apparent crossing over *within* a gene that determines eye color. At

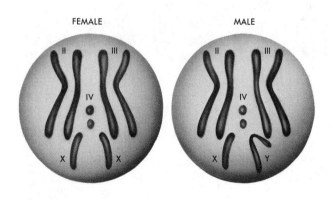

Figure 1-4
The chromosomes of male and female *Drosophila*.

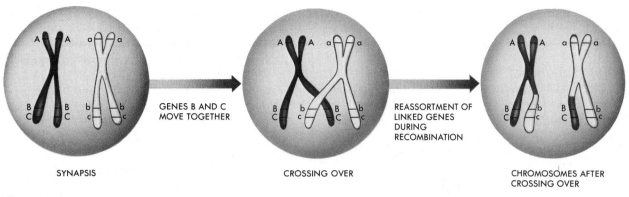

GENES B AND C
MOVE TOGETHER

REASSORTMENT OF
LINKED GENES
DURING
RECOMBINATION

SYNAPSIS

CROSSING OVER

CHROMOSOMES AFTER
CROSSING OVER

Figure 1-5
Reassortment of genes by crossing over.

that time there was no way of knowing that the two mutants being examined contained mutations in the same gene. The mutations might have been in adjacent genes that were functionally related and that therefore produced seemingly identical changes (such as white eyes). Could crossing over cause rearrangements within genes themselves? The answer had to await the development of systems that employed many, many mutations in the same gene, and that allowed millions—not merely thousands—of progeny from the same genetic cross to be examined.

The One-Gene-One-Protein Hypothesis

Geneticists first focused on finding new hereditary traits and mapping their respective genes to precise chromosomal locations. The traits they studied were obvious morphological ones like eye color and wing shape, markers whose chemical basis was (and often still is) totally obscure. How most mutations lead to altered phenotypes necessarily remained unknown. As early as 1909, however, the English physician Archibald Garrod noted that several human hereditary traits were metabolic diseases characterized by the failure of known chemical reactions to take place. (Today one of the best understood of such genetic diseases is phenylketonuria, in which the amino acid phenylalanine cannot be converted to the related amino acid tyrosine. This "error" leads to a buildup in the blood of the toxic intermediary metabolite phenylpyruvate.) Garrod hypothesized that such metabolic diseases, which he called "inborn errors of metabolism," were due to the ab-

sence of specific enzymes that were synthesized under the direction of the wild-type genes. If Garrod's hunch was correct, then for every enzyme, and perhaps for each protein that a cell possesses, there must exist a corresponding gene.

Because so few details of cellular metabolisms were then understood, some 30 years had to pass before definitive experiments could be done to prove the one-gene-one-enzyme relationship. These studies were done in the early 1940s at Stanford University by the geneticist George Beadle and the biochemist Edward Tatum on the mold *Neurospora.* This microorganism normally grows on a simple diet of glucose and inorganic ions. However, exposure to x-rays and ultraviolet light, agents that were by then known to vastly increase the rate at which mutant genes arise, led to production of mutant *Neurospora* strains that multiplied only when their normal diets were supplemented with additional food molecules (growth factors) that were specific to each mutant cell. Some of these mutant *Neurospora* cells required specific amino acids like arginine or cysteine, whereas others required a particular vitamin or one of the purine or pyrimidine building blocks of the nucleic acids. In each case, the specific metabolic requirements were afterwards shown to be due to the absence of one of the enzymes involved in the specific metabolic pathway that led to the synthesis of the missing growth factors. A few years later, the one-gene-one-enzyme concept was extended in general when the molecular defect that causes the human disease sickle-cell anemia was shown by Linus Pauling at the California Institute of Technology to be due to chemically altered (mutant) hemoglobin molecules.

READING LIST

Books

Alberts, B., D. Bray, J. Lewis, M. Raff, K. Roberts, and J. D. Watson. *Molecular Biology of the Cell.* Garland, New York, 1983.

Fruton, J. S. *Molecules and Life.* Wiley-Interscience, New York, 1972.

Lehninger, A. *Principles of Biochemistry.* Worth, New York, 1982.

Stryer, L. *Biochemistry,* 2nd ed. W. H. Freeman and Company, San Francisco, 1981.

Watson, J. D. *Molecular Biology of the Gene,* 3rd ed. Benjamin-Cummings, Menlo Park, Cal., 1976.

Zubay, G. *Biochemistry.* Addison-Wesley, Reading, Mass., 1983.

Original Research Papers (Reviews)

PROTEIN STRUCTURE

Sanger, F., and H. Tuppy. "The amino acid sequence in the phenylalanyl chain of insulin." *Biochem. J.,* 49: 463–490 (1951).

Sanger, F., and E. O. P. Thompson. "The amino acid sequence in the glycl chain of insulin." *Biochem. J.,* 53: 353–374 (1953).

GENES AND THE CHROMOSOMAL THEORY OF HEREDITY

Mendel, G. English translation of Mendel's experiments in plant hybridization, reprinted in:

Peters, J. A., ed. *Classic Papers in Genetics.* Prentice-Hall, Englewood Cliffs, N.J., 1959; and Stern, C., and E. R. Sherwood, eds. *The Origin of Genetics.* W. H. Freeman and Company, San Francisco, 1966.

Stevens, N. M. "Studies in spermatogenesis with especial reference to the 'accessory' [sex] chromosome." *Carn. Inst. Wash.,* publ. 36, pp. 1–32 (1905).

Morgan, T. H. "Sex-limited inheritance in *Drosophila.*" *Science,* 32: 120–122 (1910).

Muller, H. J. "Artificial transmutation of the gene." *Science,* 46: 84–87 (1927).

Stadler, L. J. "Mutations in barley induced by x-rays and radium." *Science,* 68: 186–187 (1928).

THE ONE-GENE-ONE-PROTEIN HYPOTHESIS

Garrod, A. E. "Inborn errors of metabolism." *Lancet,* 2: 1–7, 73–79, 142–148, 214–220 (1908). (Also published as a book by Oxford University Press, London, 1909.)

Beadle, G. W., and E. L. Tatum. "Genetic control of biochemical reactions in *Neurospora.*" *Proc. Natl. Acad. Sci. USA,* 27: 499–506 (1941).

Pauling, L., H. A. Itano, S. J. Singer, and I. C. Wells. "Sickle cell anemia: A molecular disease." *Science,* 110: 543–548 (1949).

2

DNA Is the
Primary Genetic Material

The realization that genes determine the structure of proteins was a very important milestone in the development of genetics, but it did not have any immediate consequences. As long as the molecular structure of the gene was unknown, there was no way to think constructively about gene–protein relations. In fact, as recently as 1950 there was no general agreement on which class of molecules genes belonged to. Nevertheless, the best guess was that the gene was the deoxyribonucleic acid, a still poorly understood polymeric macromolecule that was just starting to be called by its abbreviation DNA.

DNA Is Sited Exclusively on Chromosomes

For many years it was hoped that as microscopes improved, it might eventually be possible to see genes sitting side by side along chromosomes. But even with the advent in the early 1940s of the first electron microscopes, which had a potential resolution over 100 times greater than that of light microscopes, there were disappointments. The first electron-microscope pictures of chromosomes showed no repeating pattern at the molecular level; this suggested a highly irregular gene structure that would not be simple to interpret. Attempts to purify chromosomes away from other cellular constituents were much more informative, although it was impossible to obtain really pure chromosomes.

Two main chromosomal components were almost invariably found: (1) deoxyribonucleic acid (DNA), and (2) a class of small, positively charged proteins known as the histones; these, being basic, neutralized the acidity of DNA. DNA had been known to be a major constituent of the nucleus (hence the name "nucleic" acid) ever since its discovery in 1869 by the Swiss scientist Frederick Miescher. In the 1920s, with the DNA-specific purple dye developed by the German chemist Robert Feulgen, DNA was found to be sited exclusively on the chromosomes. DNA therefore had the location expected for a genetic material. In contrast, the histones could apparently be ruled out as genetic components because they were absent from many sperm, which contained instead even smaller basic proteins, the protamines. But most biochemists were not inclined to focus attention on DNA. They thought it would not be nearly as specific as the proteins, of which they knew an unlimited number could be constructed by chaining together the 20 amino acids in different orders. So it was widely believed that some minor and not yet well-characterized protein component of the chromosomes might be found to be the true genetic material.

Cells Contain RNA as well as DNA

Already late in the nineteenth century it had been discovered that cells have a second kind of nucleic acid—what we now call ribonucleic acid (RNA). Unlike DNA, which is located exclusively in the nucleus, RNA is found in the cytoplasm as well as in the nucleus. Within the nucleus, RNA is concentrated in a few dense granules (nucleoli) that are attached to chromosomes.

Both DNA and RNA resemble proteins, in that they are constructed from many smaller building blocks linked end to end. However, nucleotides, the building blocks of nucleic acid, are more complex than any amino acid. Each nucleotide

contains a phosphate group, a sugar moiety, and either a purine or a pyrimidine base (flat, ring-shaped molecules containing carbon and nitrogen) (Figure 2-1). When nucleotides are linked together in large numbers, they are called polynucleotides.

Early on, the sugar component of RNA was known to be different from that of DNA. Yet it was not until the 1920s that the work of Phoebus Levine of the Rockefeller Institute revealed that the sugar of DNA is deoxyribose (hence the name deoxyribonucleic acid). Two purines and two pyrimidines are found in both DNA and RNA. The two purines, adenine and guanine, are used in both DNA and RNA; the pyrimidine cytosine is likewise found in DNA and RNA. However, the pyrimidine thymine is found only in DNA, while the structurally similar pyrimidine uracil appears in RNA (Figure 2-2).

In both DNA and RNA, the nucleotides are linked together to form very long polynucleotide chains. The linkage consists of chemical bonds running from the phosphate group of one nucleotide to the deoxyribose group of the adjacent nucleotide. Each deoxyribose (ribose) residue contains several atoms to which phosphate groups might

attach, and there was initially much difficulty in identifying the exact atoms that are bridged by the phosphate groups.

Another question that went unanswered for a long time was how the four different nucleotides are ordered along a DNA (or RNA) molecule. No methods existed for even estimating the exact amounts of the four nucleotides within DNA or RNA, and as late as the 1940s, the possibility could not be dismissed that DNA and RNA had regular repeating structures in which each base was repeated every four nucleotides along its polynucleotide chain. More interesting, however, was the alternative possibility, that there were a very large number of different DNA and RNA molecules, each with its own specific irregular sequence of bases. If this were the way these molecules were constructed, DNA and RNA could have encoded in their varying base sequences the massive amount of information needed to order the very large number of amino acid sequences found in the proteins of the living world.

A Biological Assay for Genetic Molecules Is Discovered

DNA did not begin to get the serious attention that, with hindsight, we now realize it should have had until it was shown through a biological assay to be able to alter the heredity of certain bacteria. The development of this assay was not at all premeditated, but arose in 1928 out of the studies of the English microbiologist Fred Griffith on the pathogenicity of the bacterium *Diplococcus pneumoniae,* which causes pneumonia. Griffith made the unexpected observation that heat-killed pathogenic cells, when mixed with living nonpathogenic cells, were able to transform a small percentage of the nonpathogenic cells into pathogens. In becoming pathogenic, the nonvirulent cells acquired the thick, outer polysaccharide-rich cell wall (the capsule) that somehow confers pathogenicity to cells that possess it. Griffith thus discovered the existence of an active (genetic?) substance that remained undamaged by the lethal exposure of pathogenic cells to heat, and that could later move into nonpathogenic cells and direct them to make capsules.

Griffith himself did not try seriously to identify the active principle. This task was taken up

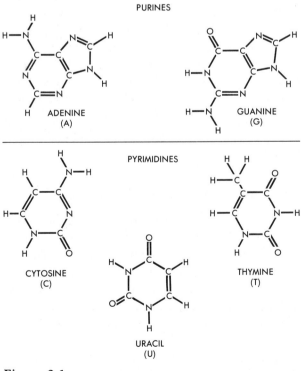

Figure 2-1
Bases of nucleic acids.

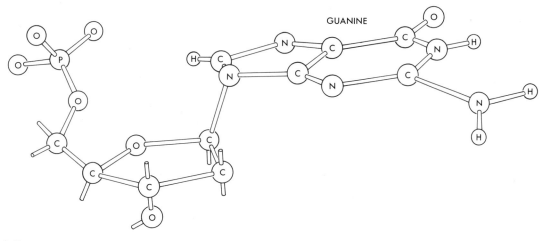

GUANINE

Figure 2-2
A DNA nucleotide. The base is attached to a deoxyribose ring that is in turn
bonded to a phosphate group. In DNA molecules, nucleotides are linked together
to form long chains by bonds running from the phosphate group of one nucleotide
to the deoxyribose group of the adjacent nucleotide. The base shown here, guanine,
can be replaced by any one of the other three DNA bases—adenine, cytosine, or
thymine.

by Oswald Avery, whose scientific career at the
Rockefeller Institute in New York had been prin-
cipally spent working out the chemistry of bacterial outer capsules. When he began working on the
"transforming factor," Avery thought it likely that
the active substance would be a complex polysac-
charide that in some way primed the synthesis of
more polysaccharides of the same kind. As his first
step, he showed that the active factor could be
extracted from heat-killed cells that were broken—a necessary precondition for the later isolation of the factor from other molecules. Then followed a decade of intensive studies that finally
ended with Avery and his younger colleagues, Ma-
clyn McCarty and Colin MacLeod, concluding that
the transforming factor was a DNA molecule. Not
only was DNA the predominant molecule in their
most purified preparation of the transforming factor, but the transforming activity was specifically
destroyed by a highly purified preparation of
DNase, a then just-discovered enzyme that specifically breaks down DNA. In contrast, the transforming activity was unaffected by exposure to enzymes that degraded proteins, or to enzymes that
degraded RNA.

Avery's experiments, first announced in 1944,
had been so carefully done that the conclusion that
DNA was the transforming factor was considered
indisputable by a majority of scientists. However,
some skeptics preferred to believe that Avery and
his colleagues had somehow missed seeing the
"genetic protein," and that DNA was required for
activity in their assay only because it functioned as
an unspecific scaffold to which the real protein
genes were fixed. Upon reflection, though, it
seems that the pinpointing of DNA should not
have been unexpected. By the time of Avery's
experiments, DNA was known to be a very large
molecule containing hundreds of nucleotides. If
the sequences of the four main nucleotides were
found to be irregular, then the number of potentially different DNA sequences would be the as-
tronomically large 4^n (n = the number of nucleotides in a chain).

The only point in question was the generality
of Avery's observation. Were all genes made of
DNA, or were there other genetic molecules that
functioned in other situations? Clearly the matter
could be quickly resolved if it proved possible to
change the heredity of other life forms through
the addition of specific DNA molecules. At that
time, though, there was no way to isolate undamaged DNA molecules from most plant and animal
sources, and it was not possible to extend transformation to other organisms. A group in France did
claim that it had been able to use DNA to change
the plumage of ducks that had grown up from eggs
into which DNA had been injected. The eggs,

however, had been obtained from a local country market and were of unclear ancestry; as a result, no one took the claims of "transduction" seriously.

Viruses Are Packaged Genetic Elements That Move from Cell to Cell

Interest in DNA had also risen as a result of its discovery in several highly purified viruses. The nature of these tiny disease-causing particles, which multiply only in living cells, was long disputed. Some scientists considered them a sort of naked gene; others preferred to think of them as the smallest form of life. Only when it became possible to purify them away from cellular debris and look at them in the electron microscope did their nature begin to be revealed. They were clearly not minute cells; rather, they lost their identity as discrete particles when they multiplied within cells. The best guess, therefore, was that they were parasites at the genetic level, and that by studying how they multiplied, definitive systems for analyzing gene structure and replication might be developed.

At this point a collection of physicists, chemists, and biologists (the "phage group") turned their attention to the growth cycle of the viruses that multiply in bacteria—the bacteriophages (*-phage* is from the Greek for "eating"). Most favored for study were a group of phages named T1, T2, T3, and λ, which multiply within the common intestinal bacterium *Escherichia coli (E. coli).* A single parental phage particle can multiply to several hundred progeny particles within roughly 20 minutes. Analysis of the genetic properties of these phages started when mutants arose during the multiplication cycle. When several independently arising mutant phages infected a single bacterium, some of the progeny phages that were produced appeared normal. Viruses were thus also capable of genetic recombination. Subsequent experiments employing many different mutants suggested that each virus particle contained several different genes linearly arranged along the viral chromosome.

By purely genetic experiments, however, scientists could not decide whether it was the DNA or one of the protein components that carried the genetic specificity. This point was not set-

tled until 1952, when at Cold Spring Harbor, New York, Alfred Hershey and Martha Chase showed that only the DNA of phages entered the host bacteria. Their surrounding protein coats remained outside and thus could be ruled out as potential genetic material (Figure 2-3).

Molecules with Complementary Sizes and Shapes Attract One Another

Even before the exact structure of DNA became known, there was speculation about the nature of the attractive forces that might bring together the appropriate monomeric precursors of proteins and nucleic acids on the surfaces of their respective master molds or templates. By far the most likely candidates were the so-called "weak bonds," as opposed to the covalent bonds that link atoms together in molecules. Enzymes are not needed to either make or break weak bonds; these bonds are spontaneously made and broken at the temperatures of living cells, and are the forces that hold molecules together when they aggregate into liquids like water, or into the rigidly defined crystals of the solid state. Weak bonds also provide the attractive forces that hold together antibodies and their respective antigens. The polypeptide chains of individual antibodies are folded in such a way that they produce cavities that are complementary in shape to the molecular groups on the surfaces of the antigens to which they bind. Such complementarity permits a large number of different weak bonds to form, and produces a correspondingly strong affinity between the antigens and their respective antibodies.

Weak bonds fall into two major classes. One class (ionic bonds and hydrogen bonds) depends in essence upon the attractive forces between atoms of opposite charges. The second class, the Van der Waals attractive forces, develops when molecules with complementary shapes come into close proximity. Both these categories of attractive forces are thus quite specific, and, particularly when several weak bonds are present simultaneously, their collective energies should be sufficient to ensure that only the correct amino acid (nucleotide) is inserted in the appropriate slot on the template surface of the genetic molecule. Given the nature of weak bonds, the product of a template-controlled polymerization process should not be a

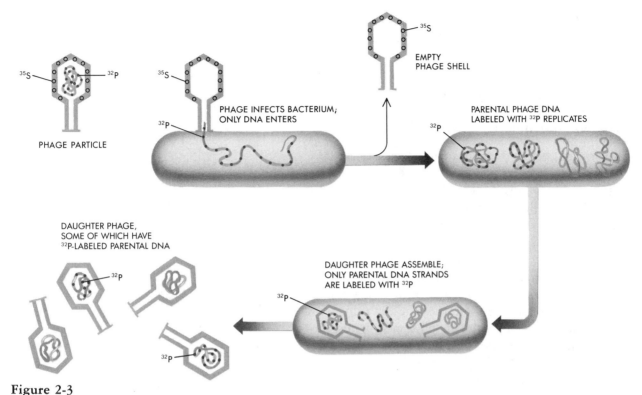

Figure 2-3
Hershey and Chase used radioactively labeled bacteriophage to demonstrate that its genetic material is DNA, not protein.

molecule identical in shape to its template, but rather one whose shape is exactly complementary in outline. The question of how a polynucleotide chain could be folded so as to be complementary in shape to another polynucleotide chain could not, however, be asked in the absence of knowledge about the covalent bonds that hold the nucleotides of DNA together.

It was thus not surprising that geneticists, impatient with the slow pace at which the fundamental chemistry of DNA was then being established, and curious about the nature of the forces leading to the pairing of homologous chromosomes prior to crossing over, asked whether there might exist still undiscovered chemical forces that were applicable only to macromolecules and that generated attractive forces between identical molecules. This hypothesis was, however, regarded as chemically unsound by serious structural chemists like Linus Pauling, who in 1940 wrote a brief statement on that subject with the theoretical physicist Max Delbrück. They dismissed as silliness the concept of

the attraction of like molecules, and stated that gene duplication would necessarily involve the synthesis of molecules with complementary outlines.

The Diameter of DNA Is Established

Although in the late 1930s Swedish physical chemists had already obtained evidence that DNA was asymmetrical from its behavior in solution, direct measurements of its size became possible only when the electron microscope came into general use in the years following the end of World War II. All carefully prepared samples showed extremely elongated molecules many thousands of angstroms ($\text{Å} = 10^{-10}$ meter) in length and approximately 20 Å thick. All the molecules were unbranched, which confirmed the highly regular backbone structure proposed by organic chemists (see below). From the length and the fact that each nucleotide base was just over 3 Å thick, it was clear that most DNA molecules were composed of

many thousands of nucleotides, and were very possibly much larger than any other natural polymeric molecules.

The Nucleotides of DNA and RNA Are Linked Together by Regular 5'–3' Phosphodiester Bonds

The knowledge that the transforming factor was DNA became general just as World War II was ending, when scientists, then dissociating themselves from military research, began looking around for new problems to take on. At that time Alexander Todd, already one of England's most effective chemists, decided to focus on the chemistry of complex nucleotides related to those found in DNA. By the early 1950s, his large research group at the Chemical Laboratories of Cambridge University established the precise phosphate–ester linkages that bound the nucleotides together. Their results were appealingly simple. These linkages were always the same, with the phosphate group connecting the 5' carbon atom of one deoxyribose residue to the 3' carbon atom of the successive nucleotide (Figure 2-4). No traces of any unusual bonds were found, and Todd's group concluded that the polynucleotide chains of DNA, like the polypeptide chains of proteins, are strictly linear molecules.

It took longer to settle the question of the nature of the linkages within RNA; its structure was not discovered until two years later, in 1955. Like DNA, RNA was found to have a highly regular backbone that employs only 5'-3' phosphodiester links to hold together its component nucleotides.

The Composition of Bases of DNA from Different Organisms Varies Greatly

Chromatographic separation methods that allowed exact measurement of the proportion of the various amino acids in proteins were first applied successfully in England in the early 1940s. Soon afterwards, these methods were extended to determining the amounts of the purine and pyrimidine bases in nucleic acids. The bases of DNA were first quantitatively analyzed by Edwin Chargaff in his laboratory at the College of Physicians and Surgeons of Columbia University. By 1951 it was not

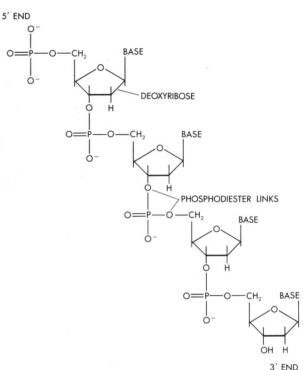

Figure 2-4
Regular phosphodiester bonds between sugar and phosphate groups form the backbone of DNA.

only clear that the four bases were not present in equal numbers, but it was also apparent that their amounts could vary greatly among distantly related organisms. Chargaff also noted that the amounts of the four bases did not vary independently but that for all the species he looked at, the amounts of the purine adenosine (A) were very close, if not identical, to the amounts of the pyrimidine thymine (T). Similarly, the amounts of the second purine, guanine (G), were always very similar if not identical to those of the second pyrimidine, cytosine (C). The number of purine groups in DNA was thus approximately equal to the number of pyrimidines.

It was not immediately known whether there was any deep significance to the equivalence of the purines and pyrimidines; much less whether the A = T and G = C relationships (Chargaff's rules) were important. To start with, the separation methods were imperfect and it was impossible to state with certainty that the ratios of these components in DNA were exactly equal. There was, moreover, the report that the bacterial virus (phage) T2 lacked any cytosine. If this was true,

then the equivalence of purines and pyrimidines that Chargaff had noticed might not be general enough to be meaningful. Soon afterwards, in 1952, the novel nature of the phage T2 DNA was explained. Although cytosine was indeed absent, there was present instead a modified cytosinelike base, and most importantly, its amount was equal to that of guanine. The significance of the A = T and G = C observations, however, remained as unclear as before, and Chargaff was initially known primarily for his demonstration that DNA molecules from different species could have dramatically different base compositions. Because all DNA molecules were not the same, most likely there existed, even within the same cell, a large number of different DNA molecules each having its own unique nucleotide sequence.

DNA Has a Highly Regular Shape

In one sense DNA chains are very regular: They contain repeating sugar– (deoxyribose–) phosphate residues that are always linked together by exactly the same chemical bonds. These identical repeating groups form the "backbone" of DNA. On the other hand, DNA's four different bases can be attached in any order along the backbone, and this variability gives DNA molecules a high degree of individuality (or specificity). So, depending on which part of a DNA molecule we focus upon, we may view it as regular or as irregular. More important, however, is how DNA "views" itself. Does its chain fold up into a regular configuration dominated by its regular backbone? If so, the configuration would most likely be a helical one in which all the sugar–phosphate groups would have identical chemical environments. On the other hand, if the chemistry of the bases dominates the DNA structure, we would fear that no two chains would have identical three-dimensional configurations. In this case, the task of figuring out how each of these differently shaped molecules could serve as a template for the formation of another DNA chain would have been beyond our reach.

The only direct way to examine the three-dimensional structure of DNA was to see how it diffracts (bends) x-rays. Dry DNA has the appearance of irregular white fluffs of cotton, but it becomes highly tacky when it takes on water, and

it can then be drawn out into thin fibers. Within these fibers, the individual long, thin DNA molecules line up parallel to one another. In structural analysis, such fibers are placed in the path of an x-ray beam and the pattern of the diffracting rays is recorded on photographic film. DNA was first examined this way in 1938 by the Englishman William Astbury, using material prepared in Sweden by E. Hammerstein, who had developed procedures that allowed DNA of very high molecular weight to be isolated from thymus glands. Astbury found that DNA did indeed yield a distinctive diffraction pattern, and so the individual DNA molecules must have had some preferred orientation. However, the individual diffraction spots were not sharp like those produced by crystalline material, and the possibility remained that DNA chains never assumed a precise configuration common to all chains. After World War II, Astbury resumed taking x-ray pictures of DNA and obtained some patterns that were considerably better defined, and from which he proposed that the individual purine and pyrimidine bases were stacked perpendicular to the long axis of their molecules as if they were a pile of pennies. His better diffraction patterns still remained far from crystalline, though, and the precise structure of the DNA remained in question.

Then in 1950 the physicist Maurice Wilkins, who was working at Kings College, London, with DNA that had been carefully prepared in Bern by the Swiss chemist R. Signer, obtained a truly crystalline diffraction pattern. The individual DNA molecules that came together to form the crystalline fibers must have been very similar in form, or they would not have been able to pack together so regularly. It thus became certain that DNA does have a precise structure, the solution of which might hopefully begin to reveal the manner in which DNA functions as a template.

The Fundamental Unit of DNA Consists of Two Intertwined Polynucleotide Chains (the Double Helix)

The data obtained from the fiber diffraction pattern of a molecule as complicated as DNA cannot by itself provide sufficient information to reveal the molecular structure. Inspection of such pat-

terns, however, often provides key parameters that strongly demarcate the outlines of the molecule under investigation. This proved to be the case with DNA, where the key x-ray patterns turned out to be not those obtained from the crystalline DNA fibers, but ones obtained from the less ordered aggregates that form when DNA fibers are exposed to a higher relative humidity and take up more water. These paracrystalline patterns were first seen by Rosalind Franklin, a colleague of Wilkins, working also in London. Her pictures revealed a dominant crosslike pattern, the telltale mark of a helix. Thus, despite the presence of an irregular sequence of bases, the sugar–phosphate backbone of DNA nevertheless assumed a helical configuration. A separate nucleotide was found every 3.4 Å along its fiber axis, with 10 nucleotides, or 34 Å, being required for every turn of the helix.

A very important inference came out of the routine measurement of the diameter of the helix. Given the measured density of DNA, the estimated 20-Å diameter of DNA was far too large for a DNA molecule containing only one chain. The fundamental unit of DNA must consist instead of two intertwined chains; these could be further shown from the diffraction patterns to run in opposite directions. This was indeed a surprising result, for prior to the use of x-ray diffraction methods, no chemist had ever suspected that DNA was a multichained molecule.

The Double Helix Is Held Together by Hydrogen Bonds Between Base Pairs

How the two chains are held together in the DNA molecule could not be ascertained from the x-ray data alone. The final clues came in the spring of 1953, when James Watson and Francis Crick, then working at the Cavendish Laboratory of Cambridge University, built three-dimensional models of DNA to look for the energetically most favorable configurations compatible with the helical parameters provided by the x-ray data. This approach quickly led them to the conclusion that the sugar–phosphate backbones are on the outside of the DNA molecule and that the purine and pyrimidine bases are on the inside, oriented in such a way that they can form hydrogen bonds to bases on opposing chains. The exact hydrogen-bonding pattern used in DNA suddenly emerged

with the realization from model building that if a purine on one chain is always hydrogen-bonded to a pyrimidine on the other chain, the area occupied by the paired bases would always be the same throughout the length of the DNA molecule. Seen from the outside, DNA is a very regular-appearing structure, despite the irregular sequence of bases on any one chain. Equally important, the two purines adenine and guanine do not unselectively bond to the two pyrimidines thymine and cytosine. Adenine (A) can pair only with thymine (T), while guanine (G) can bond only with cytosine (C) (Figure 2-5). Each of these base pairs possesses a symmetry that permits it to be inserted into the double helix in two ways (A=T and T=A; G≡C and C≡G); thus along any given DNA chain, all four bases can exist in all possible permutations of sequence (Figure 2-6).

Because of this specific base pairing, if the sequence of one chain (for example, TCGCAT) is known, that of its partner (AGCGTA) is also known. The opposing sequences are referred to as complementary, and the corresponding polynucleotide partners as complementary chains. De-

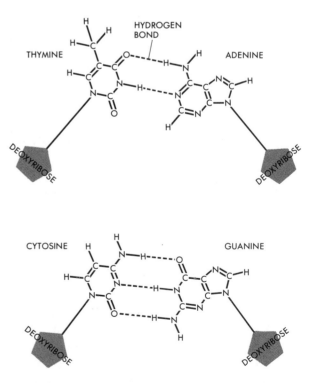

Figure 2-5
Hydrogen bonding between the adenine–thymine and guanine–cytosine base pairs.

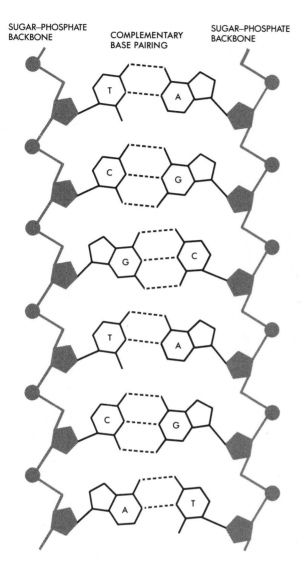

SUGAR–PHOSPHATE BACKBONE   COMPLEMENTARY BASE PAIRING   SUGAR–PHOSPHATE BACKBONE

Figure 2-6
The base pairing of two DNA chains.

The Complementary Nature of DNA Is at the Heart of Its Capacity for Self-Replication

Before we knew what genes looked like, it was almost impossible to speculate wisely about how they could be exactly duplicated prior to cell division. Any proposal had to be general. As we mentioned earlier (page 15), Linus Pauling and Max Delbrück suggested in 1940 that the surface of the gene somehow acts as a positive mold, or template, for the formation of a molecule of complementary (negative) shape, in much the same way that material can be molded around a piece of sculpture for the purpose of making a cast. The complementary-shaped negative could then serve as the template for the formation of its own complement, thereby producing an identical copy of the original mold.

Thus the realization that the two chains of DNA had complementary shapes caused great excitement. It was promptly proposed that the two strands of the double helix should be regarded as a pair of positive and negative templates, each specifying its complement and thereby capable of generating two daughter DNA molecules with sequences identical to those of the parental double helix (Figure 2-7, page 20). If this was indeed the way DNA duplicates, the process could be confirmed by proof that the parental strands separate before duplicating themselves and that each daughter molecule contains one of the parental chains.

The elucidation of the structure of DNA thus accomplished in one decisive act far more than we could ever have prudently predicted. Before the double helix was found, we hoped that determination of the correct structure of DNA would provide a firm foundation upon which a model of DNA replication could be erected. Instead, discovery of the double helix brought with it as a gigantic bonus a powerful clue, if not the total answer, to the long-desired goal of understanding gene duplication.

Of course, the possibility still existed that the self-complementary structure of the double helix had nothing to do with how it was synthesized. Virtually no one, however, really believed this alternative. The finding of the first highly plausible example of a template had to be an occasion for joy, not for scepticism. The time had also passed for any further questioning about what genes were. The name of the game was DNA.

spite the relative weakness of the hydrogen bonds holding the base pairs together, each DNA molecule contains so many base pairs that the complementary chains never spontaneously separate under physiological conditions. If, however, DNA is exposed to near-boiling temperatures, so many base pairs fall apart that the double helix separates into its two complementary chains (this process is called denaturation). The existence of the double helix provides a structural chemical explanation for Chargaff's rules: A = T and G = C. Only with complementary base pairing could all the backbone sugar–phosphate groups have identical orientations and permit DNA to have the same structure with any sequence of bases.

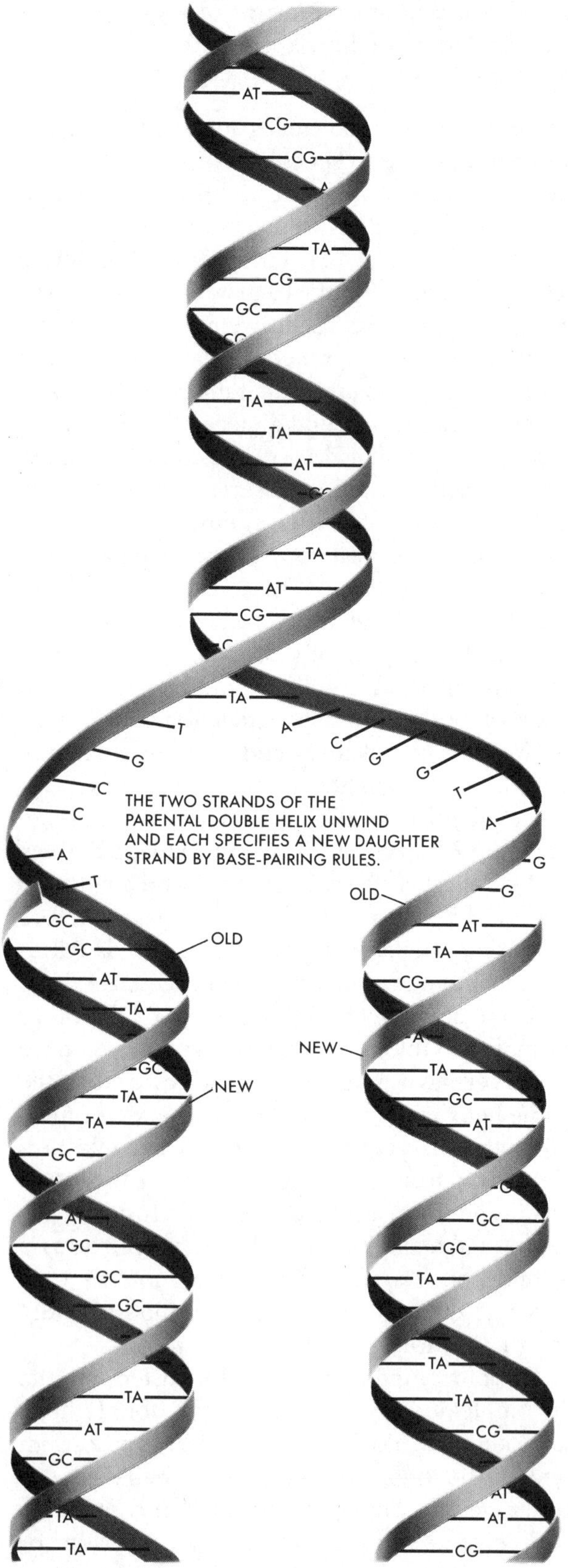

THE TWO STRANDS OF THE
PARENTAL DOUBLE HELIX UNWIND
AND EACH SPECIFIES A NEW DAUGHTER
STRAND BY BASE-PAIRING RULES.

Figure 2-7
Identical daughter double helices are
generated through the semiconservative
replication of DNA.

Proof of Strand Separation During DNA Replication

It took some five years before there was firm evidence of strand separation. Proof came from the experiments of Matthew Meselson and Franklin Stahl at the California Institute of Technology. They had the clever idea of using density differences to separate parental DNA molecules from daughter molecules. They first grew cultures of the bacterium *E. coli* in a medium highly enriched in the heavy isotopes ^{13}C and ^{15}N. By virtue of its isotopic content, the DNA in these bacteria was much heavier than the normal light DNA coming from cells grown in the presence of the far more abundant natural isotopes ^{12}C and ^{14}N. Because of its greater density, the heavier DNA could be clearly separated from the light DNA by high-speed centrifugation in cesium chloride.

When heavy-DNA-containing cells were transferred to a normal "light" medium and allowed to multiply for one generation, all the heavy DNA was replaced by DNA of a density that was halfway between heavy and light. The disappearance of the heavy DNA indicated that DNA replication is not a conservative process in which the complementary strands of the double helix stay together. Instead, its replacement by the hybrid-density DNA implied a semiconservative replication process, in which the two heavy parental strands separate to serve as templates for complementary light strands, and each daughter molecule has one heavy (parent) strand and one light strand.

Whether the complementary strands completely separate before replication starts was not immediately known. Later, abundant electron-microscope evidence of Y-shaped replication forks indicated that strand separation and replication go hand in hand. As soon as a section of double helix begins to separate for replication, the resulting single-stranded regions are quickly used as templates and become new double-helical regions.

DNA Molecules Can Be Renatured as well as Denatured

Under physiological conditions, the two strands of the double helix almost never come apart spontaneously. If, however, double helices are exposed to near-boiling temperatures or to extremes of pH (pH < 3 or pH > 10), they quickly fall apart (are

denatured) into their component single strands. At first, denaturation was regarded as essentially irreversible, but by 1960 Julius Marmur, Paul Doty, and their coworkers at Harvard showed that the complementary single strands recombine to form native double helices when they are kept for several hours at subdenaturing conditions (approximately 65°C). Such annealing "renaturation" events are very specific and only produce perfect double helices when the sequences of the bases of the two combining strands are exactly complementary. Imperfect double helices, however, can be formed between nearly complementary molecules at less stringent (lower) annealing temperatures. So by observing the extent of such imperfect renaturation events, the genetic relationships among DNA from different species can be determined.

Renaturation also can be induced to occur between DNA and RNA chains with complementary sequences. It was through the preparation of such DNA–RNA hybrid double helices that in 1961 Ben Hall and Sol Spiegelman obtained definitive proof that DNA functions in protein synthesis by serving as the template for the formation of complementary RNA chains (Chapter 3).

G–C Base Pairs Fall Apart Less Easily Than Their A–T Equivalents

Three strong hydrogen bonds link guanine to cytosine, whereas two strong hydrogen bonds hold together the adenine–thymine base pairs. Double helices containing a prevalence of G–C base pairs are thus more stable (that is, they denature at higher temperatures) than helices in which A–T base pairs predominate. In fact, the proportion of AT to GC within a DNA specimen can be directly ascertained by measuring the temperature at which half the DNA molecules fall apart into their component single strands.

Palindromes Promote Intrastrand Hydrogen Bonding

Not only do base pairs form between bases on opposing strands, but they can also form between bases of single chains that by chance (or design?) have nearby inverted repetitious sequences (palindromes) that allow the formation of hydrogen-bonded hairpin loops (Figure 2-8). The possibility thus exists that the momentary denaturation of palindromic regions often leads to the formation of semistable cruciform loops able to interact with specific DNA-binding proteins.

5-Methylcytosine Can Replace Cytosine in DNA

In many higher plant and animal DNAs a significant fraction of the cytosine residues exist in a modified form in which a methyl group is attached

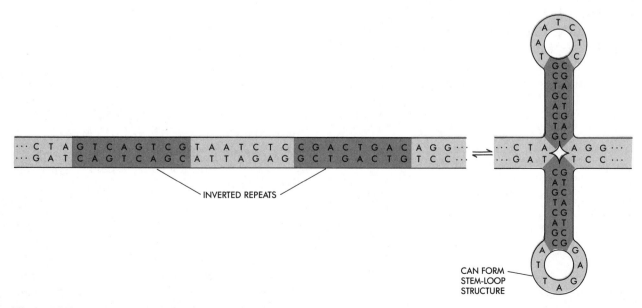

Figure 2-8
A segment of DNA containing an inverted repeat can locally denature and then renature to form stem-loop structures.

to the 5 carbon atom of the pyrimidine ring (5-methylcytosine) (Figure 2-9). Such methyl groups do not affect the way their respective molecules can hydrogen-bond, and the base pairs formed by 5-methylcytosine with guanine are equivalent in strength to those formed by cytosine. In eukaryotic DNA, cytosine residues that contain methyl groups are always located next to guanine residues on the same chain [(5')CG(3')], and DNA molecules that have relatively many (5')CG(3') segments are more methylated than those in which the (5')CG(3') dinucleotide is relatively rare. At first it was uncertain how 5-methyl-C arose, but now it is clear that methyl groups are added after their respective DNA chains are synthesized. 5-Methyl-C is particularly common in higher plants, where some 20 percent of the total cytosine residues become methylated. Why 5-methyl-C exists was initially obscure; only recently has evidence begun to accumulate that methylation of C may play a key role in the control of DNA function (page 179).

Chromosomes Contain Single DNA Molecules

When the double helix was found, it was believed that many distinct DNA molecules were used to construct all but the smallest chromosome. This picture suddenly changed with the realization that long DNA molecules are inherently fragile and easily break into much smaller fragments. When, by the late 1950s, great care was taken to prevent shearing, DNA molecules containing as many as 200,000 base pairs were immediately seen.

Now the best guess is that the chromosome of *E. coli* contains a single DNA molecule made up

– 0.3 ANTIRESTRICTION	
– 0.7 PROTEIN KINASE	**EARLY GENES**
– 1 RNA POLYMERASE	
– 1.2 REPLICATION	
– 1.3 DNA LIGASE	
– 2 ANTI-RNA POLYMERASE	
– 2.5 DNA-BINDING PROTEIN	
– 3 ENDONUCLEASE	
– 3.5 AMIDASE (LYSOZYME)	
– 4 PRIMASE	**GENES FOR DNA METABOLISM**
– 5 DNA POLYMERASE	
– 5.7 GROWTH ON LAMBDA	
– 6 EXONUCLEASE	
– 7 HOST RANGE	
– 7.3 HOST RANGE	
– 8 HEAD–TAIL	
– 9 HEAD ASSEMBLY	
– 10 CAPSID PROTEIN	
– 11 TAIL	
– 12 TAIL	
– 13 CORE	
– 14 CORE	**GENES FOR VIRION STRUCTURE AND ASSEMBLY**
– 15 CORE	
– 16 CORE	
– 17 TAIL FIBER	
– 18 DNA MATURATION	
– 19 DNA MATURATION	

Figure 2-10
A genetic and physical map of bacteriophage T7 DNA. The positions of the terminal repetition (the three black boxes) and the T7 genes are drawn to scale according to their positions in the nucleotide sequence (Dunn and Studier, 1983).

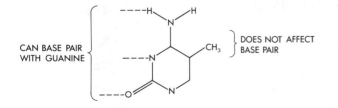

Figure 2-9
5-Methylcytosine.

of more than 4 million base pairs. Likewise, one DNA molecule is thought to exist within each of the even larger chromosomes of higher plants and animals; these chromosomes contain on the average some 20 times more DNA than the *E. coli* chromosome. A chromosome is thus properly defined as a single, genetically specific DNA molecule to which are attached a large number of positively charged (protective?) structural proteins (histones, for example), as well as other proteins whose functions have yet to be determined.

Viruses Are Sources of Homogeneous DNA Molecules

Until recently, the only DNA molecules that had been seriously studied were isolated from DNA viruses, each of which contain a single DNA molecule. Such DNA may be either linear or circular in shape, with replication generally starting at a unique internal site and moving away bidirectionally until the duplication process is completed. The various DNA phages, particularly those that multiply in *E. coli,* were the favored source of DNA in early studies, both because they are readily grown in large amounts and because many have relatively small DNA molecules that do not easily break in solution. The well-studied linear DNA of phage T7, for example, consists of approximately 40,000 (39,936) base pairs, along which 50 genes have been mapped (Figure 2-10). From DNA sequence analysis (page 61), the number of amino acids in each of the polypeptide products of these genes has been determined. Over 92 percent of the DNA's base pairs are used to specify these products, so the individual genes are very close to one another.

Phage λ DNA Can Insert Itself into a Specific Site Along the *E. coli* Chromosome

At first it was generally believed that after its entry into a bacterial cell, phage DNA always initiated a lytic multiplication cycle that resulted in the generation of hundreds of progeny phage within each infected cell. Later it became clear, largely through the work of André Lwoff at the Institute Pasteur in Paris, that certain phage DNAs have the alternate possibility of being inserted in a "prophage" form into the chromosomes of their bacterial host, where they become effectively indistinguishable from normal bacterial genes. By now the best known of such "lysogenic" phages is phage λ, whose linear DNA molecule has a molecular weight of approximately 50 million daltons. Before λ DNA inserts itself into the *E. coli* chromosome, it circularizes by the base pairing of the complementary single-stranded tails that exist at its two ends. The resulting circular λ DNA molecule then recombines into the *E. coli* chromosome by a crossover event with a specific group of bases along the *E. coli* chromosome (Figure 2-11). The

resulting linear "prophage" remains virtually inert genetically until it is provoked by a signal (usually some form of damage to the host chromosome, such as exposure to UV light) that leads to a reverse crossover event that releases λ DNA from the *E. coli* chromosome, allowing it to initiate a lytic cycle of multiplication.

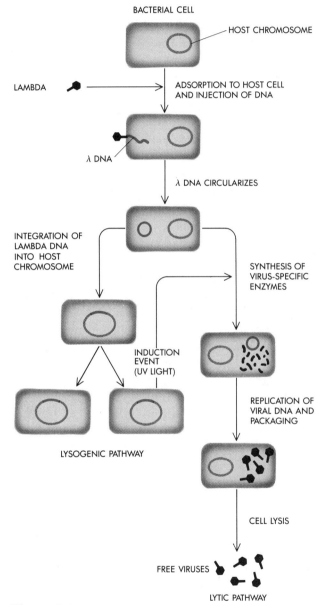

Figure 2-11
Insertion of λ phage DNA into the *E. coli* chromosome. The λ DNA circularizes by the base pairing of the complementary single-stranded tails at its ends (cohesive ends or "cos" sites). The circular molecules then recombine into the *E. coli* chromosome by a crossover event with specific bases in the chromosome.

Phage λ is by now very well known genetically. Not only have virtually all its approximately 61 genes been mapped (Figure 2-12), but also the complete sequence of the 48,513 base pairs along its chromosomes has just been determined through Frederick Sanger's application of his powerful new method for sequencing DNA (page 63).

Abnormal Transducing Phages Provide Unique Segments of Bacterial Chromosomes

The chromosomes of *E. coli*—and perhaps of all bacteria—are circular, with bidirectional DNA replication always initiated at a specific site. They are much too long to visualize in their entirety in the electron microscope, and without genetic tricks there would be no method to select any specific section to study. Even the most careful isolation procedures necessarily shear bacterial chromosomes into tens of pieces with no two fragments having the same ends. Luckily, by 1965 some high-powered bacterial genetics changed this bleak picture. Careful genetic examination of certain phages revealed that a small percentage genetically recombined their DNA with that of their host bacterial cells, to yield abnormal phages in which fragments of bacterial DNA were inserted in the phage chromosomes. The phages carrying these hybrid chromosomes are called trans-

ducing phages, and they can program still other strains of bacteria to manufacture proteins they normally cannot make. With genetic tricks, specific transducing phages carrying from one to several desired bacterial genes can be isolated. As we shall relate further on (page 82), most incisive use has been made of a defective phage that carries the the *E. coli* gene involved in the breakdown of the sugar lactose to the simpler sugars glucose and galactose.

Plasmids Are Autonomously Replicating Minichromosomes

In addition to their main chromosomes (with 4 million base pairs), many bacteria possess large numbers of tiny circular DNA molecules that may contain only several thousand base pairs. These minichromosomes, called plasmids, were first noticed as genetic elements that were not linked to the main chromosome and that carried genes that conveyed resistance to antibiotics such as the tetracyclines or kanamycin (Figure 2-13). That these genes were found on plasmids as opposed to main-chromosomal DNA was not a matter of chance. Antibiotic resistance requires relatively large amounts of the enzymes that chemically neutralize the antibiotics. By being located on plasmids, their respective genes are present in much higher copy

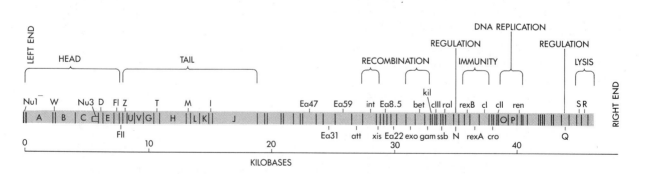

Figure 2-12
The genetic, physical, and functional structure of bacteriophage λ. The λ map coordinate system is shown on the middle line. Genes and an indication of their functions are shown above the line (Daniels, Sanger, and Coulson, 1983).

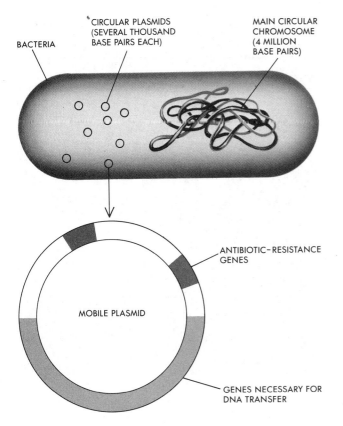

Figure 2-13
Plasmids are small, autonomously replicating
bacterial DNA molecules containing genes
conveying resistance to specific antibiotics.
Most plasmids are mobile, but nonmobilizable
plasmids also exist.

numbers than they would be if they were located
on the main chromosome.

Certain plasmids called episomes have the
ability to move on and off the main chromosomal
elements. How episomes jump on or off chromosomes was long a mystery. Now it is clear that this
capacity often reflects the possession of mobile genetic elements whose movements are accomplished through the fusion of two independently
replicating DNA units (replicons) (Chapter 11).
Plasmids that can integrate into the bacterial chromosome can be transferred from one bacterium to
another when the cells mate and a copy of the
"male" chromosome is transferred to the "female" cell. Some plasmids, however, are unable to
integrate into the bacterial chromosome, so they

cannot be transferred from cell to cell during mating. They are called nonmobilizable, and once a
gene is on such a plasmid it cannot easily move.

Because plasmid DNA is so much smaller
than even highly fragmented chromosomal DNA,
it is easily separable, and highly purified plasmid
DNA is readily obtained. In the laboratory, when
plasmid DNA is added to plasmid-free bacteria in
the presence of Ca^{++}, the DNA is taken up to
yield bacteria that will soon contain many copies of
the plasmid. In general, for reasons as yet unclear,
a given bacterial cell usually harbors only one form
of plasmid.

The number of copies of a plasmid in a host
cell depends upon the genetic constitution of the
plasmid and cell. So-called "relaxed-control" plasmids may multiply until each cell has on the average 10 to 200 copies of the plasmid. In contrast,
"stringent-control" plasmids replicate at about the
same rate as the cell's main chromosome and are
present in only one or a few copies per cell. The
relaxed plasmids are the ones used for recombinant DNA research, as we shall describe later.

Circular DNA Molecules May Be Supercoiled

As long as a DNA molecule has a linear form, its
conformation is not closely controlled by the exact
rotation of successive nucleotides around the helical axis. The original x-ray photographs of DNA
suggested an approximate 36° rotation between
successive nucleotides within the "B" form of
DNA—the form that exists in highly hydrated
fibers and that is thus presumed to be the form that
DNA takes up in solution. At that time (1953) it
was considered largely irrelevant whether DNA's
rotation angle was, say, 34.5° as opposed to 36°.
But it was later discovered that the two ends of a
linear DNA may become covalently bound to
each other, and the rotation angle became a much
more crucial parameter. With circular DNA, any
appreciable change in angle of twist between successive nucleotides leads the molecules to adopt
supercoiled configurations. In negative (positive)
supercoiling, the number of negative (positive)
supercoils exactly equals the number of added
(subtracted) twists of the double helix brought

about by an increased (decreased) rotation angle. Negative supercoiling can lead to the creation (only in AT-rich regions) of small sections of non-double-helical DNA in which the individual chains lie apart and so become susceptible to cleavage by specific DNA-cutting enzymes that attack only single-stranded as opposed to double-helical DNA (Figure 2-14).

Supercoiled DNA (whether negatively or positively twisted) adopts a more compact configuration than its "relaxed" equivalents. It thus moves faster than "relaxed" DNA when sedimented in an ultracentrifuge or subjected to gel electrophoresis (page 60) by an electrical field (Figure 2-15). At first, supercoiling was regarded as an almost accidental feature of the conditions used to study DNA in the laboratory. With time, however, it has become clear that in cells the de-

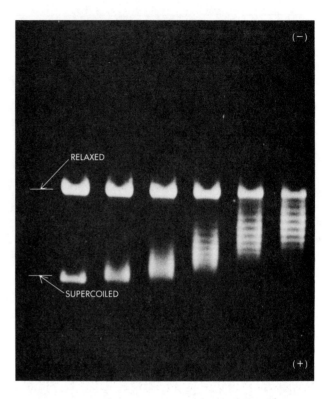

Figure 2-15
Agarose gel electrophoresis can separate DNA molecules with different amounts of supercoiling. Completely supercoiled DNA has the greatest mobility in agarose gels; molecules with progressively fewer superhelical turns migrate progressively more slowly. (Courtesy of James C. Wang.)

gree of supercoiling is strongly controlled by specific enzymes (topoisomerases and gyrases) that are capable of either adding or subtracting supercoiled twists in DNA. Now it is suspected that most circular DNA molecules (such as plasmid DNA) within cells are negatively supercoiled, with the resulting sections of separation between strands conceivably serving as control regions affecting the way DNA functions.

Most Double Helices Are Right-Handed, but Under Special Conditions Certain DNA Nucleotide Sequences Lead to Left-Handed Helices

The original model of the double helix was right-handed (the chains turn to the right as they move upward) as opposed to left-handed. Right-handed helices seemed more likely because their stereochemical configurations appeared more stable

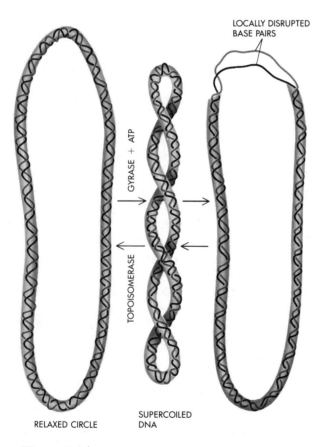

Figure 2-14
A relaxed circular DNA molecule can be twisted into a negatively supercoiled molecule by the action of DNA gyrase. The reverse reaction is catalyzed by topoisomerase ("nicking-closing" enzyme). The strain in the negatively supercoiled form can be relieved by local disruption of the double helix to produce single-stranded regions.

than those of DNA chains twisted to the left. Rigorous proof that most double helices are in fact right-handed had to await the existence of truly crystalline, short double-helical DNA segments made from chemically synthesized oligonucleotides. Such structures can be rigorously solved by x-ray diffraction techniques. Chemically synthesized DNA did not effectively become available until the late 1970s (page 63), and it was only in 1979 that rigorous proof was obtained that the solution form (the B form) of most DNA molecules is right-handed. Almost ironically, the first chemically synthesized double helix to be solved was left-handed. Its component GCGCGC chains formed left-handed double helices (Z-form DNA) in which successive GC units adopted a configuration very different from that adopted by successive nucleotides in B-form

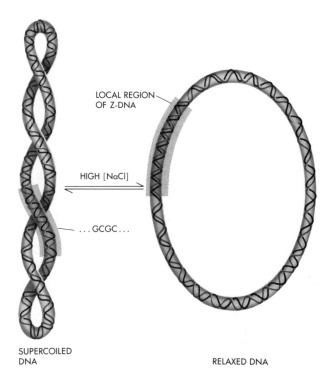

Figure 2-17
Local regions of left-handed (Z-) DNA can relieve the strain of a supertwisted DNA molecule in the same way that local single-stranded regions can.

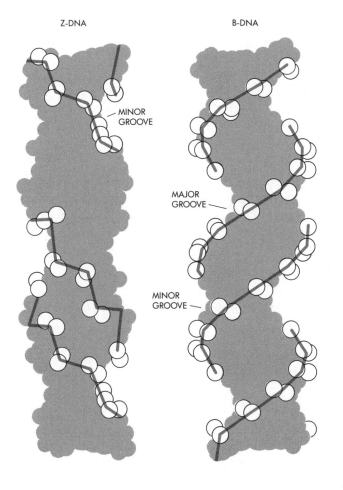

Figure 2-16
Left-handed (Z-) DNA forms a zigzag helix.

DNA (Figure 2-16). Soon afterwards, the crystal structure of several double helices formed by oligonucleotides containing all four bases revealed B-form right-handed DNA.

Double helices composed of . . . GCGC . . . sequences twist in the left-handed direction only in high concentrations of salt; in low salt concentrations, they exist in the right-handed B form. Interestingly, the methylation of C residues in . . . GCGC . . . chains stabilizes the left-handed form, permitting its existence under salt conditions similar to those *in vivo*. Whether significant amounts of biologically relevant left-handed DNA will be found in cells remains to be seen. One possible biological role for left-handed DNA might be based on the fact that small regions of Z-DNA relieve the superhelical strain in a negatively supercoiled DNA molecule, in the same way that small regions of single-stranded DNA can. These local regions of Z-DNA may bind to regulatory proteins that are specific for Z-DNA (Figure 2-17).

READING LIST

Books

Broda, P. *Plasmids.* W. H. Freeman and Company, San Francisco, 1979.

Cairns, J., G. S. Stent, and J. D. Watson, eds. *Phage and the Origins of Molecular Biology.* Cold Spring Harbor Laboratory, Cold Spring Harbor, N.Y., 1966.

Campbell, A. *Episomes.* Harper and Row, New York, 1963.

Hendrix, R., J. Roberts, F. W. Stahl, and R. Weisberg, eds. *Lambda II.* Cold Spring Harbor Laboratory, Cold Spring Harbor, N.Y., 1983.

Hershey, A. D., ed. *The Bacteriophage Lambda.* Cold Spring Harbor Laboratory, Cold Spring Harbor, N.Y., 1971.

Judson, H. F. *The Eighth Day of Creation.* Simon and Schuster, New York, 1979.

Kornberg, A. *DNA Replication.* W. H. Freeman and Company, San Francisco, 1980.

Olby, R. *The Path to the Double Helix.* University of Washington Press, Seattle, 1974.

Portugal, F. H., and J. S. Cohen. *A Century of DNA: A History of the Discovery of the Structure and Function of the Genetic Substance.* MIT Press, Cambridge, Mass., 1977. (Paperback edition, 1980.)

Stent, G. *Molecular Biology of Bacterial Viruses.* W. H. Freeman and Company, San Francisco, 1963.

Stent, G. S., and R. Calendar. *Molecular Genetics: An Introductory Narrative,* 2nd ed. W. H. Freeman and Company, San Francisco, 1978.

Watson, J. D. *The Double Helix.* Atheneum, New York, 1968. (Text and trade paperback editions.) New American Library, New York, 1969. (Paperback.)

Watson, J. D. *The Double Helix: A Norton Critical Edition.* G. S. Stent, ed. Norton, New York, 1980.

Original Research Papers (Reviews)

DNA AS THE PRIMARY GENETIC MATERIAL

Avery, O. T., C. M. MacLeod, and M. MacCarty. "Studies on the chemical nature of the substance inducing transformation of pneumococcal types." *J. Exp. Med.,* 79: 137–158 (1944).

Hershey, A. D., and M. Chase. "Independent functions of viral protein and nucleic acid in growth of bacteriophage." *J. Gen. Physiol.,* 36: 39–56 (1952).

POLYNUCLEOTIDE CHEMISTRY AND ELECTRONMICROSCOPY

Williams, R. C. "Electronmicroscopy of sodium desoxyribonucleate by use of a new freeze-drying method." *Biochim. Biophys. Acta,* 9: 237–239 (1952).

Brown, D. M., and A. R. Todd. "Nucleotides, part X. Some observations on structure and chemical behaviour of the nucleic acids." *J. Chem. Soc.,* pt. 1: 52–58 (1952).

Dekker, C. A., A. M. Michaelson, and A. R. Todd. "Nucleotides, part XIX. Pyrimidine deoxyribonucleoside diphosphates." *J. Chem. Soc.,* pt. 1: 947–951 (1953).

Chargaff, E. "Structure and function of nucleic acids as cell constituents." *Fed. Proc.,* 10: 654–659 (1951).

Wyatt, G. R., and S. S. Cohen. "The bases of the nucleic acids of some bacterial and animal viruses: The occurrence of 5-hydroxymethylcytosine." *Biochem. J.,* 55: 774–782 (1953).

THE DOUBLE HELIX

Watson, J. D., and F. H. C. Crick. "Molecular structure of nucleic acids: A structure for deoxyribose nucleic acid." *Nature,* 171: 737–738 (1953)

———. "Genetical implications of the structure of deoxyribonucleic acid." *Nature,* 171: 964–967 (1953).

———. "The structure of DNA." *Cold Spring Harbor Symp. Quant. Biol.,* 18: 123–131 (1953).

Crick, F. H. C., and J. D. Watson. "The complementary structure of deoxyribonucleic acid." *Proc. Roy. Soc., A,* 223: 80–96 (1954).

Franklin, R. E., and R. G. Gosling. "Molecular configuration in sodium thymonucleate." *Nature,* 171: 740–741 (1953).

Wilkins, M. H. F., A. R. Stokes, and H. R. Wilson. "Molecular structure of deoxypentose nucleic acids." *Nature,* 171: 738–740 (1953).

REPLICATION OF DNA

Pauling, L., and M. Delbrück. "The nature of the intermolecular forces operative in biological processes." *Science,* 92: 77–79 (1940).

Meselson, M., and F. W. Stahl. "The replication of DNA in *Escherichia coli.*" *Proc. Natl. Acad. Sci. USA,* 44: 671–682 (1958).

DNA DENATURATION AND RENATURATION

Marmur, J., and L. Lane. "Strand separation and specific recombination in deoxyribonucleic acids: Biological studies." *Proc. Natl. Acad. Sci. USA,* 46: 453–461 (1960).

Schildkraut, C. L., J. Marmur, and P. Doty. "The formation of hybrid DNA molecules, and their use in studies of DNA homologies." *J. Mol. Biol.,* 3: 595–617 (1961).

Doty, P., J. Marmur, J. Eigner, and C. Schildkraut. "Strand separation and specific recombination in deoxyribonucleic acids: Physical chemical studies." *Proc. Natl. Acad. Sci. USA,* 46: 461–476 (1960).

Hall, B. D., and S. Spiegelman. "Sequence complementarity of T2-DNA and T2-specific RNA." *Proc. Natl. Acad. Sci. USA,* 47: 137–146 (1961).

5-METHYLCYTOSINE AND GENE REGULATION

Wyatt, G. R. "The purine and pyrimidine composition of deoxypentose nucleic acids." *Biochem. J.,* 48: 584–590 (1951).

Doskočil, J., and F. Šorm. "Distribution of 5-methylcytosine in pyrimidine sequences of deoxyribonucleic acids." *Biochim. Biophys. Acta,* 55: 953–959 (1962).

Meselson, M., R. Yuan, and J. Heywood. "Restriction and modification of DNA." *Ann. Rev. Biochem.,* 41: 447–466 (1972).

Riggs, A. D. "X inactivation, differentiation, and DNA methylation." *Cytogenet. Cell. Genet.,* 14: 9–25 (1975).

Sager, R., and R. Kitchin. "Selective silencing of eukaryotic DNA." *Science,* 189: 426–433 (1975).

Holliday, R., and J. E. Pugh. "DNA modification mechanisms and gene activity during development." *Science,* 187: 226–232 (1975).

Razin, A., and A. D. Riggs. "DNA methylation and gene function." *Science,* 210: 604–610 (1980).

Gruenbaum, Y., H. Cedar, and A. Razin. "Substrate and sequence specificity of eukaryotic DNA methylase." *Nature,* 295: 620–622 (1982).

CHROMOSOMAL AND PLASMID DNA

Cairns, J. "The bacterial chromosome and its manner of replication as seen by autoradiography." *J. Mol. Biol.,* 4: 407–409 (1963).

Marmur, J., R. Rownd, S. Falkow, L. S. Baron, C. Schildkraut, and P. Doty. "The nature of intergeneric episomal infection." *Proc. Natl. Acad. Sci. USA,* 47: 972–979 (1961).

Watanabe, T. "Infectious heredity of multiple drug resistance in bacteria." *Bact. Rev.,* 27: 87–115 (1963).

Cohen, S. N., and C. A. Miller. "Multiple molecular species of circular R-factor DNA isolated from *Escherichia coli.*" *Nature,* 224: 1273–1277 (1969).

LYTIC, LYSOGENIC, AND TRANSDUCING BACTERIOPHAGES

Lwoff, A. "Lysogeny." *Bact. Rev.,* 17: 269–337 (1953).

Matsushiro, A. "Specialized transduction of tryptophan markers in *Escherichia coli* K12 by bacteriophage φ 80." *Virology,* 19: 475–482 (1963).

Daniels, D., J. L. Schroeder, F. R. Blattner, W. Szybalski, F. Sanger, A. R. Coulson, G. F. Hong, D. F. Hill, and G. B. Petersen. "Complete annotated lambda sequence, Appendix 2." In R. W. Hendrix, J. W. Roberts, F. W. Stahl, and R. A. Weisberg, eds., *Lambda II.* Cold Spring Harbor Laboratory, Cold Spring Harbor, N.Y., 1983 (in press).

Daniels, D., F. Sanger, and A. R. Coulson. "Features of bacteriophage lambda: Analysis of the complete nucleotide sequence." *Cold Spring Harbor Symp. Quant. Biol.,* 47: 1009–1024 (1983).

Dunn, J. J., and F. W. Studier. "The complete nucleotide sequence of bacteriophage T7 DNA, and the locations of T7 genetic elements." *J. Mol. Biol.* (1983) (in press).

SUPERCOILING

Vinograd, J., J. Lebowitz, R. Radloff, R. Watson, and P. Laipis. "The twisted circular form of polyoma DNA." *Proc. Natl. Acad. Sci. USA,* 53: 1104–1111 (1965).

Wang, J. C. "Interaction between DNA and an *Escherichia coli* protein W." *J. Mol. Biol.,* 55: 523–533 (1971).

Champoux, J. J., and R. Dulbecco. "An activity from mammalian cells that untwists superhelical DNA—A possible swivel for DNA replication." *Proc. Natl. Acad. Sci. USA,* 69: 143–146 (1972).

Gellert, M., K. Mizuuchi, M. H. O'Dea, and H. A. Nash. "DNA gyrase: An enzyme that introduces superhelical turns into DNA." *Proc. Natl. Acad. Sci. USA,* 73: 3872–3876 (1976).

Cozzarelli, N. R. "DNA gyrase and the supercoiling of DNA." *Science,* 207: 953–960 (1980).

Gellert, M., R. Menzel, K. Mizuuchi, M. H. O'Dea, and D. I. Friedman. "Regulation of DNA supercoiling in *Escherichia coli.*" *Cold Spring Harbor Symp. Quant. Biol.,* 47: 901–905 (1983).

Wang, J. C., L. J. Peck, and K. Becherer. "DNA supercoiling and its effects on DNA structure and function." *Cold Spring Harbor Symp. Quant. Biol.,* 47: 251–257 (1983).

LEFT-HANDED DNA

Pohl, F. M., and T. M. Jovin. "Salt-induced co-operative conformation change of a synthetic DNA: Equi-

librium and kinetic studies with poly(dG-dC)." *J. Mol. Biol.,* 67: 375–396 (1972).

Wang, A. H.-J., G. J. Quibley, F. J. Kolpak, J. L. Crawford, J. H. van Boom, G. van der Marel, and A. Rich. "Molecular structure of a left-handed DNA fragment at atomic resolution." *Nature,* 282: 680–686 (1979).

Wing, R. M., H. R. Drew, T. Takano, C. Broka, S. Tanaka, K. Itakura, and R. E. Dickerson. "Crystal structure analysis of a complete turn of B-DNA." *Nature,* 287: 755–758 (1980).

Dickerson, R. E., H. R. Drew, B. N. Conner, R. M. Wing, A. V. Fratini, and M. L. Kopka. "The anatomy of A-, B-, and Z-DNA." *Science,* 216: 475–485 (1982).

Peck, L. J., A. Nordheim, A. Rich, and J. C. Wang. "Flipping of cloned d(pCpG)n · d(pCpG)n DNA sequences from a right- to a left-handed helical structure by salt, Co(III), or negative supercoiling." *Proc. Natl. Acad. Sci. USA,* 79: 4560–4564 (1982).

Behe, M., and G. Felsenfeld. "Effects of methylation on a synthetic polynucleotide: The B-Z transition in poly(dG-m^{5}dC) · poly(dG-m^{5}dC)." *Proc. Natl. Acad. Sci. USA,* 78: 1619–1623 (1981).

3

Elucidation of the Genetic Code

Once the double helix was identified, it became possible to speculate more precisely on the one-gene-one-protein relationship. First of all, the genetic information in DNA had to be conveyed solely by the linear sequences of the four bases (A, T, G, and C) in the DNA alphabet. Therefore gene mutations had to represent changes in the sequence of bases, either through the substitution of one base pair for another, or through addition or deletion of one or many base pairs (Figure 3-1). Mutant proteins in turn clearly represented changes in the amino acid sequence, the simplest mutants being proteins in which one amino acid was replaced by another.

Identifying the Amino Acid Replacement in a Mutant Hemoglobin Molecule

The first experiments showing mutant proteins bearing single amino acid replacements focused on the sickle hemoglobin molecules produced in humans suffering from the genetic disease sickle-cell anemia. Working in Cambridge, England, in 1957, Vernon Ingram analyzed both the α and the β chains that make up the $\alpha_2\beta_2$ form of adult hemoglobin molecules. No changes were found in the sickle α chains, but each sickle β chain differed from the normal wild-type hemoglobin β chain through a specific amino acid substitution (glutamic acid → valine) that had occurred at a unique site (position 6) on the β chain. This discovery hinted that many mutations might represent single base-pair changes, as opposed to more drastic alterations in the base sequence. Probing deeper into the gene-protein relationship in sickle-cell anemia was not, however, possible at that time, because there was no conceivable way to isolate the DNA coding for the respective hemoglobin chain.

Development of Fine-Structure Genetics

The great speed at which the basic facts of molecular genetics emerged following the discovery of the double helix was possible only because of a

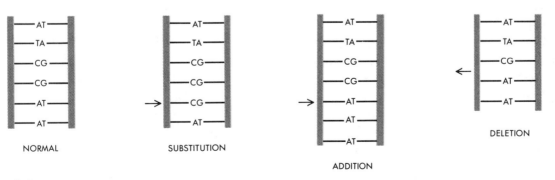

Figure 3-1
The three mechanisms of mutation: substitution, addition, and deletion of a base pair in a DNA strand.

prior decision taken collectively in the mid-1940s by a small group of younger scientists interested in the nature of the gene. This decision was to focus, whenever possible, on genetic experiments with the bacterium *Escherichia coli (E. coli)* and its various phages. Though some *E. coli* strains cause disease, other totally nonpathogenic strains (for example, B and K12) had become well adapted for laboratory use. They grew in culture extremely well, dividing as frequently as once every 20 minutes. They thus provided almost perfect systems for studying the organization of genes in the simplest known cells, and also served as ideal vehicles for observing more closely the nature of viruses.

The first rigorous genetic experiments with *E. coli* and its phages were undertaken by Max Delbrück and Salvador Luria. In the early 1940s they observed that some *E. coli* cells mutated to become resistant to specific phages, and they made accurate measurements of the spontaneous mutation rates. Soon afterwards the first phage mutants were isolated by Alfred Hershey; he showed that they recombined genetically. Yet it was not until the structure of DNA was known that the genetic structure of a single phage was thoroughly explored. The decisive experiments were done at Purdue University by Seymour Benzer, who isolated many hundreds of mutations within the *r*II gene of T4 to test whether genetic recombination (crossing over) occurs within the genes themselves. He soon found this to be the case; in fact, he discovered that *most* crossing over takes place within, rather than between, genes.

Shortly afterwards, the same conclusion was reached from genetic studies of *E. coli* itself. Successful crosses between mutant bacteria were first done at Yale University in 1946 by Joshua Lederberg and Edward Tatum. After they mixed together pairs of different *E. coli* mutants, each bearing several different mutational deficiencies, they obtained progeny bacteria that lacked the nutritional requirements of either of their respective parents. Lederberg and Tatum soon established a tentative genetic map, and the stage was set for a growing number of other geneticists to join Lederberg in exploiting the fact that very large numbers of progeny bacteria could be obtained quickly from a single genetic cross. Genetics could thus be studied much more easily with bacteria than with any higher organism. Though most of the early

research went to ordering the various genes along the single *E. coli* chromosome and establishing the existence of two different sexes, attention later turned to the structure of single bacterial genes. The situation proved to be similar to that in the *r*II gene of bacteriophage T4, with the mutations in each such bacterial gene mapping in a strictly linear order. This was to be expected if the mutable sites were the successive base pairs of the DNA molecules.

The Gene and Its Polypeptide Products Are Colinear

The gene could now be precisely defined as the collection of adjacent nucleotides that specify the amino acid sequences of the cellular polypeptide chains. Simplicity argued that the corresponding nucleotide and amino acid sequences would be colinear, and this hypothesis was soon confirmed by correlation of the relative locations of mutations in a gene with the locations of changes in its polypeptide products. The best early data were obtained at Stanford University by Charles Yanofsky, who studied mutations in the *E. coli* gene coding for tryptophan synthetase, an enzyme needed to make the amino acid tryptophan. He demonstrated very convincingly that the relative position of each amino acid replacement matched the relative position of its respective mutation along the genetic map (Figure 3-2). The molecular processes underlying colinearity, however, were not at all obvious, because the 20 different amino acids far exceeded the number of different nucleotides in DNA. A one-to-one correspondence between nucleotides and amino acids could not exist. Instead, groups of nucleotides must somehow specify (code for) each amino acid.

RNA Carries Information from DNA to the Cytoplasmic Sites of Protein Synthesis

A direct template role for DNA in the ordering of amino acids in proteins was known to be impossible, because almost all DNA is located on the chromosomes in the nucleus, whereas most, if not all, cell protein synthesis occurs in the cytoplasm. The genetic information of DNA (the nucleotide

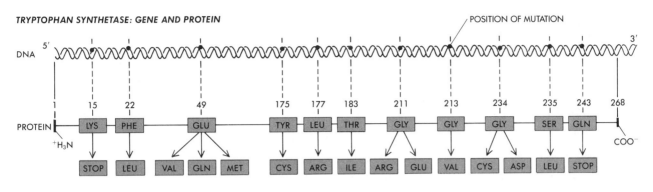

Figure 3-2
The order of mutations in a gene coding for tryptophan synthetase is the same as the order of amino acid changes in the gene's polypeptide product. The dots on the DNA strand indicate the positions of mutations; the numbers below indicate the positions of the changes in the amino acids. The amino acids that appear in these positions in normal chains are shown in the boxes below the numbers, and the amino acids resulting from the mutations appear in the bottom row of boxes. (Mutations occurring in the same position on different chains can produce different amino acid changes, as happened here at the 49th, 211th, and 234th amino acids.) Evidence of this kind showed that a gene and its polypeptide product are colinear.

sequence) thus had to be transferred to an intermediate molecule, which would then move into the cytoplasm, where it would order the amino acids. Speculation that this intermediate molecule was RNA became serious as soon as the double helix was discovered. For one thing, the cytoplasm of cells that made large numbers of proteins always contained large amounts of RNA. Even more importantly, the sugar–phosphate backbones of DNA and RNA were known to be quite similar, and it was easy to imagine the synthesis of single RNA chains upon single-stranded DNA templates to yield unstable hybrid molecules in which one strand was DNA and the other strand was RNA (Figure 3-3). Here it is important to note that the

unique base of RNA, uracil (U), is chemically very similar to thymine (T) in that it specifically base-pairs to adenine. The relationship between DNA, RNA, and protein, as conceived in 1953, was thus:

$$\text{DNA} \xrightarrow{\text{TRANSCRIPTION}} \text{RNA} \xrightarrow{\text{TRANSLATION}} \text{PROTEIN}$$

REPLICATION

where single DNA chains serve as templates for either complementary DNA molecules (in the process of DNA replication) or complementary RNA molecules (in the process of transcription). In turn, the RNA molecules serve as the templates that order the amino acids within the polypeptide

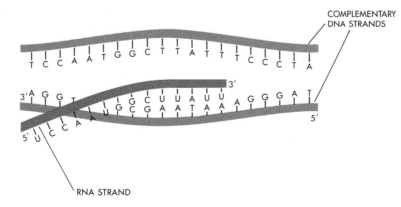
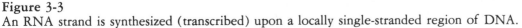

Figure 3-3
An RNA strand is synthesized (transcribed) upon a locally single-stranded region of DNA.

chains of proteins during the process of translation, so named because the nucleotide language of nucleic acids is translated into the amino acid language of proteins.

How Do Amino Acids Line Up on RNA Templates?

The groups of nucleotides that code for an amino acid are called codons. From the beginning of codon research it seemed likely that most, if not all, codons were composed of three adjacent nucleotides. Groups of two nucleotides could be arranged in only 16 different permutations ($4 \times 4 = 16$), 4 too few to code for the 20 different amino acids. Groups of three (AAA, AAC, AAU, . . .), however, could result in 64 independent permutations ($4 \times 4 \times 4 = 64$), many more than are logically needed to specify all the amino acids. So there was speculation about whether the codons might overlap in such a way that given bases would help to specify more than one amino acid. If that were true, then there would be restrictions governing which amino acids could be linked together. The first known amino acid sequences were eagerly scanned by the physicist George Gamow, to see whether some amino acids never occurred next to each other. By 1957 it was clear that no such restrictions of sequence existed and that the successive codons along an RNA chain did not overlap.

Initially it was considered possible that the sequence of amino acids was determined by the manner in which specific amino acids fit into cavities along the surfaces of RNA molecules; however, no obvious complementarity was found to exist between the specific portions (the side groups) of many amino acids and the purine and pyrimidine bases of RNA. The side groups of the amino acids leucine and valine, for example, cannot form any hydrogen bonds, and unless they were somehow modified they would not be expected to be attracted to any RNA template. This dilemma led Francis Crick to propose by early 1955 that many, if not all, amino acids had to first be attached to some form of adapter molecules before they could chemically bind to an RNA template. Testing the adapter hypothesis, however, was impossible until techniques were developed for dissecting biochemically the process by which amino acids become incorporated into growing polypeptides.

Roles of Enzymes and Templates in the Synthesis of Nucleic Acids and Proteins

The initial reaction of geneticists to the double helix was pure delight at seeing how it could function as a template. In a real sense their 50-year quest was over. In contrast, the biochemists, who were then working out how enzymes participated in the synthesis of the nucleotides and amino acids, saw their role as really just beginning. They realized that making phosphodiester and peptide bonds would also require specific enzymes. The discovery of such enzymes would demand finding conditions in which, say, DNA, RNA, or protein is made in extracts of disrupted cells. The first such experiments were difficult, because although most cells contain the enzymes to make proteins and nucleic acids, cells frequently also possess active enzymes that can break these molecules down. In the first successful experiments, only very small amounts of DNA, RNA, and protein were made. For example, test-tube-made DNA could be detected only by using radioactively labeled precursors (such as ^{14}C thymine) to distinguish it from the much larger amounts of unlabeled DNA that preexisted in the cells used to make the "cell-free extracts." However, the possession of an active extract in which, say, DNA was made allowed further experimentation with extracts that had been fractionated into various components, to discover both the exact precursors and the nature of the enzyme(s) needed for the assembly process. Such experiments would hopefully also reveal the chemical identities of the templates required in the synthetic process.

By 1960, both DNA and RNA had been successfully synthesized in highly purified cell-free extracts, and the nature of their immediate precursors was firmly established. In both cases, the precursors were nucleoside triphosphates, nucleotides containing three adjacent phosphate groups. Two of the phosphate groups are split off when adjacent nucleotides are linked together, and the energy present in the broken bonds is used to make the phosphodiester links of the sugar–phos-

phate backbone. The enzymes that directly make the phosphodiester bonds are called polymerases (because of the polymeric nature of the nucleic acids). The enzymes that make DNA are called DNA polymerases, and those that make RNA are known as RNA polymerases.

DNA polymerases catalyze the formation of DNA only in the presence of preexisting DNA templates, and the newly made DNA chains contain sequences complementary to their templates. For RNA polymerase to make RNA, DNA must be present; its template role was likewise shown by finding complementarity between DNA templates and RNA products.

Proteins Are Synthesized from the N-Terminus to the C-Terminus

One end of a polypeptide has an amino acid with a free amino group, while the other end bears an amino acid with a free carboxyl group. In 1961 Howard Dintzis at Johns Hopkins University showed that polypeptide chains grow by stepwise addition of single amino acids, starting with the amino-terminal amino acid and finishing with the carboxyl-terminal amino acid. Though all proteins are synthesized beginning with the amino acid methionine, one or several N-terminal amino acids are frequently cleaved away by proteolytic (protein-degrading) enzymes to produce functional polypeptide products that have different N-terminal amino acids than that of the primary translation product.

Three Forms of RNA Are Involved in Protein Synthesis

Fractionation of active cell-free extracts that had incorporated radioactively labeled amino acids into polypeptides revealed the temporary attachment of the newly made protein to ribosomes, the semispherical 200-Å-diameter cytoplasmic particles that contain RNA. All ribosomes were found to be built from two subunits of unequal size (their molecular weights, or MWs, are 1 and 2 million, respectively). Each of these ribosomal subunits contained an amount of protein that was roughly equal to the amount of RNA (Table 3-1). Until 1960 it was generally believed that ribosomal RNA molecules were the templates that ordered

Table 3-1

RNA Molecules in *E. coli*

Type	Relative Amount (%)	Mass (kilodaltons)*
Ribosomal RNA (rRNA)	80	1.2×10^3
		0.55×10^3
		3.6×10^1
Transfer RNA (tRNA)	15	2.5×10^1
Messenger RNA (mRNA)	5	Heterogeneous

*The largest and smallest rRNAs are part of the large (60S) subunit; the medium-sized rRNA belongs to the small (40S) subunit.

the amino acids. But, though the involvement of ribosomes in protein synthesis was indisputable, no plausible hypothesis could be proposed as to how ribosomal RNA chains of two fixed sizes (MWs 0.6 and 1.2 million) could specify polypeptides that showed so much variation in size; some chains contained as few as 50 amino acids, whereas others contained up to 2000.

Then, to everyone's great surprise, it was discovered that neither of the major ribosome RNA components had a template role. Instead, the true templates represented a minor fraction—2 to 5 percent—of all cellular RNA. Because they carried the specificity of the genes to the cytoplasm, these templates were named messenger RNA (mRNA). In the cytoplasm, mRNA moves over the surface of the ribosome, bringing successive codons into position for ordering their respective amino acids (Figure 3-4, page 36). Ribosomes are thus factories that by themselves have no specificity and to which any mRNA molecule can attach.

Equally important was the discovery that prior to their incorporation into protein, the amino acids are chemically linked to small RNA molecules called transfer RNA (tRNA). The amino-acid–tRNA complexes line up next to the codons of mRNA, with the actual recognition and binding being mediated by tRNA components. No contacts exist between the individual amino acids and the mRNA codons. Molecules of tRNA were thus the adapters whose existence had been predicted several years before by Francis Crick.

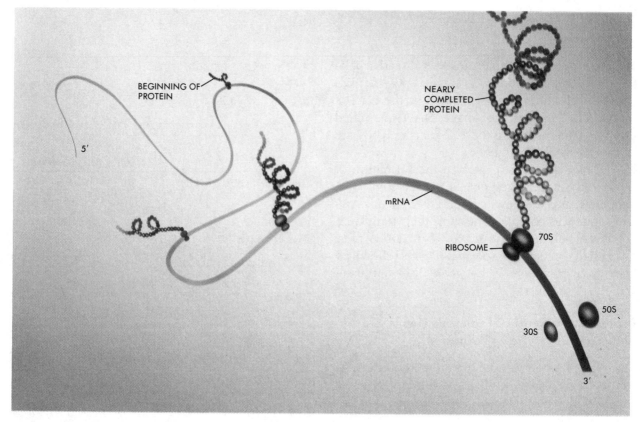

Figure 3-4
Messenger RNA carries genetic information from the DNA to the ribosomes, where it is translated into protein. The polypeptide chains are elongated as mRNA molecules move across the surfaces of the ribosomes, with the 5′ ends being translated first.

For each of the 20 different amino acids, a specific enzyme catalyzes its linkage to the 3′ end of its specific tRNA molecule. The binding of tRNA to mRNA is mediated by sets of internal nucleotides that have sequences complementary to their respective codons on the mRNA. These tRNA sequences are called anticodons (Figures 3-5 and 3-6).

Genetic Evidence That Codons Contain Three Bases

An exhaustive genetic study of a large number of phage T4 mutants that contained additions or deletions of single base pairs led Sydney Brenner and Francis Crick in 1961 to the major statement that each codon contains three bases. In Cambridge, England, they made genetic crosses to show that the addition or subtraction of either one or two base pairs invariably led to highly abnormal nonfunctional proteins. In contrast, if three base pairs were either added or subtracted, the resulting proteins frequently were totally active. They con-

cluded, as we now know correctly, that the genetic code is read in stepwise groups of three base pairs. If one or two bases are added or deleted, the resulting reading frame is upset, leading to the use of a completely new collection of codons that invariably code for amino acid sequences that make no functional sense. In contrast, when groups of three base pairs are inserted or deleted, the resulting protein, now containing one more or one less amino acid, remains otherwise unchanged and often retains full biological activity.

RNA Chains Are Both Synthesized and Translated in a 5′-to-3′ Direction

Every nucleic acid chain has a direction defined by the orientation of its sugar–phosphate backbone. The end terminating with the 5′ carbon atom is called the 5′ end, while the end terminating with the 3′ carbon atom is called the 3′ end. All RNA chains, as well as DNA chains, grow in the 5′-to-3′ direction. Translation occurs in the same direction, and as we show later (Chapter 4), in bacteria

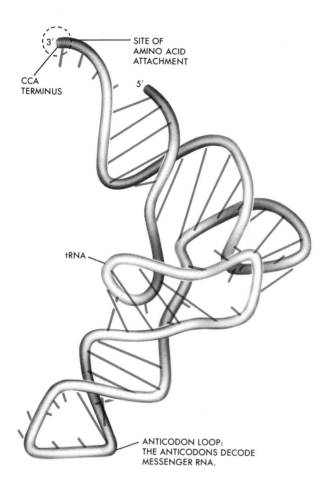

Figure 3-5
The structure of a transfer RNA molecule. Base pairing within the single-stranded molecule gives it its distinctive shape. The anticodon loop is the portion that decodes messenger RNA. An amino acid attaches to the CCA bases at the 3' end of the chain.

RNA chains can begin to be translated long before their synthesis is complete. The end of a gene at which transcription begins is called the 5' end, reflecting the 5'-to-3' order of its mRNA product. Correspondingly, the end of the gene at which transcription ceases is called the 3' end.

Synthetic mRNA Is Used to Make the Codon Assignments

The realization that ribosomes are by themselves unspecific and only become programmed to make specific RNA by binding mRNA molecules led Marshall Nirenberg and Heinrich Matthaei in 1961 to do an historic experiment. They used as mRNA an enzymatically made regular polynu-

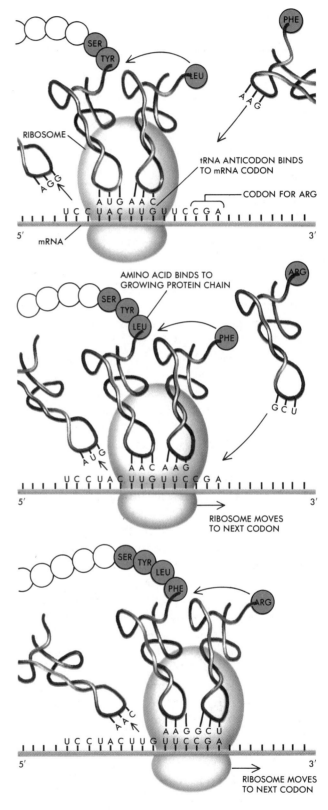

Figure 3-6
At the ribosome, the codons of a messenger RNA molecule base-pair with the anticodons of transfer RNAs, which are charged with amino acids.

cleotide poly U (UUUUUU . . .). When poly U was added to cell extracts containing ribosome molecules depleted of normal mRNA, they observed that only polyphenylalanine was synthesized. UUU thus coded for the amino acid phenylalanine. Soon poly A (AAAAAA . . .) was found to code for strings of lysine residues, while poly C (CCCCCC . . .) yielded polypeptides containing only proline. Over the next several years synthetic polynucleotides containing random mixtures of two or more nucleotides were used to tentatively decode many other codons.

The Genetic Code Is Fully Deciphered by June 1966

Most of the remaining still unidentified codons were established when H. Gobind Khorana found ways to make codons by using repeating copolymers (for example, GUGUGU . . . , AAGAAG . . . , and GUUGUU . . .). By 1966 the search for the genetic code was over, and it had been unambiguously shown that (1) all codons contain three successive nucleotides, (2) many amino acids are specified by more than one codon (the so-called degeneracy of the code), and (3) 61 of the 64 possible combinations of the three bases are used to code for specific amino acids (Table 3-2). The three combinations that do not specify any amino acid (UAA, UAG, UGA) were all found to code for stop signals that indicate chain termination.

The finding of stop codons at first created the expectation that specific start codons might also exist, especially since it was becoming more and more certain that all proteins begin with the amino acid methionine. But there is only one methionine codon (AUG), and it codes for internally located methionine as well as initiator methionine. AUGs that are used to start polypeptides are all closely preceded by a purine-rich sequence (for example, AGGA) that may help to position the starting AUG opposite the ribosomal cavity containing the initiating amino-acid–tRNA complex.

"Wobble" Frequently Permits Single tRNA Species to Recognize Multiple Codons

Initially it seemed highly probable that anticodons bind codons by means of three A—U and/or G—C hydrogen bonds that are identical to the A—T and G—C bonds that bind the two strands of the

Table 3-2

The Genetic Code

First Position (5' end)	Second Position				Third Position (3' end)
	U	C	A	G	
U	PHE	SER	TYR	CYS	U
	PHE	SER	TYR	CYS	C
	LEU	SER	Stop	Stop	A
	LEU	SER	Stop	TRP	G
C	LEU	PRO	HIS	ARG	U
	LEU	PRO	HIS	ARG	C
	LEU	PRO	GLN	ARG	A
	LEU	PRO	GLN	ARG	G
A	ILE	THR	ASN	SER	U
	ILE	THR	ASN	SER	C
	ILE	THR	LYS	ARG	A
	MET	THR	LYS	ARG	G
G	VAL	ALA	ASP	GLY	U
	VAL	ALA	ASP	GLY	C
	VAL	ALA	GLU	GLY	A
	VAL	ALA	GLU	GLY	G

Note: Given the position of the bases in a codon, it is possible to find the corresponding amino acid. For example, the codon (5')AUG(3') on mRNA specifies methionine, whereas CAU specifies histidine. UAA, UAG, and UGA are termination signals. AUG is part of the initiation signal, and it codes for internal methionines as well.

double helix together. But soon experimental results began to show that single tRNA species can bind to, say, both UUU and UUC codons. This suggested to Francis Crick that although the first two bases in a codon always pair in a DNA-like fashion, the pairing in the third position is less restrictive, due to "wobble" in the location of the third anticodon (or codon?) base. Additional types of pairing became possible, opening up the possibility that there need not always be a distinct tRNA species for each of the 61 codons corresponding to amino acids. To date, the most complete data come from yeasts that have been shown, from a combination of genetic and DNA sequence data (Chapter 12), to contain approximately 45 tRNA species. Thus, many yeast tRNA molecules have to recognize more than one codon.

Great variation exists in the relative amounts of particular tRNA species that are present in a given cell. In part, this variation reflects differ-

ences in the abundance of the amino acids the tRNAs specify. For example, the amino acids methionine and tryptophan occur relatively rarely in most proteins, and comparatively small amounts of their respective tRNAs are present. Moreover, when more than one tRNA form exists for a given amino acid, these different tRNA forms tend not to be present in equal amounts. This suggested that the more numerous tRNAs recognize the more commonly used codons for a given amino acid. This supposition was proved correct when methods for determining the precise nucleotide sequences of genes became available (Chapter 7). The rate at which an mRNA message is translated into its corresponding polypeptide chain thus may be controlled in part by whether it contains codons that are recognized by the more commonly available tRNA forms.

How Universal Is the Genetic Code?

Virtually all the experiments used to decipher the genetic code employed ribosomes and tRNA molecules from *E. coli.* It could thus be asked whether mRNA molecules are always translated into the same amino acid sequences, independent of the source of the translation machinery. At the start, the answer was thought to be yes, for it was hard to imagine how the code could change during the course of evolution. By now the initial expectations that the genetic code for chromosomal DNA would be universal have been rigorously confirmed in a large variety of organisms, ranging from the simplest prokaryotes to the most complex eukaryotes. An interesting exception, however, occurs in the genetic code used by the DNA from mitochondria. Although for many years it was believed that DNA is located only in the nucleus, by the early 1960s it had become clear that the cytoplasmic organelles, the mitochondria and the chloroplasts, both possess their own unique DNA molecules. Now it is generally believed that mitochondria and chloroplasts represent the descendents of primitive bacterial cells that became symbiotically engulfed by primitive ancestors of the present eukaryotic organisms, and increased the host's ATP-generating capacity. For the most part, the genetic code used by mitochondria is identical to that used by nuclear DNA. However, UGA, a stop codon for nuclear DNA, is read as tryptophan in mitochondria. In addition, mitochondrial AUA is read as methionine, whereas AUA is read as isoleucine in nuclear DNA.

These differences are due to the relatively small number of different tRNAs coded by mitochondrial DNA. Only 22 different tRNA species are present in mitochondria, in contrast to the more than 40 tRNAs that are available for ordinary translation. In many cases, just the first two bases in a codon are actually read, with the base in the third position playing no role in the tRNA selection process.

Average-Sized Genes Contain at Least 1200 Base Pairs

Because all codons were found to contain three base pairs, it was obvious that the number of base pairs in a gene must be at least three times the number of amino acids in its respective polypeptide. An average-sized protein of 400 amino acids was thus thought to require a section of DNA consisting of some 1200 nucleotide pairs. Because this number was found to be much smaller than the number of base pairs in even the smallest DNA molecule, it was concluded that most DNA molecules contain many genes.

Suppressor tRNAs Cause Misreading of the Genetic Code

Prior to the discovery of the double helix, suppressor genes had already been identified. These genes somehow have the potential to nullify the effects of specific mutations in a large variety of different genes. Now it is understood that many suppressor genes act by causing occasional misreadings of the genetic code. In fact, a large number of "suppressor genes" turned out to be mutant tRNAs that have anticodons complementary to one of the stop codons, but that can be charged normally with an amino acid. For example, one very common class of mutants are the so-called "nonsense" mutants in which a codon specifying an amino acid is changed to a stop codon, thus leading to the production of a truncated and usually inactive polypeptide product. The effects of such nonsense mutation can be overcome by the introduction of a suppressor tRNA that recognizes the stop codon—say, UGA—in the mutant gene and incorporates an amino acid—for instance, tyrosine—allowing nor-

mal translation to continue. In this way a normal-length, often functionally active polypeptide is produced (Figure 3-7). Such tRNA suppressors can persist in nature only if a given tRNA species has several identical genes coding for it. There exist, for example, multiple copies of the *E. coli* gene coding for the tyrosine tRNA species containing the (3')UGA(5') anticodon. So when a tyrosine-inserting suppressor tRNA arises by a mutation that lets it read the UGA stop codon, there still remain functional tyrosine tRNAs to insert tyrosine at UAU or UAC codons.

UGA-suppressor tRNAs occasionally insert amino acids at the normal sites of chain termina-tion, leading to many oversized polypeptide chains. We can assume that such events are deleterious to their respective cells, which may explain why nonsense-suppressor strains generally grow much less well than their normal equivalents.

The Signals for Starting and Stopping the Synthesis of Specific RNA Molecules Are Encoded Within DNA Sequences

The genetic information within a DNA molecule usually serves as the template for a large number of shorter RNA molecules, most of which in turn serve as templates for the synthesis of specific poly-

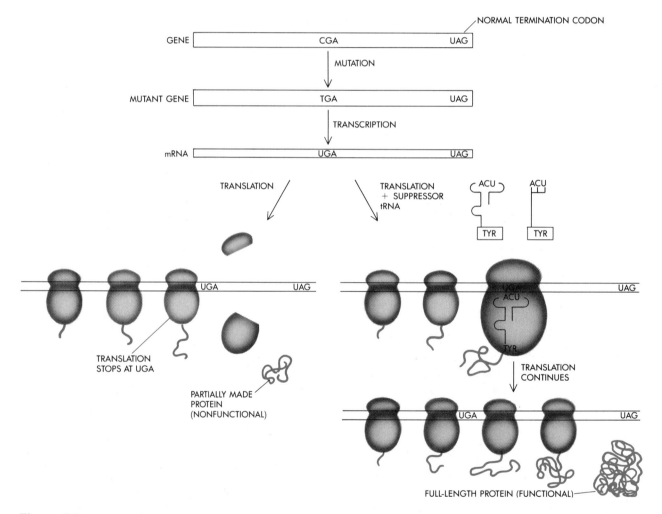

Figure 3-7
Suppressor tRNAs. A gene with a mutation that creates a translational stop codon (UGA, UAG, or UAA) is normally nonfunctional; protein synthesis will terminate prematurely and produce a partially made protein. A suppressor tRNA contains an anticodon that is complementary to one of the stop codons, but this tRNA is charged normally with an amino acid (in this case, tyrosine). *E. coli* containing such a suppressor tRNA can translate the mutant mRNA by inserting tyrosine at the mutant stop codon, and translation proceeds normally.

peptide chains. Specific nucleotide segments (often called "promoters"; page 98) are recognized by RNA polymerase molecules that start RNA synthesis (Figure 3-8a). After transcription of a functional RNA chain is finished, a second class of signals leads to the termination of RNA synthesis and the detachment of RNA polymerase molecules from their respective DNA templates. How termination is brought about is only now being worked out. RNA chains tend to end with several U residues; just before the site of termination, a nucleotide sequence capable of forming a hairpin loop is usually found (Figure 3-8b). At many termination sites, a specific RNA-chain-terminating protein (in *E. coli,* it is called "rho") plays a role.

In the transcription-initiation step, the two chains of the double helix come apart, with only one of two strands at any start site being copied into its RNA complement. The hybrid DNA–RNA double-helical sections generated as transcription proceeds have only a fleeting existence, with the newly made RNA segments quickly peeling off from the transcription complexes, allowing the just-transcribed sections to reassume their native double-helical states. In bacteria, which have no nucleus, the 5′ ends of nascent mRNA chains attach to ribosomes long before their respective chains have been completely synthesized, and as will be seen (page 48), the processes of transcription and translation can be closely coupled.

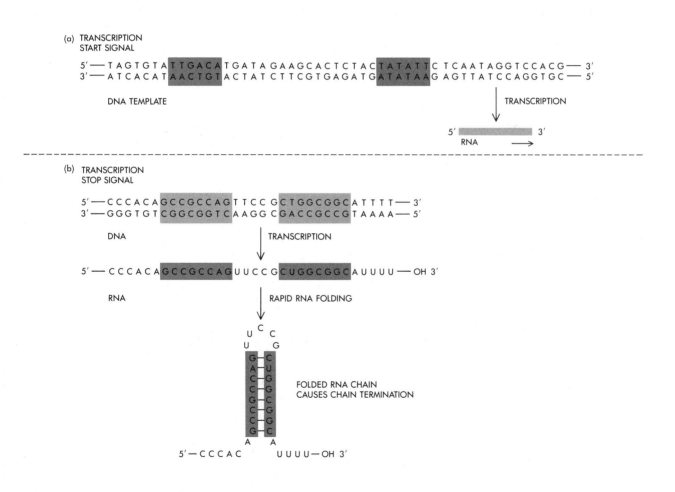

Figure 3-8
(a) RNA polymerase recognizes two blocks of sequences, one at approximately −35 and another at about −10 from the start of transcription. (b) The RNA polymerase stops after transcribing a run of U residues that follows a palindrome. The newly transcribed RNA forms a stem-loop structure that probably constitutes a termination signal for RNA polymerase.

Increasingly Accurate Systems Are Developed for the *in Vitro* Translation of Exogenously Added mRNAs

It is now more than 20 years since the first exogenously added mRNA molecules were successfully used to program the translation of functional protein molecules. In that time, increasingly reliable systems have been developed for studying the *in vitro* translation of mRNA molecules. At first only polypeptide chains of modest size (from 150 to 300 amino acids) could be made, because the early cell-free systems all contained significant amounts of contaminating ribonucleases that degraded the added mRNA templates. Since then, particularly with the development of virtually nuclease-free extracts from reticulocytes (the hemoglobin-synthesizing precursors of red blood cells), it has become possible to use mRNA preparations from many cells to routinely program the synthesis of hundreds of different proteins, including the larg-

est of the cellular polypeptides (for example, the heavy chain of myosin, which contains approximately 2000 amino acids). The proteins synthesized can be analyzed by two-dimensional gel electrophoresis (Figure 3-9). Today the presence of a given specific functional mRNA molecule can be routinely tested for by determining whether its polypeptide product is made in a cell-free protein-synthesizing extract.

Our ability to so easily translate mRNA preparations into their respective polypeptides does not mean in any sense that we understand in detail at the molecular level how proteins are made. The exact functions of the more than 60 different proteins used in constructing ribosomes remain almost totally unknown. Likewise, the functions of the two ribosomal RNA (rRNA) components of each ribosome remain as unclear as they were in 1960, when these components were found to be structural molecules as opposed to genetic information-carrying molecules.

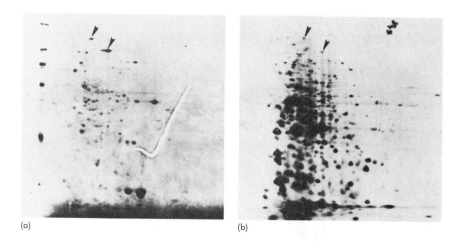

(a)   (b)

Figure 3-9
Two-dimensional gel electrophoresis of labeled proteins produced by cell-free translation of smooth-muscle mRNA. In this example, the gel was treated with the protein-staining dye Coomassie blue, photographed, and processed for fluorography, a procedure that accelerates the detection of radiolabeled proteins. (a) The stained gel. (b) The fluorogram. (Courtesy of James R. Feramisco, David M. Helfman, and G. Paul Thomas.)

READING LIST

Books

Chambliss, G., G. R. Craven, J. Davies, K. Davis, L. Kahan, and M. Nomura, eds. *Ribosomes: Structure, Function, and Genetics.* University Park Press, Baltimore, 1980.

Losick, R., and M. Chamberlin, eds. *RNA Polymerase.* Cold Spring Harbor Laboratory, Cold Spring Harbor, N.Y., 1976.

Nomura, M., A. Tissières, and P. Lengyel, eds. *Ribosomes.* Cold Spring Harbor Laboratory, Cold Spring Harbor, N.Y., 1974.

The Genetic Code. Cold Spring Harbor Symp. Quant. Biol., vol. 31. Cold Spring Harbor Laboratory, Cold Spring Harbor, N.Y., 1967.

Original Research Papers (Reviews)

FINE-STRUCTURE GENETICS

Luria, S. E., and M. Delbrück. "Mutations of bacteria from virus sensitivity to virus resistance." *Genetics,* 28: 491–511 (1943).

Hershey, A. D. "Spontaneous mutations in bacteria viruses." *Cold Spring Harbor Symp. Quant. Biol.,* 11: 67–77 (1947).

Lederberg, J., and E. L. Tatum. "Novel genotypes in mixed cultures of biochemical mutants of bacteria." *Cold Spring Harbor Symp. Quant. Biol.,* 11: 113–114 (1947).

Benzer, S. "Fine structure of a genetic region in bacteriophage." *Proc. Natl. Acad. Sci. USA,* 41: 344–354 (1955).

Ingram, V. M. "Gene mutations in human hemoglobin: The chemical difference between normal and sickle cell hemoglobin." *Nature,* 180: 326–328 (1957).

Benzer, S. "On the topology of the genetic fine structure." *Proc. Nat. Acad. Sci. USA,* 45: 1607–1620 (1959).

Yanofsky, C., B. C. Carlton, J. R. Guest, D. R. Helinski, and U. Henning. "On the colinearity of gene structure and protein structure." *Proc. Natl. Acad. Sci. USA,* 51: 266–272 (1964).

DISCOVERY OF DNA-DEPENDENT RNA SYNTHESIS

Hurwitz, J., A. Bresler, and R. Diringer. "The enzymatic incorporation of ribonucleotides into polyribonucleotides and the effect of DNA." *Biochem. Biophys. Res. Comm.,* 3: 15–19 (1960).

Weiss, S. B. "Enzymatic incorporation of ribonucleotide triphosphates into the interpolynucleotide linkages of ribonucleic acid." *Proc. Natl. Acad. Sci. USA,* 46: 1020–1030 (1960).

Stevens, A. "Incorporation of the adenine ribonucleotide into RNA by cell fractions from *E. coli* B." *Biochem. Biophys. Res. Comm.,* 3: 92–96 (1960).

MECHANISMS OF PROTEIN SYNTHESIS

Zamecnik, P. C., and E. B. Keller. "Relationship between phosphate energy donors and incorporation of labeled amino acids into proteins." *J. Biol. Chem.,* 209: 337–354 (1954).

Hoagland, M. B., E. B. Keller, and P. C. Zamecnik. "Enzymatic carboxyl activation of amino acids." *J. Biol. Chem.,* 218: 345–358 (1956).

Hoagland, M. B., M. L. Stephenson, J. F. Scott, L. I. Hecht, and P. C. Zamecnik. "A soluble ribonucleic acid intermediate in protein synthesis." *J. Biol. Chem.,* 231: 241–257 (1958).

Crick, F. H. C. "On protein synthesis. Biological replication of macromolecules." *Symp. Soc. Exp. Biol.,* 12: 138–163 (1958).

Dintzis, H. M. "Assembly of the peptide chain of hemoglobin." *Proc. Natl. Acad. Sci. USA,* 47: 247–261 (1961).

Jacob, F., and J. Monod. "Genetic regulatory mechanisms in the synthesis of proteins." *J. Mol. Biol.,* 3: 318–356 (1961).

Brenner, S., F. Jacob, and M. Meselson. "An unstable intermediate carrying information from genes to ribosomes for protein synthesis." *Nature,* 190: 576–581 (1961).

Gros, F., H. Hiatt, W. Gilbert, C. G. Kurland, R. W. Risebrough, and J. D. Watson. "Unstable ribonucleic acid revealed by pulse labelling of *Escherichia coli.*" *Nature,* 190: 581–585 (1961).

Robertus, J. D., J. E. Ladner, J. T. Finch, D. Rhodes, R. S. Brown, B. F. C. Clark, and A. Klug. "Structure of yeast phenylalanine tRNA at 3Å resolution." *Nature,* 250: 546–551 (1974).

Kim, S. H., F. L. Suddath, F. L. Quigley, A. McPherson, L. Sussman, A. H. J. Wang, N. C. Seeman, and A. Rich. "Three-dimensional tertiary structure of yeast phenylalanine transfer RNA." *Science,* 185: 435–440 (1974).

THE GENETIC CODE

Nirenberg, M. W., and J. H. Matthaei. "The dependence of cell-free protein synthesis in *E. coli* upon naturally occurring or synthetic polyribonucleotides." *Proc. Natl. Acad. Sci. USA,* 47: 1588–1602 (1961).

Crick, F. H. C., L. Barnett, S. Brenner, and R. J. Watts-Tobin. "General nature of the genetic code for proteins." *Nature,* 192: 1227–1232 (1961).

Leder, P., and M. W. Nirenberg. "RNA code words and protein synthesis II: Nucleotide sequence of a valine RNA code word." *Proc. Natl. Acad. Sci. USA,* 52: 420–427 (1964).

Nishimura, S., D. S. Jones, and H. G. Khorana. "The *in vitro* synthesis of a copolypeptide containing two amino acids in alternating sequence dependent upon a DNA-like polymer containing two nucleotides in alternating sequence." *J. Mol. Biol.,* 13: 302–324 (1965).

Crick, F. H. C. "Codon–anticodon pairing: The wobble hypothesis." *J. Mol. Biol.,* 19: 548–555 (1966).

Ikemura, T. "Correlation between the abundance of *Escherichia coli* transfer RNAs and the occurrence of the respective codons in its protein genes." *J. Mol. Biol.,* 146: 1–21 (1981).

Guthrie, C., and J. Abelson. "Organization and expression of tRNA genes in *Saccharomyces cerevisiae.*" In J. N. Strathern, E. W. Jones, and J. R. Broach, eds., *The Molecular Biology of the Yeast Saccharomyces: Metabolism and Gene Expression.* Cold Spring Harbor Laboratory, Cold Spring Harbor, N.Y., 1983, pp. 487–528.

MITOCHONDRIAL GENES

Anderson, S., A. T. Bankier, B. G. Barrell, M. H. L. deBruijn, A. R. Coulson, J. Drouin, I. C. Eperon, D. P. Nierlich, B. A. Roe, F. Sanger, P. H. Schreier, A. J. H. Smith, R. Staden, and I. G. Young. "Sequence and organization of the human mitochondrial genome." *Nature,* 290: 457–465 (1981).

Barrell, B. G., S. Anderson, A. T. Bankier, M. H. L. deBruijn, E. Chen, A. R. Coulson, J. Drouin, I. C. Eperon, D. P. Nierlich, B. A. Roe, F. Sanger, P. H. Schreier, A. J. H. Smith, R. Staden, and I. G. Young. "Different pattern of codon recognition by mammalian mitochondrial tRNAs." *Proc. Natl. Acad. Sci. USA,* 77: 3164–3166 (1980).

Bonitz, S. G., R. Berlani, G. Coruzzi, M. Li, G. Macino, F. G. Nobrega, M. P. Nobrega, B. E. Thalenfeld, and A. Tzagoloff. "Codon recognition rules in yeast mitochondria." *Proc. Natl. Acad. Sci. USA,* 77: 3167–3170 (1980).

SUPPRESSOR GENES AND THEIR SUPPRESSION BY MUTANT tRNAS

Capecchi, M. R., and G. Gussin. "Suppression in vitro: Identification of serine-sRNA as a 'nonsense' suppressor." *Science,* 149: 417–422 (1965).

Engelhardt, D. L., R. Webster, R. Wilhelm, and N. Zinder. "In vitro studies on the mechanism of suppression of a nonsense mutation." *Proc. Natl. Acad. Sci. USA,* 54: 1791–1797 (1965).

Ozeki, H., H. Inokuchi, F. Yamao, M. Kodaira, H. Sakano, T. Ikemura, and Y. Shimura. "Genetics of nonsense suppressor tRNAs in *Escherichia coli.*" In D.

Söll, J. N. Abelson, and P. R. Schimmel, eds., *Transfer RNA: Biological Aspects.* Cold Spring Harbor Laboratory, Cold Spring Harbor, N.Y., 1980, pp. 341–349.

RNA CHAIN INITIATION AND TERMINATION

Roberts, J. W. "Transcription termination and its control in *E. coli.*" In R. Losick and M. Chamberlin, eds., *RNA Polymerase.* Cold Spring Harbor Laboratory, Cold Spring Harbor, N.Y., 1976, pp. 247–271. (Review.)

Adhya, S., and M. Gottesman. "Control of transcription termination." *Ann. Rev. Biochem.,* 47: 967–996 (1978).

Melnikova, A. F., R. Beabealashvilli, and A. D. Mirzabekov. "A study of unwinding of DNA and shielding of the DNA grooves by RNA polymerase by using methylation with dimethylsulphate." *Eur. J. Biochem.,* 83: 301–309 (1978).

Rosenberg, M., and D. Court. "Regulatory sequences involved in the promotion and termination of RNA transcription." *Ann. Rev. Genet.,* 13: 319–353 (1979).

Farnham, P. J., and T. Platt. "A model for transcription termination suggested by studies on the *trp* attenuator in vitro using base analogs." *Cell,* 20: 739–748 (1980).

Farnham, P. J., and T. Platt. "Rho-independent termination: Dyad symmetry in DNA causes RNA polymerase to pause during transcription in vitro." *Nuc. Acids Res.,* 9: 563–577 (1981).

Birchmeier, C., R. Grosschedl, and M. L. Birnstiel. "Generation of authentic 3' termini of an H2A mRNA in vivo is dependent on a short inverted DNA repeat and on spacer sequences." *Cell,* 28: 739–745 (1982).

Gamper, H. B., and J. E. Hearst. "Size of the unwound region of DNA in *Escherichia coli* RNA polymerase and calf thymus RNA polymerase II ternary complexes." *Cold Spring Harbor Symp. Quant. Biol.,* 47: 447–453 (1983).

IN VITRO SYSTEMS FOR PROTEIN SYNTHESIS

Roberts, B. E., and B. M. Patterson. "Efficient translation of tobacco mosaic virus RNA and rabbit globin 9S RNA in a cell-free system from commercial wheat germ." *Proc. Natl. Acad. Sci. USA,* 70: 2330–2334 (1973).

Marcus, A., D. Efron, and D. P. Weeds. "The wheat embryo cell-free system." *Methods Enzymol.,* 30: 749–753 (1974).

Pelham, H. R. B., and R. J. Jackson. "An efficient mRNA-dependent translation system from reticulocytes lysates." *Eur. J. Biochem.,* 67: 247–256 (1976).

4

The Genetic Elements
That Control Gene Expression

Even before the basic outline of the genetic code became established, it was obvious that intricate molecular mechanisms must exist in cells to control the numbers of their respective proteins. Within *E. coli,* for example, the relative amounts of the different proteins vary enormously (from less than .01 percent to about 2 percent of the total) depending on their function, even though each protein product is coded by a single gene along the *E. coli* chromosome. *A priori* we can imagine two ways that the cell might achieve this differential synthesis. The first way would be by the evolution of molecular signals that control the rates at which specific mRNA molecules are transcribed off their DNA templates (transcriptional control). The second way would involve molecular devices for controlling the rate at which mRNA molecules, once synthesized, are translated into their polypeptide products (translational control). It makes sense for a cell not to make more mRNA molecules than it needs, and most initial attention by molecular biologists focused on whether transcriptional control existed. Here genetic analysis of the so-called "induced" enzymes of bacteria presented the first key insight, from which has emerged definitive understanding at the molecular level about how mRNA synthesis is regulated. Only recently has translational control been equally appreciated as a major factor in cellular existence.

Repressors Control Inducible
Enzyme Synthesis

Bacteria generally exist in environments that change rapidly, and some of the enzymes that they may need at one moment may be useless or even counterproductive soon afterwards. Similarly, they may suddenly require an enzyme whose presence was previously unnecessary. One way for bacteria to meet these challenges would be for them to carry on simultaneously the synthesis of all such enzymes, whether or not the enzymes' substrates were present. This would be wasteful, however, and in fact bacteria do not operate this way. Instead, many bacterial genes are constructed to function at highly variable rates, so that they make mRNA molecules at appreciable levels only when their genes receive signals from outside to go into action.

The compounds that transmit these signals are called inducers. For example, *E. coli* cells normally make the enzyme β-galactosidase at high rates only when its inducer, lactose, is present. (For lactose to function as a food source, it must be cleaved into the simpler sugars glucose and galactose, and β-galactosidase is the enzyme that catalyzes this splitting.) The presence of lactose greatly increases the rate at which RNA polymerase can bind to the beginning of the gene that codes for β-galactosidase and initiate the synthesis of the respective mRNA. In turn, the higher amounts of β-galactosidase mRNA lead to correspondingly more β-galactosidase.

Genetic analysis proved crucial to working out the molecular details of this adaptive phenomenon, with the essential clues emerging from the study of mutants that were unable to vary the amount of β-galactosidase. The key findings were that whether lactose was present or absent, some mutants made maximum amounts of enzyme while other mutants produced only traces. Such results led Jacques Monod and François Jacob of the

Institut Pasteur in Paris to postulate: (1) that a specific repressor molecule exists that binds near the beginning of the β-galactosidase gene at a specific site called the operator and that, by binding to the operator site on the DNA, sterically prevents RNA polymerase from commencing synthesis of β-gal mRNA; and (2) that lactose acts as an inducer which, by binding to the repressor, prevents the repressor from binding to the operator. In the presence of lactose, the repressor is inactivated and the mRNA is made. Upon removal of lactose, the repressor regains its ability to bind to the operator DNA and switch off the lactose gene (Figure 4-1).

Bacterial Genes with Related Functions Are Organized into Operons

The *E. coli* gene for β-galactosidase is located adjacent to two additional genes involved in lactose metabolism. One gene codes for lactose permease, a protein that facilitates the specific entry of lactose into bacteria, while the second codes for thiogalactosidase transacetylase, an enzyme that may help to remove lactoselike compounds that β-galactosidase cannot split into useful metabolites. The same mRNA that codes for β-galactosidase codes for the permease and the acetylase; thus when lactose is added to *E. coli* cells, the relative amounts of all three proteins rise coordinately. The collections of adjacent genes that are transcribed into single mRNA molecules are called operons. Some operons are large; for example, the eleven proteins involved in the synthesis of the amino acid histidine are all translated off one extremely large mRNA molecule containing over 10,000 nucleotides.

Promoters Are the Start Signals for RNA Synthesis

The site where RNA polymerase binds to the beginning of an operon is called the promoter. The

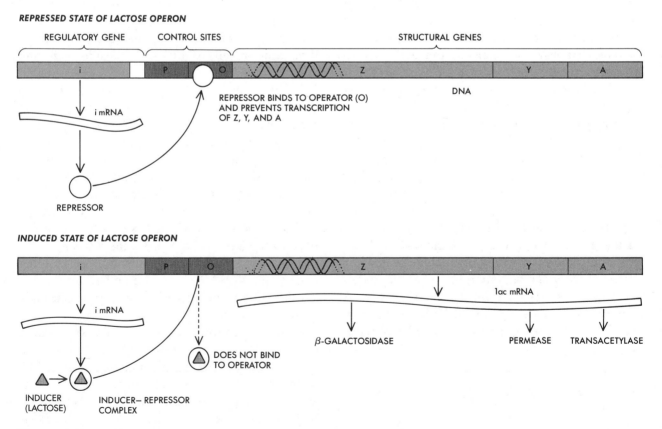

Figure 4-1
Repressors and inducers control the functioning of the genes belonging to the lactose (lac) operon.

maximum rate of transcription from a particular segment of DNA depends on the sequence of bases in its promoter, and initiation frequency can vary by a factor of at least 1000 (Figure 4-2). Two highly conserved separate sets of nucleotide blocks make up the *E. coli* promoter. One set, consisting of six base pairs (bp), is centered several base pairs upstream of the mRNA start site (at about −10), whereas the second block, of about ten base pairs, is centered some 25 nucleotides further upstream (at about −35). The initial step in *E. coli* transcription is believed to be the recognition and binding of an RNA polymerase molecule to the −35 region. Subsequently, the −10 region is thought to melt (open up) into its component single strands, allowing transcription to begin at the +1 position. Both blocks were initially identified by the existence of point mutations that blocked RNA synthesis. These mutations affect only the synthesis of the mRNA molecule immediately downstream, so promoters are examples of "cis"-acting control elements. In contrast, repressors are not limited in their binding to the DNA molecule that carries their genetic information, so they have been called "trans"-acting control elements.

The operator sequence to which a repressor binds is always close to the promoter for the operon being controlled (Figure 4-1). In some way the binding of a repressor blocks the binding of RNA polymerase to the nearby promoter. The control of a promoter by a repressor is thus an example of negative control. Promoters can also be under the control of positive effectors that increase the rate at which mRNA chains are made. Positive control elements most likely act by helping to open up the two chains of the double helix at the promoter site, thereby facilitating the binding of RNA polymerase.

Constitutive Synthesis of Repressor Molecules

Each repressor is coded by a specific gene. The gene for the lactose repressor lies immediately in front of the operon. In other cases, however, the repressor gene is widely separated from the operon genes on which it acts. The rate at which repressors are made is normally unchanging; such invariant synthesis is known as constitutive synthesis. The exact rate of this constitutive synthesis is a function of the structure of the promoter of the repressor gene. Normally the promoters of repressor genes function at very low rates, leading to the presence of only a few repressor mRNA molecules in the average cell. However, there exist promoter mutants that allow much higher rates of repressor mRNA synthesis and, correspondingly, much

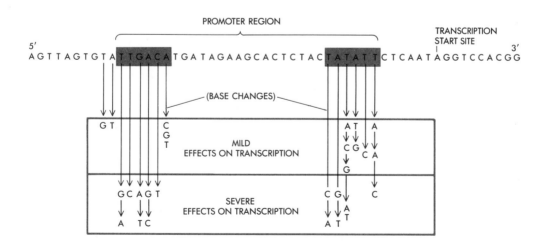

Figure 4-2
Specific DNA sequences are important for efficient transcription of *E. coli* genes by RNA polymerase. The boxed sequences at approximately −35 and −10 are highly conserved in all *E. coli* promoters. Point mutations in these regions have noticeable effects upon transcription efficiency.

higher numbers of repressor molecules per cell. Even in the presence of high levels of inducer, such mutant cells make smaller-than-usual amounts of the induced proteins.

Repressors Are Isolated and Identified

Because the genetic studies leading to the postulation of repressors were so complete, the role of repressors as key bacterial control elements seemed almost inescapable. Final proof, though, had to await the development of biochemical procedures by which individual repressors could be isolated, chemically identified, and shown to bind specifically to their respective operators. These steps depended on the development of genetic techniques for increasing the number of repressor molecules per cell beyond the few copies normally present. As long as repressors were only available in the amounts that occurred in nature, there was no effective way to isolate them.

Two tricks were used in conquering the lactose repressor. The first was the genetic manipulation of the *E. coli* genome. The gene coding for the lac repressor and the beginning of the lac operon was attached to the phage λ chromosome. When such phages multiply in *E. coli,* several hundred copies of the β-lac chromosome are produced, as well as correspondingly large numbers of lac-repressor mRNA molecules. In this way the amounts of repressor per cell were amplified about tenfold. Still further enrichment came from constructing the β-lac strains with mutant promoters, which overproduced lac-repressor mRNA by another factor of ten. The resulting amounts of repressor became sufficient to allow Walter Gilbert and Benno Müller-Hill to demonstrate at Harvard University in 1966 that the lactose repressor is a protein of MW 38,600 and that it has two specific binding sites—one for lactoselike compounds, the other for DNA containing the lactose operator. Virtually simultaneously, Mark Ptashne, also at Harvard, isolated from phage λ the repressor that controls the rates at which several classes of λ-specific mRNA are made. This repressor is a 26,000-MW polypeptide chain, and it likewise binds only to its specific operator.

Synthesis of the classical λ repressor is itself under the control of a second λ-specific repressor-like protein called "cro." Cro is a relatively small molecule composed of two identical polypeptide chains with 66 amino acids each. It acts by binding to the promoter of the gene coding for the λ repressor and thus turning off the synthesis of the repressor RNA. Recently cro has been crystallized and its exact three-dimensional structure worked out by x-ray crystallography. At roughly the same time, the three-dimensional structure of the DNA-binding fragment of the λ repressor has been elucidated, and so has the structure of the DNA-binding protein CAP (catabolite activator protein). How each of these proteins binds to DNA is at this moment not yet firmly established, but the correct molecular interactions are soon likely to be known.

Positive Regulation of Gene Transcription

More recently, several positive regulators of RNA polymerase binding, and hence gene expression, have been isolated from *E. coli* and also shown to be proteins. The best-understood positive regulator protein signals to the appropriate genes that glucose is not available as a food source. When glucose is absent, the amounts of the intracellular regulator cyclic AMP (cAMP) build up. This cAMP then binds to the DNA-binding protein known as CAP. The resulting cAMP–CAP complexes, by binding to the respective promoters, help to activate operons whose enzymes can break down alternative sugars such as lactose or galactose.

In bacteria, the regulation of DNA functioning is thus controlled in part by the binding of specific regulatory proteins to control sequences situated at the beginnings of their various genes (operons).

Attenuation

The expression of many bacterial operons involved in amino acid biosynthesis is also influenced by a process called "attenuation." This phenomenon was discovered through the elegant experiments by Charles Yanofsky and his coworkers on the tryptophan (trp) operon of *E. coli.* This by now exhaustively studied operon consists of a transcriptional control region and five structural genes that encode the enzymes involved in the last steps of tryptophan biosynthesis (Figure 4-3). The

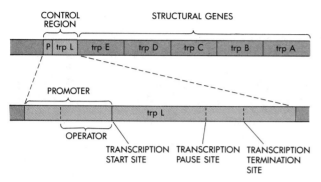

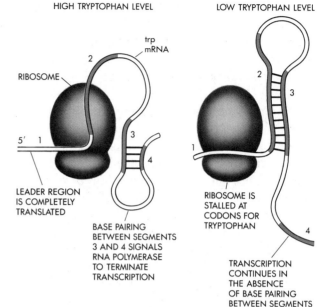

Figure 4-3
The *E. coli* tryptophan (trp) operon. Control elements at the beginning of the operon are highlighted.

Figure 4-4
A model for attenuation at the trp operon. The 5′ end of the trp operon mRNA is rich in tryptophan codons. When tryptophan is available, normal translation of this leader sequence occurs. As this happens, the next segment of the trp mRNA forms a stem-loop structure that apparently does not allow RNA polymerase to continue the transcription of the remainder of the trp operon. Note that this "attenuated" state is the *normal* situation; it is only when the tryptophan level drops that the attenuation is relieved. This is believed to occur when a ribosome is stalled trying to translate the leader sequence; the resulting formation of a different stem-loop structure in the next segment of trp mRNA allows the RNA polymerase to continue with the transcription of the remainder of the trp operon.

rate at which transcription of trp mRNA begins is controlled by a tryptophan-activated repressor molecule that can block the access of RNA polymerase to the trp promoter. Once started, however, transcription does not necessarily extend to the end of the operon to produce the very long full-length trp mRNA. Instead, incomplete transcription ("attenuation") frequently occurs to produce a relatively short 162-base mRNA molecule that codes for a correspondingly small trp L (leader) protein. Although the function (if any) of the leader protein has not been discovered, it has the very interesting property of being rich in the amino acid tryptophan, whose sole codon is UGG.

Whether or not attenuation takes place is a function of the exact folding pattern (the secondary structure) of the nascent mRNA chain; this structure in turn depends on the extent to which ribosomes have translated the beginning of the message. When tryptophan is readily available, this first section of the trp mRNA is translated normally; in this situation, the trp mRNA forms a secondary structure causing premature termination of transcription. If, however, tryptophan is in short supply, translation of trp mRNA is stalled at the UGG codons at the beginning of the message. The stalling enables the next segment of RNA transcript to form an alternative secondary structure that permits the RNA polymerase molecule to proceed with transcription (Figure 4-4). Tryptophan starvation therefore increases expression of the trp operon in two ways: by removing the trp-activated repressor (thus increasing transcription

by a factor of 70), and by relieving the normal attenuation (increasing transcription another 8- to 10-fold). Overall, then, the expression of the operon can be increased approximately 600-fold.

Translational Control

The expression of proteins in *E. coli* is also controlled at the level of translation. Efficient initiation of translation depends upon the existence of a group of six to eight purine-rich nucleotides just

upstream from the AUG initiation codon. The existence of this "ribosome-binding site" was first
pointed out in 1974 by John Shine and Larry Dalgarno in Canberra, Australia. They noted its close
complementarity in sequence to the 3' end of the
smaller (16S) of the two rRNA molecules found
in bacterial ribosomes. Apparently the initial positioning of mRNA upon the smaller ribosomal
subunit requires base pairing between the 16S
rRNA chain and the ribosome-binding (Shine–
Dalgarno) sequences. In general, the most efficiently translated mRNAs have their ribosome-
binding sequences centered eight nucleotides
upstream from the initiation codon. Mutations in
this region, including mutations that place this sequence closer to or further away from the AUG
start codon, can greatly lower the translational efficiency of their respective mRNAs. It is important
to note that possession of an appropriately sited
Shine–Dalgarno sequence does not guarantee initiation of protein synthesis. Many such sequences
are effectively buried in secondary-structure loops
and have no way of interacting with 16S rRNA.

Recent evidence shows that the efficiency with
which given Shine–Dalgarno sequences work can
be modulated by proteins that bind to them and
block their availability. The best-understood example involves the ribosomal proteins (r-proteins)
of *E. coli* (Figure 4-5). When the rate of r-protein
synthesis exceeds the rate at which rRNA is made,
free r-proteins accumulate and certain "key" ones
bind to the Shine–Dalgarno sequences on the r-
protein mRNA molecules. In this way ribosomal
proteins are not synthesized faster than they can be
used in making ribosomes.

Different sets of r-proteins are contained on a
number of different operons in the *E. coli* genome.
Each operon encodes its own "key" protein,
which inhibits the expression of the entire operon.
Key r-proteins also bind to rRNA early during
ribosome assembly. The sequences on rRNA that
bind these proteins and the sequences on the r-
protein mRNA that interact with key proteins are
quite similar (Figure 4-6). Translational control
thus represents competition between rRNA and
r-protein mRNA for the binding of these key proteins. When r-protein mRNAs are rendered untranslatable by the binding of key proteins, they
are degraded more rapidly than usual.

Once translation has started, its rate is determined by the availability of the various tRNA spe-

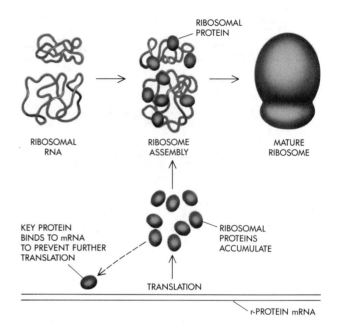

Figure 4-5
Ribosomal protein levels control the
translation of ribosomal protein mRNAs.
When the rate of r-protein synthesis exceeds
the rate of rRNA synthesis, free r-proteins
accumulate. Some of them bind to the
Shine–Dalgarno sequences on the r-protein
mRNA and prevent further translation. This
mechanism ensures that r-proteins are not
synthesized faster than they can be used in
making ribosomes.

cies corresponding to the specific codons employed in the mRNA molecules. As we discussed
in Chapter 3, different tRNA molecules are present in quite different amounts, with those present
in larger quantities generally corresponding to
more commonly used codons. Messages having a
high proportion of rarely used codons are thus
translated more slowly than those containing
widely used codons.

Early Difficulties in Probing Gene Regulation in Higher Plants and Animals

By the late 1960s the question was asked whether
the genes of higher cells are also regulated by
specific DNA-binding proteins. In particular,
would such proteins be the key to understanding
the mysteries of embryology, through which fertilized eggs divide and differentiate, eventually giving rise to the highly specific cell types that make
up our tissues and organs? The only way to pro-

16S rRNA

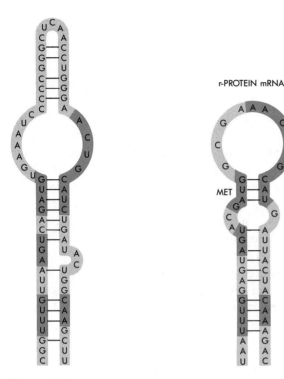

r-PROTEIN mRNA

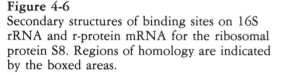

Figure 4-6
Secondary structures of binding sites on 16S rRNA and r-protein mRNA for the ribosomal protein S8. Regions of homology are indicated by the boxed areas.

the DNA level, however, came to naught. There was no way of finding the particular sections of DNA to which the hormone–receptor complexes were binding, and of distinguishing specific binding from nonspecific binding. Moreover, although mutant cells were found in which hormones did not function normally, there was no simple way to map the relevant genes. We could not—and still cannot—make conventional genetic crosses between higher cells growing in culture. As a result, even if operators and repressors existed in higher cells, the possibility of genetically demonstrating them was slim, if not nonexistent. And as long as it remained impossible to isolate from the large chromosomes of higher organisms the specific DNA segments that coded for particular proteins, direct tests for the presence of specific DNA-binding proteins were also out of the question. As we shall describe later, the great importance of recombinant DNA techniques has stemmed from the fact that they allow specific fragments of any DNA to be isolated.

Purification of *Xenopus* Ribosomal RNA Genes

Even before the advent of recombinant DNA, several perceptive observations by Don Brown in Baltimore and by Max Birnsteil in Edinburgh led to the isolation, in virtually pure form, of the genes coding for the ribosomal RNAs (18S, 28S, and 5S RNAs) of the toad *Xenopus.* This was possible because, first of all, these genes undergo enormous amplification in the very large *Xenopus* oocytes, to the point where they can represent almost 70 percent of the total nuclear DNA; and secondly, because their base composition is significantly different from the rest of the nuclear DNA, allowing further purification on density (CsCl) gradients. 5S DNA was purified by exploitation of its high AT content, which allows it to bind many more silver ions (Ag^{++}) than most other DNA can; Ag–DNA is, of course, much denser than DNA, and the 5S genes that migrate with the bulk of DNA in a normal CsCl gradient appear as a small satellite peak in a Ag–CsCl gradient.

The individual 18S and 28S genes were found to be linked closely together, separated by about 1000 bases of DNA. This 18S–28S DNA unit is present in a tandem array of approximately 450 copies, and each 18S–28S unit is separated from

ceed was to study single animal or plant cells growing in culture, as opposed to studying whole organisms. Virtually all serious embryologists were hoping that model cell culture systems in which the functions of single genes could be followed during differentiation would soon be developed. Merely watching proteins come and go during differentiation, however, would not by itself be a major step forward. We wanted to discover the signals that would turn a given gene on or off, and to determine how so many proteins can have their expression so exquisitely coordinated.

The only signals already identified in higher organisms were certain steroid hormones that had recently been shown to bind to specific receptor molecules in the cytoplasm. These hormone–receptor complexes then moved to the nucleus, where they bound to chromosomes and somehow turned the synthesis of many specific proteins on or off. All attempts to pursue this phenomenon at

the next unit by about 5 kilobases of "spacer" DNA. The 18S and 28S rRNAs are produced from a larger (40S) RNA transcript, which is processed to produce the mature ribosomal RNAs. The 5S genes, which are not linked to the 18S and 28S genes, were found to be present in 10,000 to 20,000 copies, also in tandem arrays separated by spacers that varied in length.

The purified 5S genes were faithfully transcribed when microinjected into *Xenopus* oocyte nuclei, indicating that the information necessary for accurate initiation and termination of transcription is contained on the purified genes (in other words, the genes have cis-acting controlling elements). These studies on purified ribosomal RNA genes were the prototype of similar experiments on a large number of other eukaryotic genes that became available only after the advent of recombinant DNA technology.

Eukaryotic mRNAs Have Caps and Tails

With most eukaryotic DNA so difficult to get a handle on, attention initially focused primarily on eukaryotic mRNA—in particular, on mRNAs that code for the more abundant cellular molecules like hemoglobin, the chicken egg protein ovalbumin, and the immunoglobulin chains made by antibody-producing cells. Most eukaryotic mRNAs, unlike their prokaryotic equivalents, were found to have long runs of A (poly-A tails) at their 3' ends. These tails do not come from sequences encoded in the DNA but are added after transcription has terminated. Subsequently it was found that the 5' ends of eukaryotic mRNAs are blocked by the addition of m⁷Gppp caps (7-methylguanosine residues joined to the mRNAs by triphosphate linkages) that are added during the synthesis of the primary transcript (Figure 4-7). No obvious expla-

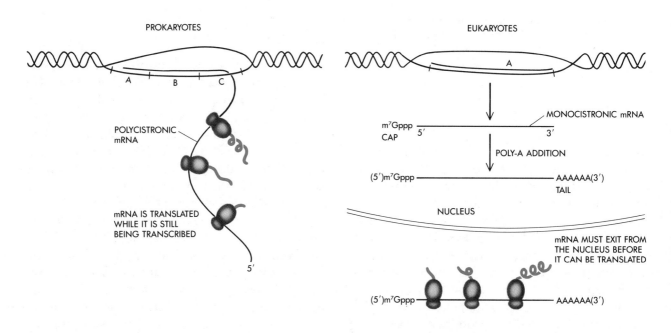

Figure 4-7
Synthesis and translation of prokaryotic vs. eukaryotic mRNAs. Prokaryotic mRNAs are often polycistronic; that is, they contain information for more than one protein. The 5' end of the message is being translated (and probably degraded) while the 3' end is still being transcribed. Eukaryotic mRNAs are monocistronic; one mRNA codes for only one protein. During transcription, methyl caps are added at the 5' end, and after transcription, a poly-A tail is added at the 3' end. The mRNA must then be transported through the nuclear membrane into the cytoplasm, where it can be translated.

nation could be given for why eukaryotic mRNAs have their ends so protected; until more convincing explanations arise, we can assume that the caps and tails give eukaryotic mRNA added protection against attack by RNA-degrading enzymes.

Three Kinds of RNA Polymerases in Eukaryotes

The successful synthesis of eukaryotic RNA in the test tube led to the realization that all eukaryotic cells contain three different RNA polymerases, each with a distinct functional role. Ribosomal RNA (rRNA) is transcribed off rDNA genes by the enzyme RNA polymerase I. The synthesis of messenger RNA is catalyzed by RNA polymerase II, with the subsequent addition of the poly-A tails carried out by the enzyme poly-A synthetase. Transfer RNA and a variety of smaller nuclear and cellular RNAs with still undetermined functions are made by RNA polymerase III. This multiplicity of RNA polymerases in eukaryotes contrasts with the single form of RNA polymerase that is responsible for all RNA synthesis in prokaryotes. The existence of three different types of eukaryotic RNA polymerases suggests the existence of three different types of promoters that can be regulated independently.

Eukaryotic DNA Is Organized into Nucleosomes

Also distinctive of eukaryotic cells is the way the DNA molecules are tightly complexed with proteins to form chromatin, nucleoprotein fibers with a beaded appearance. The key to the beadlike organization of chromatin is its histone proteins, of which there are five main classes: H1, H2A, H2B, H3, and H4. They come together to form "nucleosome" units containing an octomeric core composed of two molecules each of the H2A, H2B, H3, and H4 components, around which some 160 base pairs of DNA are wound. Connecting the nucleosome units are short stretches of linker DNA, where histone H1 is found. Also found within chromatin are a large variety of different DNA-binding proteins. None of their functions are known, but it seems likely that some have specific roles in regulating gene expression.

The basic nucleosome structure must clearly be modified as transcription passes through individual units. As soon as chromatin began to be examined seriously with the electron microscope, observers claimed that especially "active" chromatin (for example, that making ribosomal RNA) lacked the normal beaded appearance. Until more individual genes could be examined, however, the molecular nature of active chromatin remained unknown.

Animal Viruses Are Model Systems for Gene Expression in Higher Cells

Although for a long time the chromosomal DNA of higher cells seemed virtually impossible to study, strong arguments were made that some real answers might come from investigating how DNA viruses multiply in animal cells. Many of these viruses grew well in cultured cells and could easily be labeled with radioactive isotopes. Moreover, some—like the monkey virus SV40 and the mouse virus polyoma—had remarkably small circular DNA molecules containing maybe as few as 5000 base pairs, coding for less than ten proteins. They thus resembled in genetic simplicity the simple bacterial viruses from which so much had been learned. These viruses were also attractive because several were found to transform certain normal cells into cancerous cells capable of forming tumors in appropriate animals. Each of these viruses thus might contain one or more genes coding for proteins capable of transforming a normal cell into its cancerous equivalent. By studying such viruses, biochemists hoped to learn some of the principles of gene expression in higher cells, as well as a few basic facts about the origin and nature of cancer cells.

Therefore, in the late 1960s, deeper understanding of the DNA tumor viruses became the goal of an increasing number of scientists, many of whom had decided to switch their research from bacteria and bacteriophages to cells of higher animals. By 1970, the SV40 and polyoma viruses were both shown to have a life cycle that is neatly divisible into early and late phases. During the early phase of the SV40 life cycle, the cell makes mRNA coding for a viral protein that accumulates in the cell nucleus. There this viral protein plays a necessary role in SV40 DNA replication. When it was discovered that this protein was the major

SV40-coded product that was consistently present in cells made cancerous by the virus, the protein was considered the key to the cancer-causing ability of SV40. Since its discovery, therefore, this protein has been known as the T (tumor) protein. The late stage of the SV40 life cycle is marked by the synthesis of SV40 DNA as well as the synthesis of mRNA, which codes the structural proteins of the virus particle's outer shell.

How to confirm these facts at the molecular level, however, was not obvious, because there was as yet no way to relate the SV40 or polyoma virus mRNAs and the corresponding proteins to specific sections of the SV40 and polyoma DNAs. The circular forms of the viral DNAs at first seemed to preclude finding any specific reference point (such as an end) from which, with the electron microscope, the binding sites of the early and late mRNA could be mapped.

The RNA Tumor Viruses Replicate by Means of a Double-Stranded DNA Intermediate

Equally inaccessible at the molecular level were the genomes of the various RNA tumor viruses. They had suddenly become much more interesting, through the independent discoveries in 1970 by Howard Temin and S. Mizutani and by David Baltimore that their replication involves the reverse transcription of the infecting RNA genomes into complementary DNA strands. The enzyme involved, reverse transcriptase, is coded for by the respective viral genomes, which incorporate it into the infectious viral particles so that it is able to act the moment the infecting RNA chromosome enters an appropriate host cell. Soon after its synthesis, the complementary DNA strand acts as a template to make the double-helical DNA representation of the infecting RNA strand. In turn, this DNA becomes integrated as a provirus into host chromosomal DNA, where the viral life cycle can be completed through transcription of the proviral DNA into an RNA strand identical to that found in the infectious virus.

Working out how all these steps occurred—and determining at the molecular level how such viruses made cells cancerous—necessarily seemed in 1970 to be tasks for the distant future. The fact that animal cells and their viruses would never be as simple to work with as bacteria or their phages was something we were realizing we would have to live with. Moreover, there was no economically feasible way to grow animal cells on the same large scale as bacteria. Though considerable progress had been made in learning how the DNA and RNA tumor viruses give rise to cancer, it seemed impossible that we would ever attain a really precise understanding of the crucial events at the molecular level—as long as the amounts of viral nucleic acids and proteins that could be obtained were severely limited, and as long as there were no methods for isolating specific segments of cellular DNA.

READING LIST

Books

Beckwith, J. R., and D. Zipser, eds. *The Lactose Operon.* Cold Spring Harbor Laboratory, Cold Spring Harbor, N.Y., 1970.

Lewin, B. *Gene Expression,* vol. 1: *Bacterial Genomes.* Wiley, New York, 1974.

Lewin, B. *Gene Expression,* vol. 2: *Eucaryotic Chromosomes,* 2nd ed. Wiley, New York, 1980.

Lewin, B. *Gene Expression,* vol. 3: *Plasmids and Phages.* Wiley, New York, 1977.

Miller, J. H., and W. S. Reznikoff, eds. *The Operon,* 2nd ed. Cold Spring Harbor Laboratory, Cold Spring Harbor, N.Y., 1980.

Tooze, J., ed. *Molecular Biology of Tumour Viruses.* Cold Spring Harbor Laboratory, Cold Spring Harbor, N.Y., 1973.

Chromatin. Cold Spring Harbor Symp. Quant. Biol., vol. 42. Cold Spring Harbor Laboratory, Cold Spring Harbor, N.Y., 1978.

Structures of DNA. Cold Spring Harbor Symp. Quant. Biol., vol. 47. Cold Spring Harbor Laboratory, Cold Spring Harbor, N.Y., 1983.

Original Research Papers (Reviews)

REPRESSORS AND OPERATORS

Jacob, F., and J. Monod. "Genetic regulatory mechanisms in the synthesis of proteins." *J. Mol. Biol.,* 3: 318–356 (1961).

Beckwith, J. R., and W. R. Signer. "Transposition of the *lac* region of *E. coli.* I. Inversion of the *lac* operon and transduction of *lac* by φ80." *J. Mol. Biol.,* 19: 254–265 (1966).

Shapiro, J., L. MacHattie, L. Eron, G. Ihler, K. Ippen, and J. Beckwith. "Isolation of pure *lac* operon DNA." *Nature,* 224: 768–774 (1969).

Gilbert, W., and B. Müller-Hill. "Isolation of the lac repressor." *Proc. Natl. Acad. Sci. USA,* 56: 1891–1898 (1966).

Ptashne, M. "Isolation of the λ phage repressor." *Proc. Natl. Acad. Sci. USA,* 57: 306–313 (1967).

Ptashne, M., and N. Hopkins. "The operators controlled by the λ phage repressor." *Proc. Natl. Acad. Sci. USA,* 60: 1282–1287 (1968).

Wang, J., M. D. Barkley, and S. Bourgeois. "Measurements of unwinding of *lac* operator by repressor." *Nature,* 251: 247–249 (1974).

Kim, R., and S.-H. Kim. "Direct measurement of DNA unwinding angle in specific interaction between *lac* operator and repressor." *Cold Spring Harbor Symp. Quant. Biol.,* 47: 481–484 (1983).

Anderson, W. F., D. H. Ohlendorf, Y. Takeda, and B. W. Matthews. "Structure of the cro repressor from bacteriophage λ and its interaction with DNA." *Nature,* 290: 754–758 (1981).

Ptashne, M., A. D. Johnson, and C. O. Pabo. "A genetic switch in a bacterial virus." *Sci. Am.,* 247(5): 128–140 (1982).

Pabo, C. O., and M. Lewis. "The operator-binding domain of λ repressor: Structure and DNA recognition." *Nature,* 298: 443–447 (1982).

Steitz, T. A., D. H. Ohlendorf, D. B. McKay, W. F. Anderson, and B. W. Matthews. "Structural similarity in the DNA-binding domains of catabolite gene activator and *cro* repressor proteins." *Proc. Natl. Acad. Sci. USA,* 79: 3097–3100 (1982).

Matthews, B. W., D. H. Ohlendorf, W. F. Anderson, R. G. Fisher, and Y. Takeda. "Cro repressor protein and its interaction with DNA." *Cold Spring Harbor Symp. Quant. Biol.,* 47: 427–433 (1983).

PROMOTERS

Pribnow, D. "Bacteriophage T7 early promoters: Nucleotide sequences of two RNA polymerase binding sites." *J. Mol. Biol.,* 99: 419–443 (1975).

Gilbert, W. "Starting and stopping sequences for the RNA polymerase." In R. Losick and M. J. Chamberlin, eds., *RNA Polymerase.* Cold Spring Harbor Laboratory, Cold Spring Harbor, N.Y., 1976, pp. 193–205. (Review.)

Wang, J. C., J. H. Jacobsen, and J.-M. Saucier. "Physiochemical studies on interactions between DNA and RNA polymerase. Unwinding of the DNA helix by *E. coli* RNA polymerase." *Nuc. Acids Res.,* 4: 1225–1241 (1977).

Chamberlin, M. J., W. C. Nierman, J. Wiggs, and N. Neff. "A quantitative assay for bacterial RNA polymerases." *J. Biol. Chem.,* 254: 10061–10069 (1979).

McClure, W. R. "Rate-limiting steps in RNA chain initiation." *Proc. Natl. Acad. Sci. USA,* 77: 5634–5638 (1980).

Youderian, P., S. Bouvier, and M. Susskind. "Sequence determinants of promoter activity." *Cell,* 30: 843–853 (1982).

Ackerson, J. W., and J. D. Gralla. "In vivo expression of *lac* promoter variants with altered −10, −35, and spacer sequences." *Cold Spring Harbor Symp. Quant. Biol.,* 47: 473–476 (1983).

ATTENUATORS

Bertrand, K., I. Korn, F. Lee, T. Platt, C. L. Squires, C. Squires, and C. Yanofsky. "New features of the regulation of the tryptophan operon." *Science,* 189: 22–26 (1975).

Oxender, D. L., G. Zurawski, and C. Yanofsky. "Attenuation in the *Escherichia coli* tryptophan operon: The role of RNA secondary structure involving the Trp codon region." *Proc. Natl. Acad. Sci. USA,* 76: 5524–5528 (1979).

Yanofsky, C. "Attenuation in the control of expression of bacterial operons." *Nature,* 289: 751–758 (1981).

Yanofsky, C., and R. Kolter. "Attenuation in amino acid biosynthetic operons." *Ann. Rev. Genet.,* 16: 113–134 (1982).

POSITIVE CONTROL

Zubay, G., D. Schwartz, and J. Beckwith. "Mechanism of activation of catabolite-sensitive genes: A positive control system." *Proc. Natl. Acad. Sci. USA,* 66: 104–110 (1970).

Epstein, W., L. B. Rothman, and J. Hesse. "Adenosine 3':5'-cyclic monophosphate as mediator of catabolite repression in *Escherichia coli.*" *Proc. Natl. Acad. Sci. USA,* 72: 2300–2304 (1975).

Simpson, R. B. "Interaction of the cAMP receptor protein with the lac promoter." *Nuc. Acids Res.,* 8: 759–766 (1980).

McKay, D. B., I. T. Weber, and T. A. Steitz. "Structure of catabolite gene activator protein at 2.9Å resolution: Incorporation of amino acid sequence and interactions with cyclic-Amp." *J. Biol. Chem.,* 257: 9518–9524 (1982).

TRANSLATIONAL CONTROL

Shine, J., and L. Dalgarno. "The 3'-terminal sequence of *Escherichia coli* 16S ribosomal RNA: Complementarity to nonsense triplets and ribosome binding sites." *Proc. Natl. Acad. Sci. USA,* 71: 1342–1346 (1974).

Gold, L., D. Pribnow, T. Schneider, S. Shinedling, B. S. Singer, and G. Stormo. "Translational initiation in prokaryotes." *Ann. Rev. Microbiol.,* 35: 365–403 (1981).

Nomura, M., J. L. Yates, D. Dean, and L. E. Post. "Feedback regulation of ribosomal protein gene expression in *Escherichia coli:* Structural homology of ribosomal RNA and ribosomal protein mRNA." *Proc. Natl. Acad. Sci. USA,* 77: 7084–7088 (1980).

Nomura, M., D. Dean, and J. L. Yates. "Feedback regulation of ribosomal protein synthesis in *Escherichia coli.*" *TIBS,* 7: 92–95 (1982).

ISOLATION OF RIBOSOMAL RNA GENES

Weinberg, R. A., and S. Penman. "Processing of 45 S nucleolar RNA." *J. Mol. Biol.,* 47: 169–178 (1970).

Birnstiel, M. L., M. Chipchase, and J. Spiers. "The ribosomal RNA cistrons." *Prog. Nuc. Acid Res. Mol. Biol.,* 11: 351–389 (1971). (Review.)

Brown, D. D., and K. Sugimoto. "The structure and evolution of ribosomal and 5 S DNAs in *Xenopus laevis* and *Xenopus mulleri.*" Cold Spring *Harbor Symp. Quant. Biol.,* 38: 501–505 (1974).

Brown, D.D., and J.B. Gurdon. "High-fidelity transcription of 5S DNA injected into *Xenopus* oocytes." *Proc. Natl. Acad. Sci. USA,* 74: 2064–2068 (1977).

CAPS AND POLY-A TAILS

Lim, L., and E. S. Canellakis. "Adenine-rich polymer associated with rabbit reticulocyte messenger RNA." *Nature,* 227: 710–712 (1970).

Darnell, J. E., R. Wall, and R. J. Tushinski. "An adenylic acid-rich sequence in messenger RNA of HeLa cells and its possible relationship to reiterated sites in DNA." *Proc. Natl. Acad. Sci. USA,* 68: 1321–1325 (1971).

Lee, S. Y., J. Mendecki, and G. Brawerman. "A polynucleotide segment rich in adenylic acid in the rapidly-labeled polyribosomal RNA component of mouse sarcoma 180 ascites cells." *Proc. Natl. Acad. Sci. USA,* 68: 1331–1335 (1971).

Edmonds, M., M. H. Vaughan Jr., and H. Nakazato. "Polyadenylic acid sequences in the heterogeneous nuclear RNA and rapidly-labeled polyribosomal RNA of HeLa cells: Possible evidence for a precursor relationship." *Proc. Natl. Acad. Sci. USA,* 68: 1336–1340 (1971).

Aviv, H., and P. Leder. "Purification of biologically active globin messenger RNA by chromatography on oligothymidylic acid–cellulose." *Proc. Natl. Acad. Sci. USA,* 69: 1408–1412 (1972).

Shatkin, A. J. "Capping of eucaryotic mRNAs." *Cell,* 9: 645–653 (1976).

NUCLEOSOMES

Olins, A. L., and D. E. Olins. "Spheroid chromatin units (v bodies)." *Science,* 183: 330–332 (1974).

Kornberg, R. "Chromatin structure: A repeating unit of histones and DNA." *Science,* 184: 868–871 (1974).

McGee, J. D., and G. Felsenfeld. "Nucleosome structure." *Ann. Rev. Biochem.,* 49: 1115–1155 (1980).

TUMOR VIRUSES

Black, P. W., W. P. Rowe, H. C. Turner, and R. J. Huebner. "A specific complement-fixing antigen present in SV40 tumor and transformed cells." *Proc. Natl. Acad. Sci. USA,* 50: 1148–1156 (1963).

Benjamin, T. L. "Virus-specific RNA in cells productively infected or transformed by polyoma virus." *J. Mol. Biol.,* 16: 359–373 (1966).

Sambrook, J., H. Westphal, P. R. Srinivasan, and R. Dulbecco. "The integrated state of viral DNA in SV40-transformed cells." *Proc. Natl. Acad. Sci. USA,* 60: 1288–1295 (1968).

Baltimore, D. "Viral RNA-dependent DNA polymerase." *Nature,* 226: 1209–1211 (1970).

Temin, H. M., and S. Mizutani. "Viral RNA-dependent DNA polymerase." *Nature,* 226: 1211–1213 (1970).

5

Methods of
Creating Recombinant DNA Molecules

Given the length of even the smallest DNA molecules, the isolation of the first repressors signified to many biologists more the end of an era rather than the beginning of a new cycle of important conceptual advances. Unless some radically new tricks emerged to manipulate DNA, there could be no immediate bright future for eukaryotic molecular biologists. So, to avoid possibly marking time, several distinguished contributors to our primary knowledge on the storage of genetic information in DNA left molecular genetics to start up new careers in neurobiology. What they did not and could not have foreseen was the very rapid development over the next several years of the enzymological and chemical techniques that gave rise to recombinant DNA, and the consequent period of scientific excitement and achievement that has seen few if any parallels in the history of biological research.

Nucleic Acid Sequencing Methods Are Developed

For deeper insights about the organization of DNA, methods had to be developed to reveal exact nucleotide sequences—first, of selected regions of a gene; then of an entire gene; and finally of an entire chromosome. The first nucleic acid sequences to be established were not of DNA, but were of the relatively small tRNA molecules that contain 75 to 80 nucleotides. By 1964, the sequence of the yeast alanine tRNA molecule was worked out. To do this, Robert Holley and his colleagues at Cornell University had to find specific enzymes that broke the tRNA chains reproducibly into smaller and smaller discrete fragments, until they could be sequenced directly by simple stepwise degradation procedures.

With each passing year these methodologies

greatly improved, and by 1975 the complete sequence of the RNA chromosome* of the single-stranded RNA phage MS2 was worked out in Walter Fiers' laboratory in Ghent. For the first time the precise way in which a simple chromosome was put together could be visualized. And for the first time the exact codons that specify the amino acids of the three proteins coded by the three genes of phage MS2 became known, as well as the stop codons that signal chain termination. Few nucleotides separated the three genes, but unexpectedly long untranslated regions (of 129 and 174 bases, respectively) existed at the two ends. Here, as on all other messenger-RNA-like molecules, the two physical ends never acted as start or stop signals.

Direct sequencing of any DNA molecule was not then possible because there was no way to cut DNA at specific points to produce discrete reproducible fragments having unique sequences. The available deoxyribonucleases (DNases) all cut DNA into hopelessly heterogeneous collections of small fragments whose order within the original DNA could never be deciphered.

Restriction Enzymes Make Sequence-Specific Cuts in DNA

All the nucleases, the enzymes that were first found to break the phosphodiester bonds of nucleic acids, showed very little sequence dependency; the most specific was the T1 RNase, which

---

* RNA replaces DNA as the genetic material in many viruses (for example, tobacco mosaic, influenza, polio, certain RNA phages). Replication follows the same pattern used for DNA, with single RNA chains serving as templates to make chains with complementary sequences. The specific replication enzymes are coded on the viral RNA chromosomes and called replicases. Equally important, one of the complementary partners can also function as mRNA by combining with ribosomes and coding directly for the amino acids of the viral proteins.

was found to cut only next to guanine residues. Highly preferred sites of cleavage on certain RNAs were discovered, but these reflected the way single-stranded RNA molecules fold into complex three-dimensional arrangements rather than any tendency of the enzymes to cut within specific base sequences. The prevailing opinion was that highly specific nucleases would never be found, and that therefore the isolation of discrete DNA fragments, even from viral DNA, would not be possible. The only grounds for thinking differently were observations, beginning as early as 1953, that when DNA molecules from one strain of *E. coli* were introduced into a different *E. coli* strain (for example, *E. coli* strain B vs. *E. coli* strain C), they rarely functioned genetically. Instead, the foreign DNAs were almost always quickly fragmented into smaller pieces. Quite infrequently, the infecting DNA molecule would not be broken down because it had somehow become modified so that it and all its descendants could now multiply on the new bacterial strain. In 1966 chemical analysis of a small viral DNA modified in such a way that it could survive in a different strain of *E. coli* revealed the presence of one to several methylated bases not present in the unmodified DNA. Methylated bases are not inserted as such into growing DNA chains; they arise through the enzymatically catalyzed addition of methyl groups to newly synthesized DNA chains.

The stage was thus set in the late 1960s for Stewart Linn and Werner Arber, working in Geneva, to find in extracts of cells of *E. coli* strain B both a specific modification enzyme that methylated unmethylated DNA and a "restriction" nuclease that broke down unmethylated DNA. Over the next several years, the discovery of restriction nucleases and their companion modification methylases in two other *E. coli* strains opened up the possibility that many site-specific nucleases might exist. None of these early *E. coli* restriction enzymes lived up to their finders' first hopes, however, because although the enzymes recognized specific unmethylated sites, they cleaved the DNA at random locations far removed from these sites.

Soon specific restriction nucleases that did cleave at specific sites in DNA were identified. The first was discovered in 1970 by Hamilton Smith of Johns Hopkins University, who followed up his accidental finding that the bacterium *Haemophilus influenzae* rapidly broke down foreign phage DNA. This degradative activity was subsequently observed in cell-free extracts and shown to be due to a true restriction nuclease, because the enzyme broke down *E. coli* DNA, whereas it failed to cut up the DNA of the *Haemophilus* cells from which it had been extracted. Highly purified *Hin*dII, as this enzyme is called, was found to bind to the following set of sequences, in which the arrows indicate the exact cleavage sites, and Py and Pu represent any pyrimidine or purine residue:

$$(5')GTPy{\downarrow}PuAC(3')$$
$$(3')CAPu{\uparrow}PyTG(5')$$

Since then, restriction enzymes that cut specific sequences have been isolated from some 230 bacterial strains, and over 91 different specific cleavage sites have been found (Table 5-1). Some of these enzymes recognize specific groups of four bases, whereas many others recognize groups of

Table 5-1

Some Restriction Enzymes and Their Cleavage Sequences

Microorganism	Abbreviation	Sequence $5' \rightarrow 3'$ $3' \rightarrow 5'$
Bacillus amyloliquefaciens H	*Bam*HI	G↓G A T C C C C T A G↑G
Brevibacterium albidum	*Bal*I	T G G↓C C A A C C↑G G T
Escherichia coli RY13	*Eco*RI	G↓A A T T C C T T A A↑G
Haemophilus aegyptius	*Hae*II	Pu G C G C↓Py Py↑C G C G Pu
Haemophilus aegyptius	*Hae*III	G G↓C C C C↑G G
Haemophilus haemolyticus	*Hha*I	G C G↓C C↑G C G
Haemophilus influenzae R$_d$	*Hin*dII	G T Py↓Pu A C C A Pu↑Py T G
Haemophilus influenzae R$_d$	*Hin*dIII	A↓A G C T T T T C G A↑A
Haemophilus parainfluenzae	*Hpa*I	G T T↓A A C C A A↑T T G
Haemophilus parainfluenzae	*Hpa*II	C↓C G G G G C↑C
Providencia stuartii 164	*Pst*I	C T G C A↓G G↑A C G T C
Streptomyces albus G	*Sal*I	G↓T C G A C C A G C T↑G
Xanthomonas oryzae	*Xor*II	C G A T C↓G G↑C T A G C

six. The ones that bind to only four bases cut many more times in a given DNA molecule than the ones that have to recognize a specific group of six. A six-base restriction sequence may not exist even once in a given viral DNA molecule. For example, the $\frac{\text{GAATTC}}{\text{CTTAAG}}$ recognition sequence of the *E. coli* RI enzyme is not present in phage T7 DNA, which is 40,000 base pairs long.

Restriction Maps Are Highly Specific

The various fragments generated when a specific viral DNA is cut by a restriction enzyme can be easily separated by using agarose gel electrophoresis (Figure 5-1). The rate at which the fragments move is a function of their lengths, with small fragments moving much faster than large fragments. Depending on the concentration of agarose, the larger fragments may hardly be able to move into the gel. Restriction fragments move unharmed through such agarose gels and can be eluted as biologically intact double helices. Staining of these gels with dyes that bind to DNA generates a series of bands, each corresponding to a restriction fragment whose molecular weight can be established by calibration with DNA molecules of known weights. Different restriction enzymes necessarily give different restriction maps for the same viral DNA molecule. In general, the most useful enzymes are the ones that have rare recognition sequences and that therefore produce small numbers of fragments that can easily be separated from one another on the agarose gels.

The first restriction map was obtained in 1971 by Daniel Nathans, a colleague of Hamilton Smith at Johns Hopkins. Nathans used the *Hin*dII enzyme to cut the circular DNA of SV40 into 11 specific fragments. The order in which these 11 fragments occurred in the SV40 DNA could be deduced by studying the patterns of fragments produced as the digestion proceeded to completion. The first cut broke the circular molecule into a linear structure that was then cut into progressively smaller fragments. By following the pattern of production of, first, the overlapping intermediate-sized fragments and, from them, the fragments of the complete digest, Nathans produced a restriction map that located the sites on the circular viral DNA that are attacked by the restriction enzyme (Figure 5-2). Repeating the experiment with other enzymes produced a more detailed map with many different restriction sites.

With this information it became possible to determine the positions of regions of biological importance on the circular viral DNA. For example, by briefly radioactively labeling replicating viral DNA and then digesting it with *Hin*dII, Nathans proved that the replication of SV40 DNA always begins in one specific *Hin*dII fragment and proceeds bidirectionally around the circular DNA molecule. Subsequently, by using other enzymes (including *Eco*RI, which cuts SV40 DNA only once), experimenters precisely located the site of the initiation of DNA replication at some 1700 base pairs away from the *Eco*RI site.

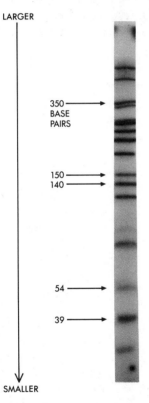

LARGER

350 → BASE PAIRS

150 →
140 →

54 →

39 →

SMALLER

Figure 5-1
Restriction fragments of DNA may be separated by electrophoresis, a process in which an electric field is used to move the DNA fragments through porous agarose gels. The smaller fragments move faster than the larger ones, so the sizes of the fragments can be estimated by comparing their positions in the gels to the positions of DNA molecules of known sizes. (Courtesy of John C. Fiddes and Howard M. Goodman.)

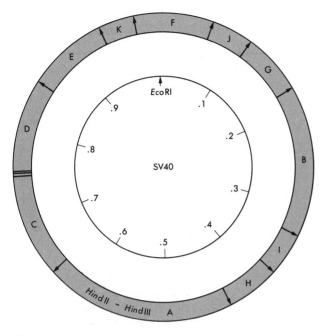

Figure 5-2
A restriction map of SV40 produced with the
enzymes *Hin*dII and *Hin*dIII.

The restriction maps and restriction fragments
were then used to identify on the viral DNA the
regions that specify the mRNAs of the viral proteins at different stages during viral replication. To
do this, radioactively labeled mRNA from infected cells was isolated at early and late times after
infection. Pure restriction fragments of the viral
DNA were prepared and denatured so that the
two DNA chains of the double helix separated.
The mRNA was then mixed with the separated
DNA strands in conditions that allowed the RNA
to form RNA–DNA double helices with DNA
strands that had a complementary base sequence.
Such RNA–DNA hybridization experiments revealed that both early and late viral RNA are
coded by continuous DNA regions, each spanning
about half of the total SV40 DNA. The promoters
of both the early and the late mRNAs are near the
origin of DNA replication, but synthesis of early
mRNA proceeds in one direction, and that of late
mRNA proceeds in the other. These techniques,
which allowed identification of the genetically significant regions of DNA, were important in themselves. But they also paved the way to the development of extremely useful new methods of DNA
sequencing and recombinant DNA techniques.

Restriction Fragments Lead to Powerful New Methods for Sequencing DNA

When the first restriction fragments became available, there was no good method of sequencing them
directly. The only realistic way to proceed was to
use RNA polymerase to synthesize their complementary RNA chains, on which the elegant new
RNA-sequencing procedures of Fred Sanger could
be employed. In the mid 1960s Sanger had stopped
sequencing proteins and turned his attention to
working out fast simple procedures for sequencing
long stretches of RNA. By employing Sanger's
procedures, Sherman Weissman at Yale University
and Walter Fiers in Ghent established, by the end
of 1976, the sequence of more than half of the over
5200 base pairs of the DNA of SV40 virus.

A breakthrough came with the advent of
methods that allowed sequencing of 100-to-500-
base-pair fragments of DNA. Sanger devised the
first of these direct DNA-sequencing methods, the
"plus-minus" method, in 1975. It is based on the
elongation of DNA chains with DNA polymerase.
With this technique the 5386-base-pair sequence
of the small DNA phage ϕX176 was quickly determined. An equally powerful method based on
the chemical degradation of DNA chains was developed at Harvard University by Allan Maxam
and Walter Gilbert in 1977 (Figure 5-3, page 62).
All the 5226 base pairs of SV40 DNA became
quickly known, as did those of the small recombinant plasmid pBR322, whose 4362 bases were
determined in less than a year by Greg Sutcliffe in
Gilbert's laboratory.

Sanger later devised a second method for sequencing DNA, and again he used enzymatic
rather than chemical techniques. Specific termina-
tors of DNA chain elongation—2′,3′-dideoxy nucleoside triphosphates—were synthesized. These
molecules can be incorporated normally into a
growing DNA chain through their 5′ triphosphate
groups. However, they cannot form phosphodiester bonds with the next incoming deoxynucleotide triphosphates (dNTPs). When a small amount
of a specific dideoxy NTP (say, ddATP) is included along with the four deoxy NTPs normally
required in the reaction mixture for DNA synthesis by DNA polymerase, the products are a series
of chains that are specifically terminated at the

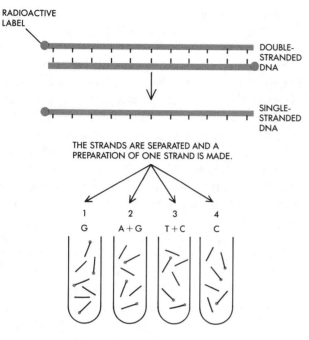

RADIOACTIVE LABEL

DOUBLE-STRANDED DNA

SINGLE-STRANDED DNA

THE STRANDS ARE SEPARATED AND A PREPARATION OF ONE STRAND IS MADE.

1	2	3	4
G	A + G	T + C	C

A CHEMICAL AGENT DESTROYS ONE OR TWO OF THE FOUR BASES AND SO CLEAVES THE STRANDS AT THOSE SITES. THE REACTION IS CONTROLLED SO THAT ONLY SOME STRANDS ARE CLEAVED AT EACH SITE, GENERATING A SET OF FRAGMENTS OF DIFFERENT SIZES.

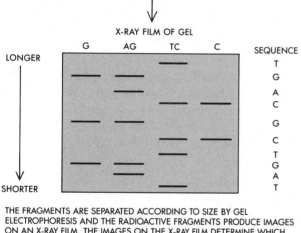

X-RAY FILM OF GEL

LONGER

SHORTER

THE FRAGMENTS ARE SEPARATED ACCORDING TO SIZE BY GEL ELECTROPHORESIS AND THE RADIOACTIVE FRAGMENTS PRODUCE IMAGES ON AN X-RAY FILM. THE IMAGES ON THE X-RAY FILM DETERMINE WHICH BASE WAS DESTROYED TO PRODUCE EACH RADIOACTIVE FRAGMENT.

SEQUENCE OF ANALYZED STRAND

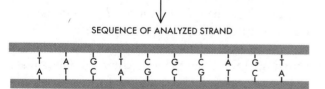

SEQUENCE OF COMPLEMENTARY STRAND

THE SEQUENCE OF THE DESTROYED BASES YIELDS THE BASE SEQUENCE OF THE ANALYZED STRAND.

Figure 5-3
The Maxam and Gilbert DNA-sequencing procedure. A segment of DNA is labeled at one end with ^{32}P. The labeled DNA is divided into four samples and each sample is treated with a chemical that specifically destroys one or two of the four bases in the DNA. The conditions of the reaction are controlled so that only a few sites are nicked in any one DNA molecule. When these nicked molecules are treated with piperidine, the DNA backbone is broken at the site at which the base had been destroyed. This generates a series of labeled fragments, the lengths of which depend on the distance of the destroyed base from the labeled end of the molecule. For instance, if there are G residues 5, 11, 15, and 22 bases away from the labeled end, then treatment of the DNA strand with chemicals that cleave at G will generate fragments 5, 11, 15, and 22 bases in length. (Such treatment will also, of course, give fragments of 6, 4, and 7 bases, and so on; these are the unlabeled fragments between the G residues.) The sets of labeled fragments obtained from each of the four reactions are run side by side on an acrylamide gel that separates DNA fragments according to size, and the gel is autoradiographed. The pattern of bands on the x-ray film is read to determine the sequence of the DNA.

dideoxy residue (Figure 5-4a). Thus four separate reactions, each containing a different dideoxy NTP, can be run, and their products displayed on an acrylamide gel (Figure 5-4b). With this technique the 5577-base-pair sequence of the phage G4, a relative of ϕX174, was determined quickly.

The exact sequence of any segment of DNA of reasonable size is now a feasible project, and already the base sequences are known for the whole genomes of the SV40 and polyoma viruses, and for the operators and promoters used in the regulation of several bacterial operons (lactose and galactose) and several eukaryotic genes. From the sequence of a gene it is a simple matter to deduce the amino acid sequence of the protein it specifies; in fact, nowadays it is often faster to determine the sequence of a protein by this indirect

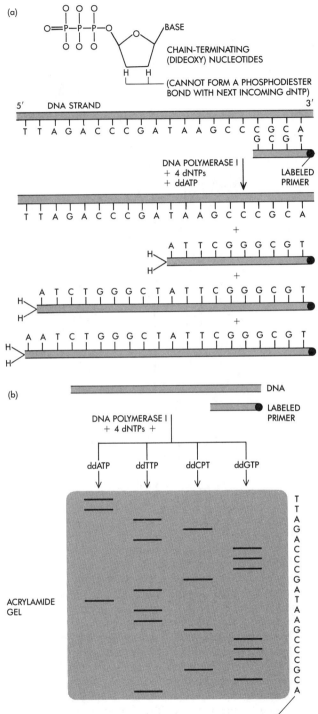

Figure 5-4

The Sanger DNA-sequencing procedure.
(a) 2',3'-Dideoxynucleotides of each of the four bases are prepared. These molecules can be incorporated into DNA by *E. coli* DNA polymerase because they have a normal 5' triphosphate; however, once incorporated into a growing DNA strand, the dideoxynucleotide (ddNTP) cannot form a phosphodiester bond with the next incoming dNTP. Growth of that particular DNA chain stops. A Sanger sequencing reaction consists of a DNA strand to be sequenced, a short labeled piece of DNA (the primer) that is complementary to the end of that strand, a carefully controlled ratio of one particular dideoxynucleotide with its normal deoxynucleotide, and the other three dNTPs. When DNA polymerase is added, normal polymerization will begin from the primer; when a ddNTP is incorporated, the growth of that chain will stop. If the correct ratio of ddNTP:dNTP is chosen, a series of labeled strands will result, the lengths of which are dependent on the location of a particular base relative to the end of the DNA.

(b) A DNA strand to be sequenced, along with labeled primer, is split into four DNA polymerase reactions, each containing one of the four ddNTPs. The resultant labeled fragments are separated by size on an acrylamide gel, autoradiography is performed, and the pattern of the fragments gives the DNA sequence.

route rather than by directly sequencing the protein. Whereas protein sequencing can take months and even years, DNA sequencing can often be accomplished in a matter of weeks.

Oligonucleotides Can Be Synthesized Chemically

The emergence of quick convenient methods for the synthesis of moderately long oligonucleotides with defined sequences has followed close upon the development of rapid sequencing methods. Until recently, it was a formidable task to put together only a few nucleotides. Now, however, sequences of 12 to 20 nucleotides can be synthesized in three or four days. Chemical synthesis is based on the ability to protect specifically (that is, to

prevent having a chemical reaction occur at) either the 5′ or the 3′ end of a mono- or oligonucleotide. This is done by hanging a large blocking group onto either the 5′ or the 3′ hydroxyl (Figure 5-5). Different blocking groups are used: Some can be removed with acid, some with base. Thus a 5′-blocked mononucleotide can be chemically condensed with a 3′-blocked molecule, resulting in a dinucleotide that is blocked at both ends. Either the 5′ or the 3′ blocking group is then removed (using either acid or base) and the dinucleotide is reacted with an appropriately unblocked mono- or dinucleotide. This cycle of condensation, removal of one or the other blocking group, and recondensation can be repeated many times until the oligonucleotide of the desired length is obtained.

Recently, it has been found possible to attach the first nucleotide to a solid support and to add nucleotides in a stepwise fashion, washing the support matrix between each step. This procedure is, of course, amenable to automation.

Eco RI Produces Fragments Containing Sticky (Cohesive) Ends

Restriction enzymes like *Hin*dII break DNA at the center of their recognition sites to produce blunt-ended fragments that are base-paired out to their ends and have no tendency to stick together (Figure 5-6). In contrast, the *Eco* RI enzyme makes staggered cuts that create short four-base single-stranded tails on the ends of each fragment (Figure 5-7). Many other restriction enzymes also make staggered cuts, and many specific tail sequences are known. Complementary single-stranded tails tend to associate by base pairing and thus are often called cohesive, or sticky, ends. For example, the linear molecules that *Eco* RI generates by cutting circular SV40 DNA often temporarily recyclize by base pairing between their tails. Fragments held together by such base pairing can be permanently rejoined by adding the enzyme DNA ligase to catalyze the formation of new phosphodiester bonds.

Base pairing occurs only between complementary base sequences, so the cohesive AATT ends produced by *Eco* RI will not, for example, pair with the AGCT ends produced by *Hin*dIII. But any two fragments (regardless of their origin) produced by the same enzyme can stick together and later be joined together permanently by the action

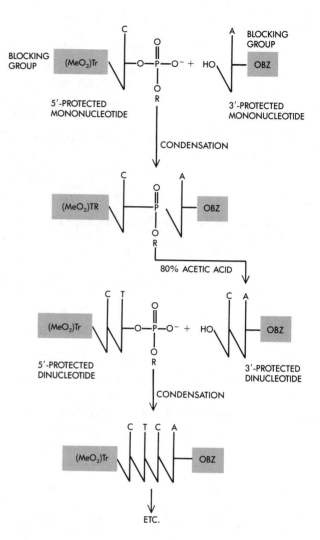

Figure 5-5
Chemical synthesis of an oligonucleotide. Blocking groups are added to the 5′ or 3′ ends of mononucleotides. These blocking groups are large organic residues that prevent the formation of a phosphodiester bond; they can be specifically removed when desired. If a 5′-protected nucleotide is condensed with a 3′-protected nucleotide, a normal phosphodiester bond will form between the two unblocked ends, resulting in a dinucleotide that is blocked at both ends. Now the blocking group at either the 5′ or the 3′ end of the dinucleotide is removed by use of either acid or base. A 5′-protected dinucleotide can be condensed with a 3′-protected mononucleotide (or dinucleotide) to form a trinucleotide (or tetranucleotide). This cycle of condensation, specific removal of the blocking group from one end, and recondensation is repeated until the oligonucleotide of desired length is synthesized.

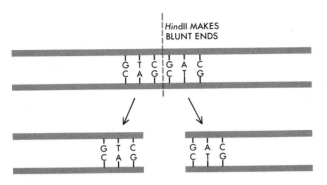

Figure 5-6
The *Hin*dII restriction enzyme cuts DNA at the center of its recognition site, leaving blunt ends.

of the enzyme DNA ligase (Figure 5-7). Such experiments were first done at Stanford in 1972 by Janet Mertz and Ron Davis, who realized that *Eco*RI in conjunction with DNA ligase would provide a general way to achieve *in vitro* site-specific genetic recombination.

Many Enzymes Are Involved in DNA Replication

The enzyme DNA ligase, which may be used to seal together restriction fragments, is but one of many enzymes that are now known to be involved in DNA replication. The first enzyme to make DNA chains *in vitro,* DNA polymerase I, was for almost ten years believed to be the main—if not the only—polymerase needed to link up deoxynucleotides into DNA chains. In 1967, however, an *E. coli* mutant was found that had almost no DNA polymerase I yet that made DNA at normal rates. Within a year, two new DNA polymerases were discovered, of which DNA polymerase III is now considered the enzyme involved in most DNA chain elongation. Much of the inherent complexity of DNA replication arises because the two chains of the double helix run in opposite directions $(5' \rightarrow 3'$ and $3' \rightarrow 5')$, and their daughter strands must likewise run in opposite directions. Yet the elongation of all individual daughter chains occurs in the $5' \rightarrow 3'$ direction. This apparent paradox was resolved by the realization that one daughter strand grows continuously in one direction while the other daughter strand is made discontinuously from smaller pieces that are individually elongated in the opposite direction. This feature necessitates both a special polymerase to fill in the gaps (most likely DNA polymerase I) and the joining enzyme DNA ligase (Figure 5-8, page 66).

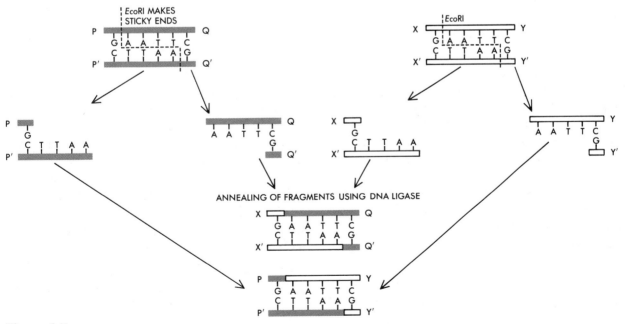

Figure 5-7
The *Eco*RI restriction enzyme makes staggered, symmetrical cuts in DNA away from the center of its recognition site, leaving cohesive or "sticky" ends. A sticky end produced by *Eco*RI digestion can anneal to any other sticky end produced by *Eco*RI cleavage.

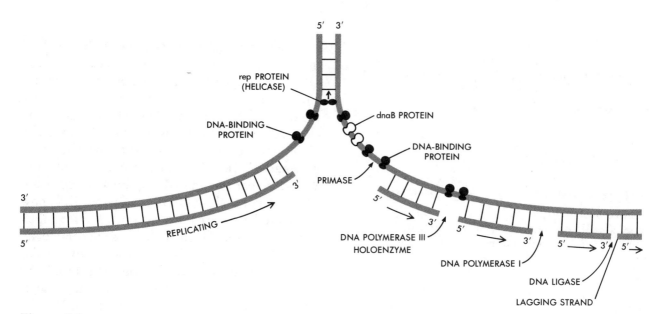

Figure 5-8
The enzymes involved in DNA replication in *E. coli.* Several enzymes have been found to be necessary for DNA replication in *E. coli.* DNA polymerase cannot initiate chains *de novo,* but requires a primer. This is provided by an RNA polymerase called a primase, which in the presence of the dnaB protein synthesizes a short stretch of RNA. DNA polymerase III can then take over and use this RNA as a primer to continue the synthesis of DNA. A protein called rep is necessary to unwind the DNA helix to allow replication. A single-stranded DNA-binding protein is also necessary to stabilize the single-stranded regions of DNA that are transiently formed during the replication process. Finally, since DNA polymerase can synthesize DNA in only the 5'-to-3' direction, one of the strands must be synthesized discontinuously (the "lagging strand"). This leads to a series of short stretches of DNA with gaps in between. These gaps are filled by the action of DNA polymerase I and sealed with DNA ligase.

In addition, other enzymes edit the DNA to remove erroneously incorporated bases or repair chains damaged by agents like ultraviolet light or x-rays. Still more proteins are needed to separate the parental strands at the replication fork, as well as to bind temporarily to single-stranded regions prior to their conversion to double helices. Also needed are several specific proteins involved in initiating DNA chains.

Most of these proteins were found over the last twenty years in the laboratory of Arthur Kornberg at Stanford University or by his former collaborators working elsewhere (Figure 5-8).

Sticky Ends May Be Enzymatically Added to Blunt-Ended DNA Molecules

The calf thymus enzyme terminal transferase, which adds nucleotides to the 3' ends of DNA chains, provides a general method for creating cohesive ends on blunt-ended DNA fragments. For example, if polydeoxy A (AAAA . . .) is added to

the two 3' ends of one double-stranded fragment, and polydeoxy T (TTTT . . .) is added to the 3' ends of another fragment, the two fragments, when mixed together, can form base pairs between their complementary tails. Appropriate enzymes can be added to fill in any single-stranded gaps, and finally DNA ligase can be used to permanently join the two fragments together (Figure 5-9). These procedures, developed at Stanford in 1971–1972 by Peter Lobban and Dale Kaiser, and by David Jackson and Paul Berg, provide a second general method for creating recombinant DNA molecules. They do, however, introduce regions

of $\dfrac{\text{AAAA} \ldots}{\text{TTTT} \ldots}$ base pairs at the junctions be-

tween the fused fragments. Such additional base sequences could affect the function of the joined molecules, and whenever possible, cohesive ends generated by restriction enzyme cuts are used to create recombinant DNA molecules.

Small Plasmids Are Vectors for the Cloning of Foreign Genes

The realization that *Eco* RI generates specific cohesive ends that can later be sealed up by DNA ligase was followed within a year by the development of the first practical method for systematically cloning specific DNA fragments, regardless of their origin. The essential trick was the random insertion of the *Eco* RI fragments of a DNA molecule into circular plasmid DNA that also had been cut with *Eco* RI. This procedure led to hybrid plasmids, which could be used to infect bacteria. Each bacterial cell acquired a recombinant plasmid carrying a specific foreign DNA fragment. In the first such experiments, carried out in early 1973 by Herbert Boyer and Stanley Cohen and their collaborators at Stanford and the University of California, San Francisco, the small *E. coli* plasmid pSC101 (Figure 5-10) was used because it contained only a single *Eco* RI recognition site and was converted by *Eco* RI into linear molecules. When such DNA was mixed with foreign DNA fragments also possessing cohesive ends generated by *Eco* RI, and DNA ligase was added, new hybrid plasmids were created (Figure 5-11, page 68). Each contained one or more pieces of foreign DNA inserted into the RI site of the plasmid.

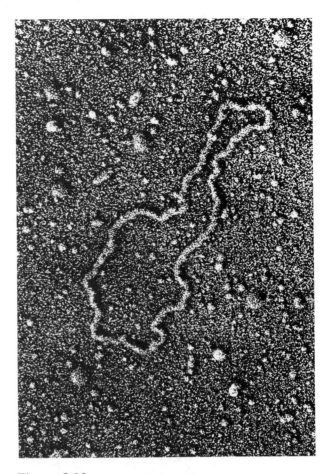

Figure 5-10
Electron micrograph of plasmid pSC101. (Courtesy of Stanley N. Cohen, Stanford University.)

In the original Boyer–Cohen use of plasmid pSC101, the foreign DNA was that of another plasmid, and the recombinant was a new plasmid containing two origins of replication. The possibility then existed of doing further experiments in which all sorts of foreign DNA, both from microbes and from higher plants and animals, would be inserted into these plasmids. For example, the *E. coli* chromosome with its 4 million base pairs contains about 500 different recognition sites for *Eco* RI. By random insertion of the *Eco* RI fragments into pSC101, it would be possible to clone all the *E. coli* genes in the form of fragments that would be easily isolatable for subsequent genetic as well as biochemical manipulation.

DNA of Higher Organisms Becomes Open to Molecular Analysis

Most important were the possibilities that recombinant DNA opened up for the analysis of DNA

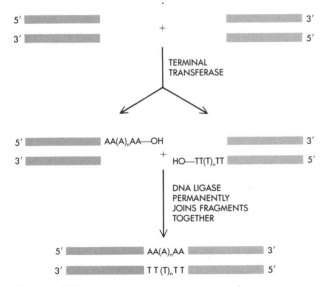

Figure 5-9
Two blunt-ended DNA fragments can be joined together by adding polydeoxy-A or polydeoxy-T tails to the ends of the fragments. The complementary tails will form base pairs, and enzymes can be used to fill in any single-stranded gaps and join the fragments permanently.

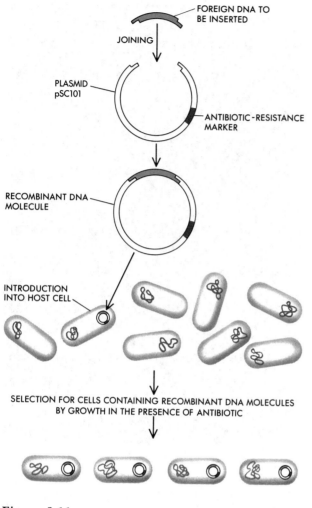

Figure 5-11
The cloning of DNA in a plasmid.

from plants and animals. Although transducing phages carrying specific parts of bacterial chromosomes had long been available, no one had succeeded in constructing transducing animal viruses carrying specific chromosomal genes such as the ones that code for hemoglobin or the muscle proteins actin and myosin. Now by randomly inserting large numbers of different restriction fragments of human DNA into the proper bacterial plasmids (this procedure is called "shotgunning" DNA), we had a high likelihood of subsequently generating one or more bacterial clones containing the recombinant plasmid carrying the specific human gene that was wanted. The moment an appropriate selection technique for a specific mRNA species (such as human hemoglobin mRNA) could be devised, finding the right clone would be no problem. Furthermore, given the

right clones, direct searches could be made for DNA-binding proteins with possible control roles. Moreover, fine-structure analysis of genes should then be virtually commonplace.

With the advent of new recombinant DNA procedures, we now had the perfect method of isolating desired DNA restriction fragments, if not the intact genomes of DNA tumor viruses. Previously, when growing small batches of tumor viruses by conventional cell culture methods, we had never been absolutely sure that we were not subjecting ourselves to some risk of cancer. If such a risk existed, it was assumed to be small because the viruses used (like SV40 or polyoma) were quite similar to common human viruses that apparently did not cause tumors. Nonetheless, it would be a relief to researchers to be able to stop growing intact tumor viruses in large batches, and it was clear that recombinant DNA procedures should be taken up by tumor virologists at maximum possible speed.

Scientists Voice Concerns About the Dangers of Unrestricted Gene Cloning

However, an earlier talk by Janet Mertz, then a student in Paul Berg's laboratory at Stanford, about the possibility of cloning SV40 in *E. coli* and its phages had already started a counterreaction. Particularly concerned was the cell biologist Robert Pollack, then working on SV40-mediated transformation of mouse cells at Cold Spring Harbor. He had already had doubts about the safety of SV40 itself, and he was even more concerned that SV40-carrying bacteria could conceivably act as a vector to transmit human cancer. After hearing Mertz talk in the summer of 1971 at Cold Spring Harbor about Berg's plans to use the currently studied terminal transferase tails (such as AAAA and TTTT) to genetically engineer SV40, Pollack telephoned Berg to ask him if he had worried about the potential biohazards of recombinant DNA. Although Berg indicated that he saw no reason to panic, he said he would consider postponing such work, and in fact later he made the decision to refrain from cloning tumor virus genomes.

These events, which took place two years before the antibiotic-resistance plasmids were employed as cloning vectors, no doubt reflected an undercurrent of concern among molecular biolo-

gists that they knew too little about tumor viruses to treat them casually. Also to be considered was the intense excitement created by the new restriction enzymes like *Eco* RI. Their mere existence was sufficient to keep Berg's laboratory, as well as that of other leading tumor virologists, very busy over the next several years without needing to employ recombinant DNA technologies immediately.

A further concern surfaced in the following year as the opinion spread that the DNA of mice, and possibly of all higher cells, harbored latent RNA tumor virus DNA. Many of the recombinant plasmids in which scientists hoped to find human antibody genes might instead contain genes capable of causing cancer. If so, should all recombinant DNA experiments using vertebrate DNA be regarded as potentially dangerous? The answers were mixed. Those hoping to soon do a particular experiment, say with human DNA, had no worries about either their own safety or that of others. Yet often these same people would question the advisability of medically oriented microbiologists making recombinant plasmids containing mixtures of many antibiotic-resistance markers, arguing that maybe they would spread out of control through human populations. And virtually everybody was united in the opinion that researchers should not have unrestricted freedom to do experiments that might have military consequences. Certain experiments *sounded* nasty, even though no one concerned knew any of the technical aspects of biological warfare that had long remained tightly classified in military secrecy.

The first debates as to whether recombinant DNA should be used in the laboratory thus preceded researchers' actual ability to move full speed ahead. At first such conversations aroused no really deep emotions, either among those who said they were concerned or among those who thought that the whole matter was academic silliness created by naive souls who had never worked with real pathogens. Almost everyone was deep into his or her experiments of the moment, and concern over experiments that would not come to

pass for at least a year or two seemed of little consequence.

But after the first *Eco* RI–pSC101 cloning experiments were announced, it became clear that it would soon be impossible to go on doing molecular genetics as if recombinant DNA did not exist. The question was whether to move ahead as fast as possible or to try to invent methods that would reassure the worriers without straight-jacketing most future explorations of recombinant DNA.

Guidelines for Recombinant DNA Research Are Proposed at the Asilomar Conference

In an attempt to decide whether any restrictions should be placed on recombinant DNA research, a gathering of more than 100 internationally respected molecular biologists was held in February of 1975 at the Asilomar Conference Center located near Monterey, California. In the absence of knowledge about whether any danger might exist, a nearly unanimous consensus emerged to restrict further cloning of DNA to organisms that had been specifically genetically disabled so that they would not grow well outside of the test tube. The use of such "safe" vectors was partly intended to reassure the public that recombinant DNA procedures would not result in the "escape" of new types of organisms that might have the potential for causing diseases or that might in some other way upset the normal balance of life. Afterwards, the "Asilomar recommendations" were considered by a special committee appointed by the National Institutes of Health (USA). In its deliberation it urged a tightening of the guidelines in a way that would effectively preclude the use of recombinant DNA procedures to study cancer viruses (cancer genes). These recommendations became codified in official governmental regulatory guidelines that took force in July 1976. They remained in effect until early in 1979. When they were relaxed, effective work on cancer viruses (genes) commenced.

READING LIST

Books

Grobstein, C. *A Double Image of the Double Helix: The Recombinant-DNA Debate.* W. H. Freeman and Company, San Francisco, 1979.

Jackson, D., and S. Stich. *The Recombinant DNA Debate.* Prentice-Hall, Englewood Cliffs, N.J., 1979.

Kornberg, A. *DNA Replication.* W. H. Freeman and Company, San Francisco, 1980.

Kornberg, A. *1982 Supplement to DNA Replication.* W. H. Freeman and Company, San Francisco, 1982.

Morgan, J., and W. J. Whelan, eds. *Recombinant DNA and Genetic Experimentation.* Pergamon, Elmsford, N.Y., 1979.

Watson, J. D., and J. Tooze. *The DNA Story.* W. H. Freeman and Company, San Francisco, 1981.

Wu, R., ed. *Recombinant DNA, Methods in Enzymology, Vol. 68.* Academic, New York, 1979.

Original Research Papers (Reviews)

RNA SEQUENCING

Holley, R. W. "The nucleotide sequence of a nucleic acid." *Sci. Am.,* 214(2): 30–39 (1966).

Fiers, W., R. Contreras, F. Duerinck, G. Haegeman, D. Iserentant, J. Merregaert, W. Min Jou, F. Molemans, A. Raeymaekers, V. Berghe, G. Volckaert, and M. Ysebaert. "Complete nucleotide sequence of bacteriophage MS2 RNA: Primary and secondary structure of replicase gene." *Nature,* 260: 500–507 (1976). Correction in *Nature,* 26a: 810.

RESTRICTION ENZYMES AND MAPS

Linn, S., and W. Arber. "Host specificity of DNA produced by *Escherichia coli,* X. *In vitro* restriction of phage fd replicative form." *Proc. Natl. Acad. Sci. USA,* 59: 1300–1306 (1968).

Meselson, M., and R. Yuan. "DNA restriction enzyme from *E. coli.*" *Nature,* 217: 1110–1114 (1968).

Smith, H. O., and K. W. Wilcox. "A restriction enzyme from *Hemophilus influenzae,* I. Purification and general properties." *J. Mol. Biol.,* 51: 379–391 (1970).

Kelly Jr., T. J., and H. O. Smith. "A restriction enzyme from *Hemophilus influenzae,* II. Base sequence of the recognition site." *J. Mol. Biol.,* 51: 393–409 (1970).

Roberts, R. J. "Restriction and modification enzymes and their recognition sequences." *Nuc. Acids Res.,* 11: r135–r167 (1983).

Danna, K., and D. Nathans. "Specific cleavage of simian virus 40 DNA by restriction endonuclease of Hemophilus influenzae." *Proc. Natl. Acad. Sci. USA,* 68: 2913–2917 (1971).

Sharp, P. A., B. Sugden, and J. Sambrook. "Detection of two restriction endonuclease activities in *Hemophilus parainfluenza* using analytical agarose–ethidium bromide electrophoresis." *Biochemistry,* 12: 3055–3062 (1973).

SEQUENCING DNA

Sanger, F., G. G. Brownlee, and B. G. Barrel. "A two-dimensional fractionation procedure for radioactive nucleotides." *J. Mol. Biol.,* 13: 373–398 (1965).

Fiers, W., F. Contreras, G. Haegeman, R. Rogers, A. Vande Voorde, H. Van Heuverswyn, J. Van Herreweghe, G. Volckaert, and M. Ysebaert. "Complete nucleotide sequence of SV40 DNA." *Nature,* 273: 113–120 (1978).

Reddy, V. B., B. Thimmappaya, R. Dhar, K. N. Subramanian, B. S. Zain, J. Pan, P. K. Ghosh, M. L. Celma, and S. M. Weissman. "The genome of simian virus 40." *Science,* 200: 494–502 (1978).

Sanger, F., and A. R. Coulson. "A rapid method for determining sequences in DNA by primed synthesis with DNA polymerase." *J. Mol. Biol.,* 94: 444–448 (1975).

Maxam, A. M., and W. Gilbert. "A new method of sequencing DNA." *Proc. Natl. Acad. Sci. USA,* 74: 560–564 (1977).

Sanger, F., S. Nicklen, and A. R. Coulson. "DNA sequencing with chain-terminating inhibitors." *Proc. Natl. Acad. Sci. USA,* 74: 5463–5467 (1977).

Sanger, F., G. M. Air, B. G. Barrel, N. L. Brown, A. R. Coulson, J. C. Fiddes, C. A. Hutchison III, P. M. Slocombe, and M. Smith. "Nucleotide sequence of bacteriophage ϕX174." *Nature,* 265: 687–695 (1977).

Sutcliffe, G. "Complete nucleotide sequence of the *E. coli* plasmid pBR322." *Cold Spring Harbor Symp. Quant. Biol.,* 43: 77–90 (1979).

OLIGONUCLEOTIDE SYNTHESIS

Heyneker, H. L., J. Shine, H. M. Goodman, H. Boyer, J. Rosenberg, R. E. Dickerson, S. A. Narang, K. Itakura, S. Linn, and A. D. Riggs. "Synthetic *lac* operator is functional *in vivo.*" *Nature,* 263: 748–752 (1976).

Gait, M. J., and R. C. Sheppard. "Rapid synthesis of oligodeoxyribonucleotides: A new solid-phase method." *Nuc. Acids Res.,* 4: 1135–1158 (1977).

Khorana, H. G. "Total synthesis of a gene." *Science,* 203: 614–625 (1979).

Itakura, K., and A. D. Riggs. "Chemical DNA synthesis and recombinant DNA studies." *Science,* 209: 1401–1405 (1980).

ENZYMOLOGY OF DNA REPLICATION

Kornberg, A. "Biologic synthesis of deoxyribonucleic acid." 1959 Nobel Prize lecture reprinted in *Science,* 131: 1503–1508 (1960).

Weiss, B., and C. C. Richardson. "Enzymatic breakage and joining of deoxyribonucleic acid, I. Repair of single-strand breaks in DNA by an enzyme system from *Escherichia coli* infected with T4 bacteriophage." *Proc. Natl. Acad. Sci. USA,* 57: 1021–1028 (1967).

Olivera, B. M., and I. R. Lehman. "Linkage of polynucleotides through phosphodiester bonds by an enzyme from *Escherichia coli.*" *Proc. Natl. Acad. Sci. USA,* 57: 1426–1433 (1967).

Zimmerman, S. B., J. W. Little, C. K. Oshinsky, and M. Gellert. "Enzymatic joining of DNA strands: A novel reaction of diphosphopyridine nucleotide." *Proc. Natl. Acad. Sci. USA,* 57: 1841–1848 (1967).

THE FIRST RECOMBINANT DNA MOLECULES

Mertz, J. E., and R. W. Davis. "Cleavage of DNA by RI restriction endonuclease generates cohesive ends." *Proc. Natl. Acad. Sci. USA,* 69: 3370–3374 (1972).

Jackson, D., R. Symons, and P. Berg. "Biochemical method for inserting new genetic information into DNA of simian virus 40: Circular SV40 DNA molecules containing lambda phage genes and the galactose operon of *Escherichia coli.*" *Proc. Natl. Acad. Sci. USA,* 69: 2904–2909 (1972).

Lobban, P., and A. D. Kaiser. "Enzymatic end-to-end joining of DNA molecules." *J. Mol. Biol.,* 79: 453–471 (1973).

Cohen, S., A. Chang, H. Boyer, and R. Helling. "Construction of biologically functional bacterial plasmids *in vitro.*" *Proc. Natl. Acad. Sci. USA,* 70: 3240–3244 (1973).

6

The Isolation of Cloned Genes

In the mid-1970s, our ability to exploit recombinant DNA methods to full potential faced several obstacles. One was the need for development of disabled hosts and vectors that would have no significant probability of surviving outside the laboratory, and that would satisfy the criteria of biological containment established in the 1976 National Institutes of Health guidelines (Chapter 5). Another, more fundamental obstacle was the need for methods that would allow the identification of bacteria that carried the cloned genes in which we were interested.

"Safe" Bacteria and Plasmid Vectors Are Developed

The release of the original NIH guidelines thus did not let recombinant DNA research take off immediately. In the first place, none of the so-called "safe" bacterial hosts and plasmid and phage vectors had yet been developed and certified as meeting the safety criteria specified by the guidelines. At Asilomar in February 1975, there was talk that only a few weeks' work would be necessary to make "safe" bacteria or phage, but it was not until late 1976 that the first of the "safe" bacteria (the EK2 category) became available. Certification of the plasmid (phage) vectors, into which DNA fragments were to be inserted, consumed more time. Even after the Recombinant DNA Advisory Committee (RAC) gave its approval to the now widely used pMB9 and pBR322 plasmids, several months passed before they were formally certified by the NIH directorate in the late spring of 1977. Not surprisingly, the RAC's approval was interpreted as a green light by sev-

eral European laboratories, who saw no reason to be held back by procedural restraints imposed in the United States.

The first "safe" E. coli K12 strain was developed in 1976 by Roy Curtiss III (University of Alabama), who named it χ1776 after the Bicentennial. Among the many features that should prevent the escape of χ1776 to the outside world is its metabolic requirement for diaminopimelic acid, an intermediate in the biosynthesis of lysine, and a substance that is not present in human intestines. χ1776 also possesses a fragile cell wall that bursts open in low salt concentrations or in the presence of even a trace of detergent. Unfortunately, those who began using χ1776 found it difficult to work with. It grew to much lower cell densities than ordinary E. coli K12 strains, and even when enough cells had been grown, introduction of recombinant DNA molecules into χ1776 cells proved difficult. Much effort was therefore spent in developing other "safe" E. coli K12 derivatives that eventually were officially certified as approved EK2 hosts.

The guidelines also made it necessary to use only plasmids that had been modified by mutational events so that they could not move by a sexual process within the human intestine from a "safe" to an "unsafe" strain of bacteria. Fortunately, it is a simple matter to construct new plasmids lacking all the genes controlling movement, or mobilization, of plasmids from cell to cell.

Why Use Drug-Resistance Plasmids?

One of the most obvious concerns about the biosafety of recombinant DNA research was that it

might contribute to the further spread of bacterial drug resistance. The extensive use and abuse of antibiotics in medicine and animal husbandry have caused many natural strains of bacteria to become resistant to the most common antibiotics. Usually this resistance depends on acquisition by the bacterial cell of a plasmid that specifies enzymes that can break down the drugs.

As vectors for recombinant DNA, plasmids with genes for antibiotic resistance offer a great advantage. When they are used together with host bacterial cells that are plasmid-free and therefore antibiotic-sensitive, the entry of the resistance plasmid carrying the recombinant DNA can be easily detected. The plasmid pBR322 is an example of one that carries genes for resistance against two antibiotics, ampicillin and tetracycline. Moreover, sites for restriction enzymes are present within these antibiotic-resistance genes (Figure 6-1). So when a piece of foreign DNA is recombined into one or the other drug-resistance gene, that gene is inactivated. This means that the successful insertion of a piece of foreign DNA into one of the antibiotic-resistance genes is easily detected: The genetic potential for that resistance is eliminated (Figure 6-2). When the plasmid is introduced into a host bacterium, it results in the bacterium acquiring the resistance specified by the second, and still intact, resistance gene.

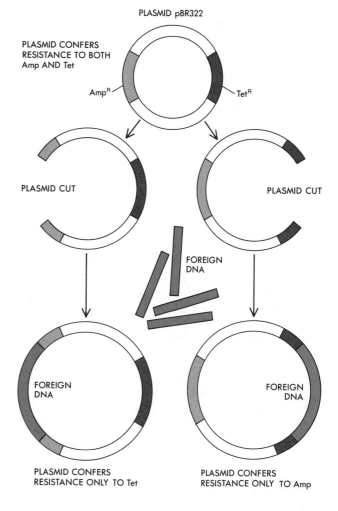

Figure 6-2
The insertion of foreign DNA into the ampicillin-resistance gene (Amp^R) or the tetracycline-resistance gene (Tet^R) of pBR322 inactivates that gene while the other drug-resistance gene remains functional.

By using plasmids carrying antibiotic-resistance genes that are already widespread in nature, by mutating the plasmids so that they cannot move spontaneously from cell to cell, and by using "safe" strains of bacteria, experiments with drug-resistance plasmids can be very useful without incurring a significant risk of contributing to the spread of antibiotic resistance. As some medical microbiologists pointed out during the recombinant DNA debate, even if these precautions were not taken, the contribution of recombinant DNA research to the spread of antibiotic resistance would have been trivial compared to that resulting from the excessive daily use of antibiotics in medicine and agriculture.

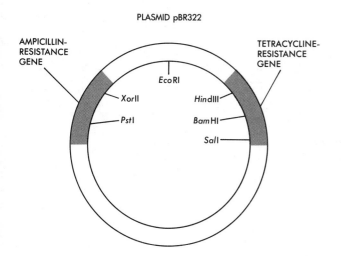

Figure 6-1
Several restriction enzymes have recognition sites within the antibiotic-resistance genes of plasmid pBR322.

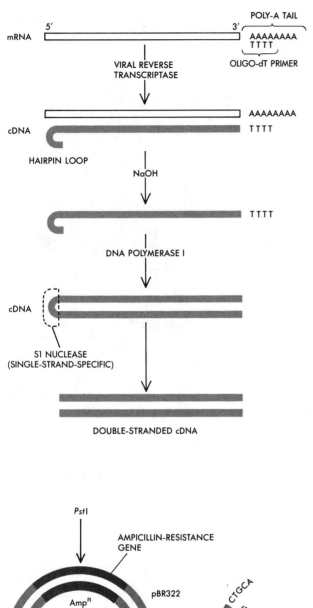

Probes for Cloned Genes

If we take the total DNA of a human cell, cut it into fragments with a suitable restriction enzyme, join the fragments to plasmid vectors, and introduce the recombinant DNAs into a population of bacteria, we have a so-called human gene "library." It is, however, a library without a proper index, and we are faced with the problem of separating the bacterium carrying the human gene we

Figure 6-3
The synthesis of double-stranded cDNA from mRNA. A short oligo-dT chain is hybridized to the poly-A tail of an mRNA strand. The oligo-T segment serves as a primer for the action of reverse transcriptase, which uses the mRNA as a template for the synthesis of a complementary DNA strand. The resulting cDNA ends in a hairpin loop. Once the mRNA strand is degraded by treatment with NaOH, the hairpin loop becomes a primer for DNA polymerase I, which completes the paired DNA strand. The loop is then cleaved by S1 nuclease to produce a double-stranded cDNA molecule.

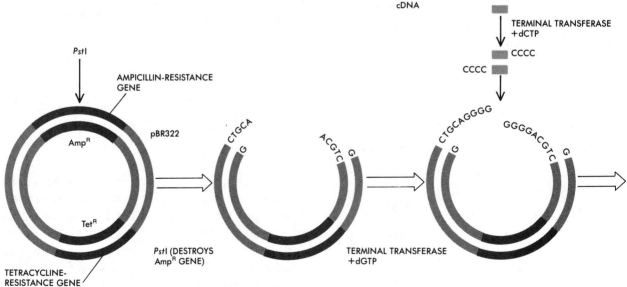

Figure 6-4
The cloning of cDNA by using homopolymeric tails. The Amp^R gene of pBR322 is cleaved by *Pst*I, and terminal transferase and dGTP are used to add poly-G tails to the 3' single-stranded ends. Terminal transferase and dCTP are used to attach poly-C tails to the 3' ends of double-stranded cDNA, and the cDNA segment is then inserted into the plasmid. Enzymes are added to fill in the remaining gaps in the two strands of the plasmid. The result is a recombinant plasmid that can be inserted into bacteria and then detected by selecting for bacteria that are resistant to tetracycline but not to ampicillin.

want from the millions of others. To overcome this "needle-in-the-haystack" dilemma, nucleic acid probes are now used to exploit the complementarity of nucleic acid sequences. Such probes are radioactively labeled mRNAs specified by the genes to be isolated. Given such mRNA, the gene library can be screened for the bacteria that have DNA complementary to the mRNA probe.

There is a snag, however. It is generally impossible to obtain totally pure mRNA probes, even for proteins that are made in large amounts in certain differentiated cells. For example, in immature red blood cells, only about half the mRNA is for hemoglobin, the protein that in the mature red blood cells accounts for well over 90 percent of the total protein. If an impure preparation of mRNA is used as a probe for a cloned gene, many false positives can arise; the less pure the mRNA, the greater that risk. Fortunately, by a circuitous route involving what amounts to cloning the mRNA itself, pure probes can be obtained.

Synthesis and Cloning of cDNA

As we have mentioned (Chapter 4, page 52), virtually every eukaryotic mRNA molecule has at its 3'

end a run of adenine nucleotide residues called a poly-A tail. Whatever its function, the poly-A tail provides a very convenient way to synthesize a strand of DNA complementary to the mRNA: If short chains of oligo-dT are mixed with the mRNA, they hybridize to the poly-A tail to provide a primer for the action of the enzyme reverse transcriptase (Figure 6-3). This enzyme, which is isolated from certain RNA tumor viruses (Chapter 4, page 54), can use RNA as a template to synthesize a DNA strand. (Its name comes from its ability to reverse the normal first step of gene expression.) The result of the reaction is an RNA–DNA hybrid; the newly synthesized DNA strand has a hairpin loop at its end, apparently as a result of the enzyme "turning the corner" and starting to copy itself. The hairpin loop may be an *in vitro* artifact, but it does provide a very convenient primer for the synthesis of the second strand of DNA. The resultant double-stranded cDNA has the hairpin loop intact; this can be cleaved by S1 nuclease, a single-strand-specific nuclease.

The double-stranded cDNA molecule so obtained is then inserted into pBR322, either by tailing with terminal transferase (Figure 6-4), as described earlier (Figure 5-9), or by attaching artificial restriction enzyme sites onto the ends of the

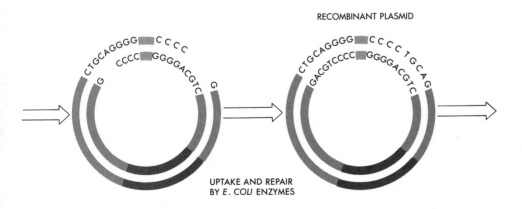

RECOMBINANT PLASMID

UPTAKE AND REPAIR BY *E. COLI* ENZYMES

SELECT FOR TETRACYCLINE-RESISTANT COLONIES

cDNA. These restriction enzyme sites, called "linkers," are eight-to-ten-base-pair oligonucleotides that are synthesized chemically (Chapter 5, page 63). The linkers are added to the double-stranded cDNA by using DNA ligase. Then the linkers are cut open with the restriction enzyme, and the cDNA, now containing sticky ends generated by the enzyme, is inserted into pBR322 that has been cleaved with the same enzyme (Figure 6-5). The resulting cDNA-containing recombinant plasmid is then introduced into the appropriate *E. coli* strain and propagated.

Identifying Specific cDNA Clones

If the set of experiments just described is done with a mixed population of mRNAs—which in practice is always the case because pure species of mRNA are not easily obtained—the cDNA also will be mixed. But the system can be arranged so that each bacterial cell receives only one cDNA recombinant molecule. Each cell and its descendant will carry only one sort of cDNA. The problem now is to discover which sort of cDNA is present in each cell.

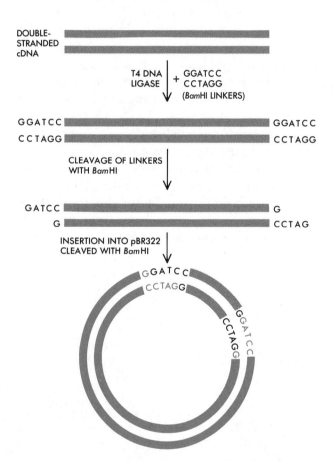

Figure 6-5
The cloning of cDNA by using *Bam* HI linkers. *Bam* HI linkers are added to the ends of a double-stranded cDNA molecule. The linkers are then cleaved with *Bam* HI, and the enzyme is also used to cleave a pBR322 molecule. The sticky ends on the cDNA and at the cleavage site in the plasmid are complementary, allowing the cDNA to be inserted into the plasmid.

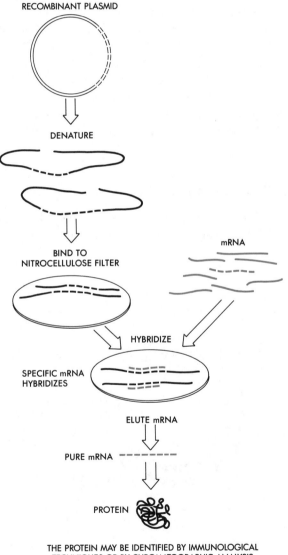

THE PROTEIN MAY BE IDENTIFIED BY IMMUNOLOGICAL TECHNIQUES OR BY CHROMATOGRAPHIC ANALYSIS

Figure 6-6
Identification of specific cDNA plasmids.

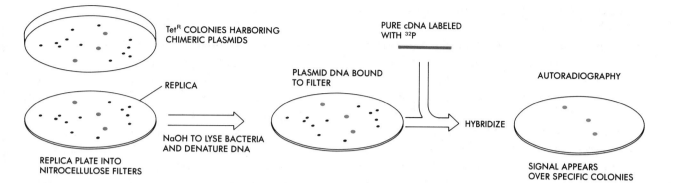

Figure 6-7
Identification of bacterial colonies harboring a specific cDNA plasmid. A cDNA library is prepared in pBR322 and used to transform *E. coli.* The resulting colonies can be "replica-plated" onto nitrocellulose filters. A sheet of nitrocellulose is placed on the plate with the bacterial colonies. When the filter is peeled off, some of the bacteria in each colony will have transferred to the filter, and the rest will remain on the plate. The pattern of colonies on the plate and on the filter will thus be identical. The nitrocellulose filter is then treated with NaOH, which both lyses the bacteria and denatures the DNA. When the filter is baked in a vacuum oven, the DNA becomes bound to the nitrocellulose. A pure cDNA or mRNA labeled with ^{32}P is then hybridized to the filter. Bacteria that harbor a recombinant plasmid containing sequences homologous to the labeled probe will hybridize to it, and give a signal following autoradiography. The position of the positive colony on the replica filter can be compared with the master plate, and that bacterial colony can be picked and expanded.

This is done by picking single cells and growing them into large homogeneous cultures. The cDNA-containing plasmids from this genetically pure, clonal population of cells are then extracted, and their DNA is denatured so that the strands separate. The separated DNA strands can then be bound to nitrocellulose filters. When mixtures of mRNAs are passed through these filters, only the mRNA molecules that are complementary to the pure cDNA on the filters are retained. Later the bound mRNA can be washed off the filters, concentrated, and added to a cell-free system that translates mRNA into protein. The protein specified by the cDNA insert can often be identified (Figure 6-6). When this is the case, we have a pure cDNA probe for a single gene and its corresponding mRNA.

Clearly, the more abundant a given mRNA species is within an mRNA preparation, the more easily its respective cDNA probe may be obtained. So, naturally, recombinant DNA methods were first used to make probes for abundant proteins such as hemoglobin and the antibodies made in myelomas, cancer cells that produce antibody. Early on, cDNA probes were also made for ovalbumin, which constitutes more than half the protein made in hormonally stimulated cells in the chicken oviduct.

Once one cDNA probe for a given protein (and gene) has been made, it can be subsequently used with the nitrocellulose filter technique to rapidly screen large numbers of bacterial colonies for other recombinant—or, as they are sometimes called, chimeric—plasmids carrying the same or closely related gene sequences (Figure 6-7).

In this way many cDNA probes for proteins such as hemoglobin have been isolated and tested to see whether they represent complete rather than partial probes. One procedure measures the sizes of the RNA–DNA heteroduplexes formed after, say, a hemoglobin mRNA preparation is mixed with denatured cDNA probes. The cDNA clones that formed heteroduplexes with the entire mRNA are then analyzed to see if the nucleotide sequences match those of the respective proteins. As expected, this is precisely the case, with the additional finding that the mRNA molecules of many genes contain more nucleotide sequences at their 5′ ends. These extra sequences specify protein leader segments that are cleaved off soon after synthesis and whose presence is therefore often unsuspected (Chapter 7, page 96).

Genomic Fragments Are Cloned in Bacteriophage λ

Once the desired cDNA plasmid probes became available, it was possible to look directly at the structures of the chromosomal genes themselves rather than at their mRNA transcripts. Only by examining the potentially much longer fragments obtainable from the chromosomes could the possible regulatory sequences outside the 5′ and 3′ ends of the direct coding sequences of a gene be examined. Many thousands of chimeric plasmids—each bearing a specific fragment of, for example, the total human, mouse, or chicken DNA—are easily prepared, and by screening sufficiently large numbers, the entire human, mouse, or chicken genome can be examined for those fragments carrying, say, hemoglobinlike sequences. Soon, however, it became obvious that plasmids bearing large chromosomal DNA insertions were not stable and tended as they replicated to give rise to smaller plasmids in which increasing amounts of the inserted DNA had been deleted. This is a consequence of the fact that the less DNA a plasmid has, the faster it can multiply. Thus, genetic segments unnecessary for plasmid multiplication invariably tend to be lost. Inserts of cDNA probes are of course subject to the same elimination pressures, but deletions only become commonplace when the inserted fragments are several thousand base pairs larger than most cDNA clones.

In contrast to their instability in plasmids, large chromosomal DNA fragments (of about 15,000 base pairs, or 15 kilobases) are essentially stable when inserted into the DNA of specially prepared strains of phage λ (Figure 6-8). Already at the 1975 Asilomar conference it had been suggested that phage λ could be mutated so that it would be unable to insert its DNA into that of host E. coli cells, and thus it would be at least as safe as, if not safer than, disabled plasmid vectors in E. coli strains with properties similar to those of χ1776. Such "safe" λ vectors exploit the fact that the entire central section of phage λ DNA is not necessary for its replication in E. coli, but only functions to ensure the integration of the phage DNA into the host bacterial chromosome during its lysogenic phase. Special λ strains have been created in which recognition sites for the restriction enzyme EcoRI are located so as to leave intact the left and right end fragments of the viral DNA that are essential for its replication. After EcoRI cutting, these end fragments, because of their relatively large sizes, can easily be purified from all the other fragments generated by EcoRI, and can later be used to make new λ-like phages containing one left fragment, one right fragment, and one foreign DNA insert in the 15-kilobase range. Most fortunately, maturation of phage λ requires that its DNA chromosome be approximately 45 kilobases long; thus the only DNAs that are constructed in vitro and that can multiply following such manipulations are chimeric mixtures of phage ends and foreign DNA of the appropriate length.

Once a "library" of λ phage carrying specific eukaryotic genes has been constructed, it is easily screened with cDNA plasmid probes, again by using the probe's radioactivity to mark out the phage colonies (plaques) bearing complementary DNA sequences. The haploid number of chromosomes of a mammalian cell contains a total of about 3×10^9 base pairs of DNA. When DNA is cloned in bacteriophage λ, each fragment contains on the average 1.5×10^4 base pairs. Screening a mere million phage plaques will effectively sample all the DNA of the mammalian cell for a given gene. Thus, for a specific cDNA probe, at worst only a few weeks may be necessary to screen a phage λ library for the respective genes.

Cosmids Allow the Cloning of Larger Segments of Foreign DNA

Cloning the eukaryotic genome in phage λ limits the size of the eukaryotic segment that can be cloned to about 15 kilobases (kb). Often this is

long enough to contain an entire gene and its immediate flanking sequences. However, as we shall see, many genes have turned out to be much larger than was anticipated; some of them contain on the order of 35 to 40 kb. Also, cloning in phage λ does not usually permit the isolation of two linked genes on one recombinant DNA molecule. A technique for cloning much larger eukaryotic fragments in *E. coli* is known as "cosmid cloning." This technique makes use of the fact that, as we mentioned in Chapter 3, λ phage contains at each end single-stranded complementary stretches of DNA, the so-called "cos" sites. During the normal life cycle of λ, hundreds of copies of λ DNA form a long chain or concatamer, each λ genome being joined to the next one in the chain through the cos sites. The λ packaging enzymes chop this concatamer into λ-sized pieces by recognizing two cos sites 35 to 45 kb apart, cleaving this unit, and packaging it into phage heads. Thus, the cos sites are all that is necessary for packaging DNA into

phage heads. The cos sites of λ phage have been cloned into the ampicillin-resistance gene of pBR322, leaving the tetracycline-resistance gene intact. Eukaryotic DNA is partially cleaved with a restriction enzyme to give relatively large pieces of DNA. This cleaved DNA is then ligated to the cloned cos-site plasmids, which have been cleaved with a similar enzyme. Hopefully, each piece of eukaryotic DNA will end up with a cos site (with an intact Tet^R gene) at each end. When λ packaging extract is added, it will recognize and cleave 35-to-45-kb pieces of eukaryotic DNA flanked by cos sites and package these molecules into phage heads. Smaller pieces of eukaryotic DNA, even if they have cos sites at each end, will not be packaged. The phage heads containing such DNA are, of course, not viable as phage; they are used to infect *E. coli* and selected for tetracycline resistance. Once inside the *E. coli,* the eukaryotic DNA–cos hybrid molecule lives as a plasmid (Figure 6-9, page 80).

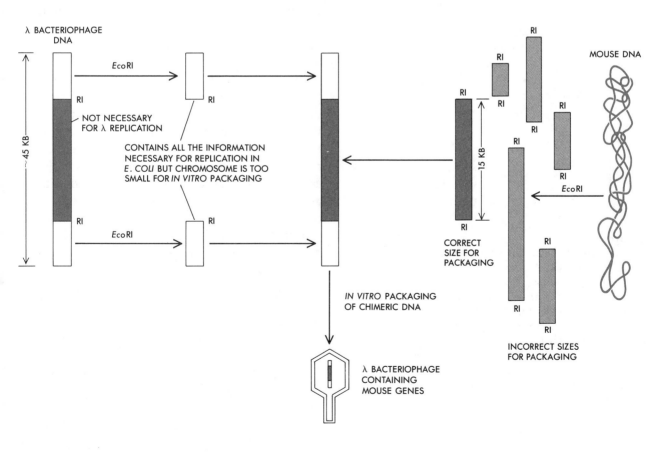

Figure 6-8
The cloning of eukaryotic genes in λ bacteriophage.

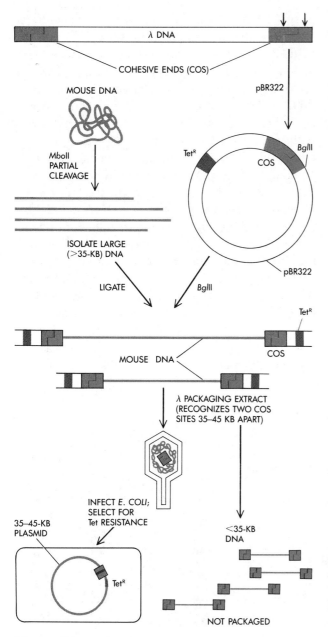

Figure 6-9
Cosmid cloning. The cos sites of λ phage are cloned in pBR322, leaving the tetracycline-resistance gene intact. Eukaryotic DNA is partially cleaved with a restriction enzyme to give large pieces of DNA. This DNA is ligated to the cloned cos-site plasmids. Lambda packaging extract is added; it will recognize and cleave 35-to-45-kb pieces of eukaryotic DNA flanked by cos sites, and package these molecules into phage heads. Smaller pieces of eukaryotic DNA will not be packaged.

The efficiency of cloning using this cosmid technology is lower than that of using phage, and cosmid libraries representing the entire genome are difficult to obtain except from organisms with relatively low DNA content (*Drosophila* is an example). However, several partial cosmid libraries can be made and then screened for the presence of a specific gene.

Chromosome Walking Is Used to Analyze Long Stretches of Eukaryotic DNA

Often we wish to analyze several hundred kilobases of contiguous information from a eukaryotic genome. It is impossible, though, to obtain this much DNA on a single phage or cosmid. However, one recombinant phage or cosmid can be used to isolate another recombinant that contains overlapping information from the genome. This technique, known as "chromosome walking," depends on isolating a small segment of DNA from one end of the first recombinant and using this piece of DNA as a probe to rescreen the phage or cosmid library in order to obtain a recombinant containing that piece of DNA and the next portion of the genome. The second recombinant is used to obtain a third, and so on, to yield a set of overlapping cloned segments (Figure 6-10). Of course, the small piece of DNA used to rescreen the library must be a single-copy element in the genome; if it is a repeated sequence, the recombinants obtained by rescreening will not necessarily represent contiguous regions of the genome. Because repetitive sequences are so ubiquitous in the genomes of most vertebrates, it is often difficult to obtain a piece of DNA that does not contain a repeated sequence. For this reason, chromosome walking has thus far been most successful in *Drosophila*, whose DNA has few repeated sequences.

An alternative to chromosome walking has been successful in higher eukaryotes. It is based on the fact that many eukaryotic proteins are encoded by several genes that are often linked (positioned near one another) in the chromosome (Chapter 7). By screening a λ library with a cDNA probe, we should be able to isolate most or all of these different genes on different λ clones. If the genes are only 10 to 15 kb apart in the genome, then a λ recombinant phage or cosmid containing one of the genes may overlap with a clone containing another gene. If enough recombinant clones are obtained, several overlapping sets can be iden-

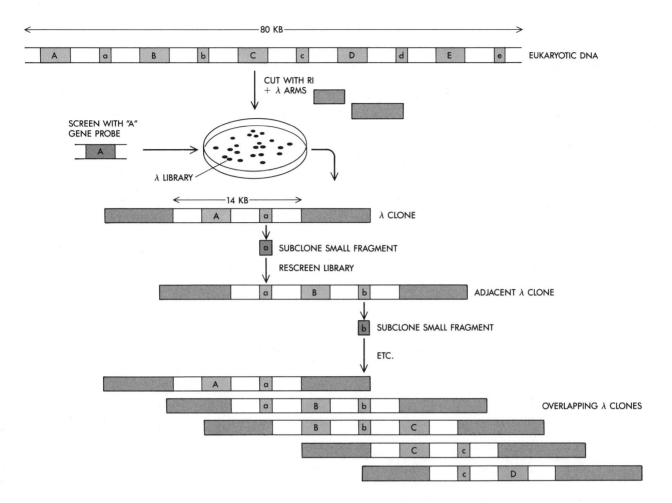

Figure 6-10
Chromosome walking. One recombinant λ phage obtained from a library of a eukaryotic genome can be used to isolate another recombinant containing the neighboring segment of eukaryotic DNA. The first recombinant is subcloned into small pieces in pBR322. A subcloned DNA fragment that is a single-copy sequence in the genome must be identified; this fragment must be different from the probe that was used to isolate the original recombinant. The subclone is used to rescreen the λ library to obtain a recombinant that has partial overlap with the first λ clone, but that also contains neighboring DNA. This second recombinant is subcloned, and a suitable fragment is used to rescreen the λ library to obtain another partially overlapping clone; and so on.

tified; these may represent 50 to 100 kb of contiguous chromosomal information. In this way a large number of the genes coding for the several classes of immunoglobulin chains have been mapped, as have many of the genes that code for proteins constituting the major histocompatibility complex (MHC) (Chapter 9).

Cloning DNA in Phage M13 Speeds Up Sanger Sequencing

Another useful cloning vehicle is the phage M13. M13 is a single-stranded phage packed in a filamentous protein coat. When a phage containing this "+" strand infects *E. coli,* the DNA replicates to form a double-stranded (+/−) intermediate, the + strands of which are repackaged into progeny virus particles. The double-stranded intermediate (the replicative form, or "RF") can be used as a cloning vector: It is small (about 7200 base pairs) and contains unique restriction sites in which DNA can easily be inserted. When a heterologous sequence is inserted into M13, only one of the strands of the insert is packaged in the phage. By cloning the insert into M13 in each of the two possible orientations, we can obtain large amounts of either one strand or the other of the

foreign DNA segment (Figure 6-11). By far the most important use of M13 cloning is to provide single-stranded template DNA for Sanger dideoxy sequence analyses (Figure 6-12) (Chapter 5). The site at which DNA is inserted into M13 is precisely known; an eight-to-ten-base oligonucleotide that is complementary to the region of M13 adjacent to the cloning site can be synthesized chemically. This oligonucleotide can serve as a primer for the dideoxy sequencing of *any* DNA fragment that is cloned into M13; it is not necessary to make a primer for each individual piece of cloned DNA

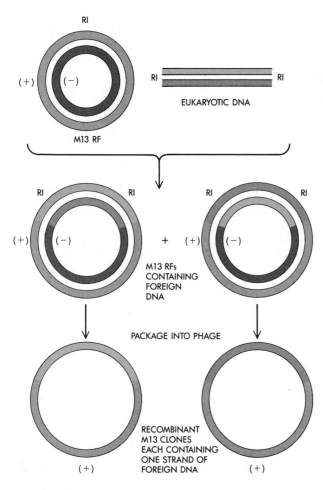

Figure 6-11
The cloning of DNA in phage M13. M13 is a single-stranded DNA phage. When it infects bacteria, it transiently goes through a double-stranded intermediate, known as the replicative form or "RF." The original strand that was in the phage is called the + strand, and the strand that is made in bacteria is called the − strand. Only the + strand is packaged into new phage coats. The RF of M13, which is essentially a plasmid, can be obtained easily. If the RF is cleaved with a restriction enzyme that cuts it once, and then ligated to a piece of foreign DNA cut with the same enzyme, the foreign DNA can be inserted into the RF in either of two different orientations, since the ends of the foreign DNA are identical. Therefore, depending on its orientation, either one strand or the other of the foreign DNA will correspond to the + strand and be packaged into phage.

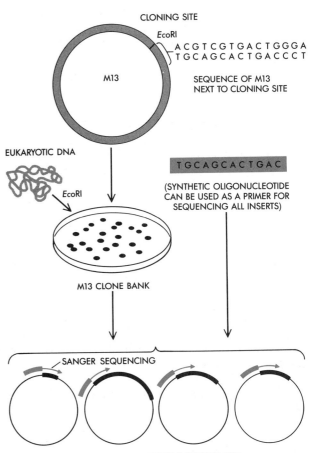

Figure 6-12
The same synthetic oligonucleotide primer can be used for Sanger sequencing of any insert in M13. All insertions of foreign DNA into the RF of M13 are made at the same site in the RF. The sequence adjacent to the foreign DNA is therefore always the same. An oligonucleotide complementary to this adjacent M13 sequence can be used as a primer to sequence any piece of foreign DNA cloned into M13.

to be sequenced. This advance allowed Sanger to sequence the entire 48,513 bases of phage λ with a speed that would have been impossible with Maxam and Gilbert sequencing, or with dideoxy sequencing if a different primer had to be made for each DNA fragment.

Southern and Northern Blotting

Given the existence of a suitable mRNA or cDNA probe, the structure of specific genes in eukaryotes may be initially analyzed without prior cloning in bacteria by using a technique developed in Edinburgh by E. M. Southern. In this "Southern blotting" method, high-molecular-weight DNA is cut with one or several restriction enzymes. The resultant fragments are separated, based on their size, by agarose gel electrophoresis. The gel is then laid onto a piece of nitrocellulose. A flow of an appropriate buffer is set up through the gel (perpendicular to the direction of electrophoresis), toward the nitrocellulose filter. This flow causes the DNA fragments to be carried out of the gel onto the filter, where they bind. Thus a "replica" of the DNA fragments in the gel is created on the nitrocellulose. A labeled probe, specific for the gene under study, is then hybridized to the filter. This probe can be a purified RNA, a cDNA made from purified RNA, or a segment of DNA that has been cloned in *E. coli*. The labeled probe will hybridize to the specific fragment (or fragments) containing complementary information. Autoradiography of the nitrocellulose filter will result in a precise and reproducible pattern of bands representing the DNA fragment or fragments that contain the gene (Figure 6-13).

Southern blotting, which was developed independently of recombinant DNA technology, has been an extremely powerful tool in analyzing the eukaryotic genome. In concert with cloning techniques, it allows the determination of (1) the presence (and number) of sites for a given restriction enzyme in a given gene, and (2) the number of copies of that gene in the genome. Most often, Southern blotting is used in preliminary analysis of a gene or gene family in order to devise the best cloning strategy.

An analogous technique for analyzing RNA has been dubbed "northern blotting." Total cellular RNA, or mRNA, is separated by size using agarose gel electrophoresis (usually in the pres-

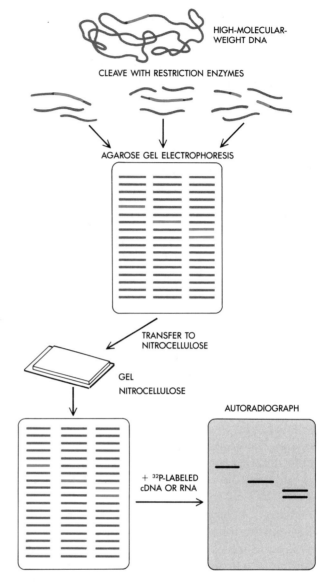

Figure 6-13
Southern blotting. Eukaryotic DNA is cleaved with one or several restriction enzymes. The cleaved DNA is separated by size using agarose gel electrophoresis. The gel is then laid on a piece of nitrocellulose, and a flow of buffer is set up through the gel onto the nitrocellulose. This causes the DNA fragments to flow out of the gel and bind to the filter. A replica of the DNA fragments in the gel is created on the filter. The filter can be hybridized to a suitable labeled probe, and specific DNA fragments that hybridize to the probe will give a signal following autoradiography.

ence of a strong denaturing agent, such as methyl mercury or formaldehyde, to prevent the formation of secondary-structure loops in the RNA). The gel is blotted either onto chemically treated paper that can covalently bind RNA, or, if formaldehyde was used as the denaturant, onto nitrocellulose. Hybridization with the appropriate labeled probe followed by autoradiography will produce bands indicating the number and size of the RNA species complementary to the probe. Northern blotting is useful as an adjunct to cDNA cloning because the size of a specific mRNA can be compared to the size of cloned cDNAs copied from that mRNA, and this reveals whether or not the cloned cDNA is indeed full-length.

Developing Procedures for Cloning Genes That Code for Less-Abundant Proteins

The most direct way to increase the concentration of a minor component in a crude mRNA preparation involves centrifuging the mRNA through a sucrose gradient. This procedure separates the mRNA into populations of different sizes, which can then be collected and tested in cell-free translational systems to determine which size class codes for the desired protein. For example, the mRNA molecules that code for the light chains of antibodies sediment at approximately 13 sedimentation units (13S), whereas those that code for the α and β chains of hemoglobin move much more slowly (9S). Further purification can be obtained by subjecting the mRNA samples enriched by centrifugation to electrophoresis in acrylamide or agarose gels under denaturing conditions, a procedure that again separates different-sized molecules, but with much higher resolution than is possible by simple centrifugation. In this way, mRNA preparations that are 80 to 90 percent pure for the antibody light chains have been obtained.

With less abundant proteins, however, the desired mRNA still remains a minor component after the foregoing procedures. Fortunately, several enrichment tricks can be used, one of which is most effective for minor proteins that are cell-specific. The rat protein α_{2u}, for example, constitutes 1 percent of the protein made in the male rat liver, but it is totally absent in the female rat liver cells. This difference reflects the fact that the synthesis of α_{2u} mRNA is stimulated by the male sex hormone testosterone and strongly inhibited by the female sex hormones, the estrogens. How

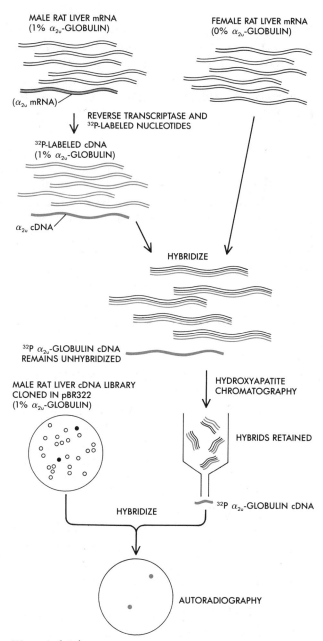

Figure 6-14
Preparation of a probe to isolate a rare cDNA. The mRNA for a protein called α_{2u}-globulin represents 1 percent of the mRNA in an adult male rat liver, but is absent from female rat liver. Male rat liver mRNA is reverse-transcribed into cDNA using labeled nucleotides. The cDNA is then hybridized with a huge excess of female liver mRNA. Most mRNA sequences in male and female liver are common, so most of the male cDNA will hybridize with an mRNA from female liver. The α_{2u} cDNA, however, will remain single-stranded and can be separated from the cDNA–mRNA hybrids by chromatography on hydroxyapatite, which binds double-stranded molecules much more avidly than single-stranded molecules. The labeled α_{2u} cDNA can then be used to screen a male rat liver cDNA library to obtain a cloned copy of the α_{2u} cDNA.

these sex hormones control mRNA synthesis remains a major unsolved problem, and the α_{2u} gene could provide an excellent system for finding the answer. When the still heterogeneous single-stranded cDNA copied from the appropriately sedimenting fraction of male rat liver mRNA is mixed under hybridizing conditions with much larger amounts of female liver mRNA, the only cDNA molecules that do not form cDNA–mRNA hybrids are the ones that code for α_{2u}. Since single-stranded nucleic acid molecules, unlike their double-helical equivalents, do not bind to hydroxyapatite, the single-stranded α_{2u} cDNA becomes much more enriched after just one passage through a hydroxyapatite column. Then it can constitute over half of the total cDNA, and the plasmid probes into which it is inserted can easily be detected (Figure 6-14).

Another trick can be used for selecting a cDNA probe when the amino acid sequence of the desired minor protein is available. This was the case for the β subunit of human chorionic gonadotropin (HCG), whose sequence of 145 amino acids had been unambiguously determined. From the amino acid sequence, the nucleotide sequences of the corresponding mRNA and cDNA were predicted. This revealed which restriction enzymes would cut the cDNA, and where. A mixed population of cloned cDNAs could then be screened with the appropriate restriction enzymes, and those that yielded fragments of the expected sizes and in the expected numbers were further analyzed. With this procedure, a cDNA probe for the HCG subunit gene was obtained.

Screening Gene Libraries with Oligonucleotide Probes

An appropriately chosen set of synthetic oligonucleotides can be used as a specific probe for a gene or an mRNA strand coding for a protein whose amino acid sequence has been determined. A sequence of five or six amino acids in the protein is used to predict all the possible mRNA sequences that may code for that stretch of the peptide. A set of complementary oligonucleotides is then synthesized. These oligonucleotides can be used directly as probes to screen a cDNA or genomic library, or as potential primers for reverse transcriptase in the synthesis of a cDNA that should be highly enriched for the specific sequence (Figure 6-15).

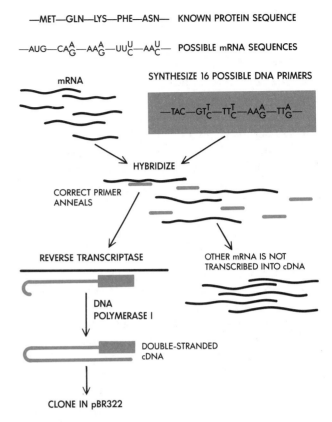

Figure 6-15
Isolation of rare cDNAs using a synthetic oligonucleotide primer. If the amino acid sequence of a protein, or part of a protein, is known, it can be used to predict the possible nucleotide sequences of its mRNA. (Because of wobble, the absolute sequence cannot be determined *a priori.*) Oligonucleotides that are complementary to all the possible mRNA sequences are synthesized chemically. When this mixture of oligonucleotides is mixed with an mRNA population containing the mRNA in question, the oligonucleotide that is complementary to the specific mRNA will anneal to it. Reverse transcriptase, as we saw above (Figure 6-3), requires a primer; normally, oligo dT is used as a primer to reverse-transcribe all poly-A-containing mRNAs. The synthetic oligonucleotide, however, will be a primer only for the mRNA to which it is complementary, and will allow its reverse transcription into cDNA. The other mRNAs in the reaction will be reverse-transcribed very inefficiently. The resultant cDNA population should be highly enriched for the specific cDNA corresponding to the primer.

Computers Match Up cDNA Classes

Because of their medical importance, many of the proteins made by human liver cells have been analyzed and their amino acid sequences determined. The sequence of the corresponding mRNAs can therefore be predicted. This information has been used at the EMBO lab in Heidelberg to identify the cloned cDNAs of several liver-specific mRNAs. Total human fetal liver mRNA was reverse-transcribed into cDNA, and the population of cDNA molecules was cloned in an M13 vector. The library of recombinant phages was first screened by hybridization for those that carry liver-specific cDNAs. The cDNA inserts in over 200 of these clones were then sequenced by the Sanger procedure. The sequences were fed into a computer that was asked to compare them to the predicted mRNA sequences of the well-characterized liver proteins. In this way, recombinant phages carrying cDNAs corresponding to parts of the mRNAs of over a dozen different human liver cell proteins have been identified. These clones are now being used as specific probes for the corresponding chromosomal genes. Recently this approach has also been used to identify cloned cDNAs for several muscle proteins.

Expression Vectors May Be Used to Isolate Specific Eukaryotic cDNAs

Specific eukaryotic cDNAs can also be isolated by looking for their expression in bacteria after cloning the cDNA in the appropriately constructed plasmid (expression vector). Such procedures often start with cDNAs made from enriched mRNA preparations. These cDNAs are then inserted into plasmids within genetic regions that can be highly expressed once the plasmid is introduced into a host bacterium. This is effected by placing the eukaryotic sequence downstream from a strong bacterial promoter. For example, when an insulin cDNA is placed in pBR322 in the gene for β-lactamase (the ampicillin-resistance gene), a mixed β-lactamase–insulin polypeptide is produced. After the enzymatic removal of most β-lactamase amino acids, biologically active insulin-like molecules are detectable (Figure 6-16).

Obligatory strong promoters like that for β-lactamase, however, do not always work well. To start with, the high-level production of foreign proteins may block the growth of their host bacteria. And, in some cases, high-level transcription from plasmid sequences interferes with the plasmid's replication and leads to the loss of plasmid sequences from the host cells. The most versatile promoters, therefore, are those that are strong but also regulatable. Several expression vectors containing such promoters have been designed and constructed.

Expression vectors that utilize the phage λ pL promoter, a regulatable promoter responsible for the synthesis of several λ genes, are very useful. In the presence of the λ repressor this promoter is blocked from working, but in the absence of control the pL is very active. The gene coding for the λ repressor can be placed on the expression vector itself, or it can be located in the host bacterial chromosome. Desired variations in the level of the repressor are achieved by using a temperature-sensitive repressor that is active at 31°C but not at 38°C. Thus, eukaryotic sequences cloned downstream from the pL promoter will not be transcribed at 31°C. High-level expression can be turned on by switching the bacteria to a higher temperature in order to inactivate the λ repressor. Using vectors containing this promoter, the rate of synthesis of a foreign protein can be brought up to 10 percent of the total bacterial protein synthesis.

The efficiency of translation of an mRNA in bacteria is critically dependent on the presence of a ribosome-binding site (the Shine–Dalgarno sequence), and on the distance between this site and the AUG initiation codon (Chapter 4, page 50). For efficient expression of eukaryotic proteins using expression vectors, the Shine–Dalgarno sequence is usually included in the vector itself. The initiation codon from the eukaryotic sequence must then be placed downstream at the correct distance from the Shine–Dalgarno site. Because this spacing may be difficult to achieve, the ATG initiation codon is sometimes included on the vector itself. This ensures that the spacing between the Shine–Dalgarno sequence and the initiation codon is optimal. This latter type of vector will result in a hybrid protein containing a few amino acids from the prokaryotic protein and the remainder from the eukaryotic insert. Such a "fusion protein" is often more stable in bacteria than the native eukaryotic protein would be, and the fusion protein can, in some cases, be treated enzymatically or chemically to release the eukaryotic peptide. The most commonly used promoters in such vectors are the lac promoter with its Shine–Dalgarno site, and a hybrid promoter consisting of the

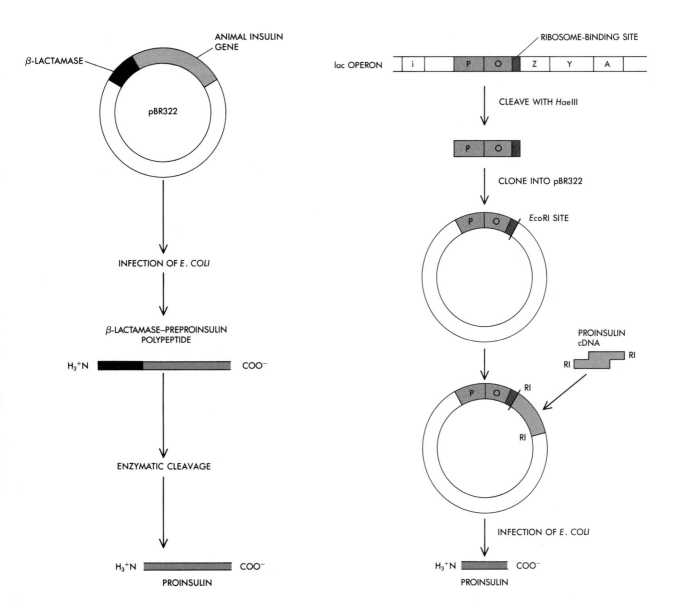

Figure 6-16
Expression of eukaryotic genes in *E. coli*. If a segment of eukaryotic DNA is cloned into pBR322 next to an active bacterial promoter, such as the promoter for β-lactamase or for β-galactosidase, the eukaryotic sequence will be transcribed into mRNA very efficiently, since the *E. coli* RNA polymerase needs only the bacterial promoter sequences to initiate transcription. Once transcription begins, it will continue until a termination signal is reached. The mRNA with eukaryotic sequences can then be translated into functional protein if appropriate ribosome-binding sites are present on the hybrid mRNA.

trp (tryptophan) promoter (which is stronger than the lac promoter) and the lac Shine–Dalgarno sequences, with or without the ATG initiation codon. Vectors containing the trp–lac fusion promoter are best cloned into a bacterial strain that contains the lac Iq repressor. This mutant "super repressor" keeps the promoter in the "off" state until an inducer, such as IPTG, is added.

Immunological Screening for the Products of Expression Vectors

The production of specific eukaryotic proteins in bacteria by using expression vectors is detected most easily by immunological tests. This approach was recently used successfully in the cloning of the cDNA coding for the β form of chicken tro-

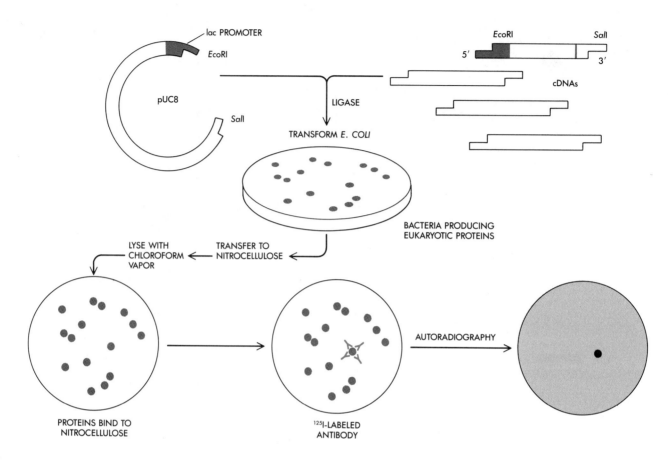

Figure 6-17
The ability to express eukaryotic proteins in *E. coli* can be used to identify and isolate specific eukaryotic cDNA clones. The cDNA for chicken tropomyosin was cloned using this method. A cDNA library was made from chicken smooth-muscle mRNA. (Tropomyosin represents about 0.5 percent of smooth-muscle mRNA.) The cDNA was cloned into pUC8 next to the lac promoter in such a way as to ensure that all the cDNAs were in the correct orientation relative to the direction of transcription from the promoter. All the resulting bacterial colonies should have produced one eukaryotic protein. Colonies that produced tropomyosin (and that therefore had the tropomyosin cDNA clone) were identified by screening the library, essentially as described in Figure 6-7, except that a labeled antitropomyosin antibody was used rather than a labeled cDNA. Colonies producing tropomyosin bound the labeled antibody and were identified by autoradiography.

pomyosin, a major component of smooth-muscle cells. In the first step, cDNA is made from total smooth-muscle cell mRNA. The double-stranded cDNAs are held together by a hairpin loop at the end coding for the amino-terminal amino acids. *Sal*I linkers are added to the 3' end of the cDNA before the hairpin loop is cleaved with S1 nuclease, and then *Eco*RI linkers are added to the 5' end. When the linkers are cut with *Eco*RI and *Sal*I, all the cDNA molecules in the population have an *Eco*RI sticky end at the 5' end and a *Sal*I sticky end at the 3' end. The resulting cDNA population is then ligated into an expression vector called pUC8, which has been cleaved with *Eco*RI and *Sal*I.

This particular vector has a strong lac promoter slightly upstream from an *Eco*RI site. Thus, all the cDNAs are inserted into pUC8 in the proper orientation for expression. The resultant bacteria colonies containing the cDNA clones are lysed with chloroform vapor, and colonies producing tropomyosin are identified by screening with an ^{125}I-labeled antibody to tropomyosin (Figure 6-17). This method and variations on it—for example, the prescreening of the mRNAs by injecting them into frog oocytes to detect fractions that code for the desired protein—should be successful in cloning any cDNA corresponding to any protein for which appropriate antibodies are available.

READING LIST

Books

Maniatis, T., E. F. Fritsch, and J. Sambrook. *Molecular Cloning: A Laboratory Manual.* Cold Spring Harbor Laboratory, Cold Spring Harbor, N.Y., 1982.

Original Research Papers (Reviews)

BACTERIAL AND PLASMID VECTORS

Bolivar, F., R. L. Rodrigues, P. J. Greene, M. C. Betlach, H. L. Heyneker, H. W. Boyer, J. Crosa, and S. Falkow. "Construction and characterization of new cloning vehicles, II: A multi-purpose cloning system." *Gene,* 2: 95–113 (1977).

Blattner, F. R., B. G. Williams, A. E. Blechl, K. Denniston-Thompson, H. E. Faber, L.-A. Furlong, D. J. Grunwald, D. O. Keifer, D. D. Moore, J. W. Schumm, E. L. Sheldon, and O. Smithies. "Charon phages: Safer derivatives of bacteriophage lambda for DNA cloning." *Science,* 196: 161–169 (1977).

Leder, P., D. Tiemeier, and L. Enquist. "EK2 derivatives of bacteriophage lambda useful in the cloning of DNA from higher organisms: The λgtWES system." *Science,* 196: 175–177 (1977).

Curtiss, R. III, M. Inoue, D. Pereira, J. C. Hsu, L. Alexander, and L. Rock. "Construction in use of safer bacterial host strains for recombinant DNA research." In W. A. Scott and R. Werner, eds., *Molecular Cloning of Recombinant DNA: Proceedings of the Miami Winter Symposia,* vol. 13. Academic, New York, 1977, pp. 99–114.

SYNTHESIS AND CLONING OF cDNA

Rougeon, F., P. Kourilsky, and B. Mach. "Insertion of a rabbit β-globin gene sequence into *E. coli* plasmid." *Nuc. Acids Res.,* 2: 2365–2378 (1975).

Rabbitts, T. H. "Bacterial cloning of plasmids carrying copies of rabbit globin messenger RNA." *Nature,* 260: 221–225 (1976).

Maniatis, T., S. G. Kee, A. Efstratiadis, and F. C. Kafatos. "Amplification and characterization of a β-globin gene synthesized *in vitro.*" *Cell,* 8: 163–182 (1976).

Scheller, R., R. Dickerson, H. Boyer, A. Riggs, and K. Itakura. "Chemical synthesis of restriction enzyme recognition sites useful for cloning." *Science,* 196: 177–180 (1977).

Seeburg, P. H., J. Shine, J. A. Martial, J. D. Baxter, and H. M. Goodman. "Nucleotide sequence and amplification in bacteria of structural gene for rat growth hormone." *Nature,* 270: 486–494 (1977).

Shine, J., P. H. Seeburg, J. A. Martial, J. D. Baxter, and H. M. Goodman. "Construction and analysis of recombinant DNA of human chorionic somatomammotropin." *Nature,* 270: 494–499 (1977).

Land, H., M. Guez, H. Hauser, W. Lindenmaier, and G. Schutz. "5′ terminal sequences of eucaryotic mRNA can be cloned with high efficiency." *Nuc. Acids Res.,* 9: 2251–2266 (1981).

Okayama, H., and P. Berg. "High-efficiency cloning of full length cDNA." *Mol. Cell. Biol.,* 2: 161–170 (1982).

IDENTIFYING SPECIFIC cDNA PROBES

Grunstein, M., and D. S. Hogness. "Colony hybridization: A method for the isolation of cloned DNAs that contain a specific gene." *Proc. Natl. Acad. Sci. USA,* 72: 3961–3965 (1975).

BACTERIOPHAGE λ LIBRARIES

Benton, W. D., and R. W. Davis. "Screening λgt recombinant clones by hybridization to single plaques *in situ.*" *Science,* 196: 180–182 (1977).

Maniatis T., R. C. Hardison, E. Lacy, J. Lauer, C. O'Connell, D. Quon, G. K. Sim, and A. Efstratiadis. "The isolation of structural genes from libraries of eucaryotic DNA." *Cell,* 15: 687–701 (1978).

COSMIDS

Hohn, B., and K. Murray. "Packaging recombinant DNA molecules into bacteriophage particles *in vitro.*" *Proc. Natl. Acad. Sci. USA,* 74: 3259–3263 (1977).

Collins, J., and B. Hohn. "Cosmids: A type of plasmid gene cloning vector that is packageable *in vitro* in bacteriophage heads." *Proc. Natl. Acad. Sci. USA,* 75: 4242–4246 (1978).

Hohn, B., and J. Collins. "A small cosmid for efficient cloning of large DNA fragments." *Gene,* 11: 291–298 (1980).

CHROMOSOME WALKS

Bender, W., P. Spierer, and D. Hogness. "Gene isolation by chromosomal walking." *J. Supra. Molec. Struc.,* 10 (suppl.): 32 (1979). (Abstract.)

Steinmetz, M., A. Winoto, K. Minard, and L. Hood. "Clusters of genes encoding mouse transplantation antigens." *Cell,* 28: 489–498 (1982).

Shimizu, A., N. Takahashi, Y. Yaoita, and T. Honjo. "Organization of the constant region gene family of the mouse immunoglobulin heavy chain." *Cell,* 28: 499–506 (1982).

SANGER SEQUENCING USING PHAGE M13

Messing, J., B. Gronenborn, B. Müller-Hill, and P. H.

Hofschneider. "Filamentous coli phage M13 as a cloning vehicle: Insertion of a *Hin*dII fragment of the *lac* regulatory region in M13 replicative form *in vitro.*" *Proc. Natl. Acad. Sci. USA,* 74: 3642–3646 (1977).

Sanger, F., A. R. Coulson, B. G. Barrell, A. J. H. Smith, and B. A. Roe. "Cloning in single-stranded bacteriophage as an aid to rapid DNA sequencing." *J. Mol. Biol.,* 143: 161–178 (1980).

Messing, J., R. Crea, and P. H. Seeburg. "A system for shotgun DNA sequencing." *Nuc. Acids Res.,* 9: 309–321 (1981).

SOUTHERN AND NORTHERN BLOTTING

Southern, E. M. "Detection of specific sequences among DNA fragments separated by gel electrophoresis." *J. Mol. Biol.,* 98: 503–517 (1975).

Alwine, J. C., D. J. Kemp, and G. R. Stark. "Method for detection of specific RNAs in agarose gels by transfer to diazobenzyloxymethyl-paper and hybridization with DNA probes." *Proc. Natl. Acad. Sci. USA,* 74: 5350–5354 (1977).

PROCEDURES FOR CLONING GENES THAT CODE FOR LESS ABUNDANT PROTEINS

Alt, F. W., R. E. Kellems, J. R. Bertino, and R. T. Schimke. "Selective multiplication of dihydrofolate reductase genes in methotrexate-resistant variants of cultured murine cells." *J. Biol. Chem.,* 252: 1357–1370 (1978).

Chan, S. J., B. E. Noyes, K. L. Agarwal, and D. F. Steiner. "Construction and selection of recombinant plasmids containing full-length complementary DNAs corresponding to rat insulins I and II." *Proc. Natl. Acad. Sci. USA,* 76: 5036–5040 (1979).

Nakanishi, S., A. Inoue, T. Kita, M. Nakamura, A. Chang, S. Cohen, and S. Numa. "Nucleotide sequence of cloned cDNA for bovine corticotropin-β-tropin precursor." *Nature,* 278: 423–427 (1979).

Fiddes, J. C., and H. M. Goodman. "The cDNA for the β-subunit of human chorionic gonadotropin suggests evolution of a gene by readthrough into the 3′-untranslated region." *Nature,* 286: 685–687 (1980).

Kurtz, D. T., and C. R. Nicodemus. "Cloning of α_{2u} globulin cDNA using a high efficiency technique for the cloning of trace messenger RNAs." *Gene,* 13: 145–152 (1981).

SCREENING GENE LIBRARIES WITH OLIGONUCLEOTIDE PROBES

Suggs, S. V., R. B. Wallace, T. Hirose, E. H. Kawashima, and K. Itakura. "Use of synthetic oligonucleotides as hybridization probes: Isolation of cloned cDNA sequences for human β_2-microglobulin." *Proc. Natl. Acad. Sci. USA,* 78: 6613–6617 (1981).

EXPRESSION VECTORS

Chang, A. C. Y., J. H. Nunberg, R. J. Kaufman, H. A. Erlich, R. T. Schimke, and S. N. Cohen. "Phenotypic expression in *E. coli* of a DNA sequence coding for mouse dihydrofolate reductase." *Nature,* 275: 617–624 (1978).

Villa-Komaroff, L., A. Efstratiadis, S. Broome, P. Lomedica, R. Tizard, S. P. Nabet, W. L. Chick, and W. Gilbert. "A bacterial clone synthesizing proinsulin." *Proc. Natl. Acad. Sci. USA,* 75: 3727–3731 (1978).

Guarente, L., G. Lauer, T. Roberts, and M. Ptashne. "Improved methods of maximizing expression of a cloned gene: A bacterium that synthesizes rabbit β-globin." *Cell,* 20: 545–553 (1980).

Gilbert, W., and L. Villa-Komaroff. "Useful proteins from recombinant bacteria." *Sci. Am.,* 242(4): 74–94 (1980).

DeBoer, H. A., L. J. Comstock, and M. Vasser. "The *tac* promoter: A functional hybrid derived from the *trp* and *lac* promoters." *Proc. Natl. Acad. Sci. USA,* 80: 21–25 (1983).

IMMUNOLOGICAL SCREENING

Broome, S., and W. Gilbert. "Immunological screening method to detect specific translation products." *Proc. Natl. Acad. Sci. USA,* 75: 2746–2749 (1978).

Helfman, D. M., J. R. Feramisco, J. C. Fiddes, G. P. Thomas, and S. Hughes. "Identification of clones that encode chicken tropomyosin by direct immunological screening of a cDNA expression library." *Proc. Natl. Acad. Sci. USA,* 80: 31–35 (1983).

The Unexpected Complexity
of Eukaryotic Genes

Examination of the structure of specific chromosomal DNA fragments began in earnest in the spring of 1977, soon after the certification of the first "safe" vectors. Then everyone's attention was focused on the regulatory sequences for mRNA synthesis; it was thought likely that these lay in front of the 5' ends of the coding segments themselves. It was taken for granted that the nucleotide sequences within the genes would be identical to the ones in the cDNA probes. Yet the first preliminary results indicated that the restriction fragments obtained from the segments of chromosomal DNA were frequently different from the ones generated from their related cDNA probes. Initially, these observations caused only puzzlement, because it seemed impossible for the cDNA (or mRNA) sequences to be other than identical to the sequences of the genes from which they were transcribed.

Split Genes Are Discovered

Then—during the 1977 Cold Spring Harbor Symposium—came the first announcements of an mRNA splicing phenomenon during adenovirus multiplication. The primary viral RNA transcripts within the nucleus of an infected cell were found to be shortened by removal of one or more internal sections to produce smaller mRNA molecules; these moved to the cytoplasm, where they served as templates for viral protein synthesis. Quickly, the generality of splicing was extended to SV40, and the question immediately arose whether splicing might also be involved in the processing of most chromosomal DNA. For several years it had

been known that many eukaryotic mRNAs were first synthesized as large precursors (pre-mRNAs) that were later processed in the nucleus to much smaller products. But until the announcement of adenovirus splicing, it had always been assumed that this processing necessarily and exclusively involved removal of long, possibly regulatory sections at the 5' and 3' ends of pre-mRNA.

So, during the summer of 1977, the patterns from restriction enzyme fragments of chromosomal DNA were reinterpreted to see whether they might indicate noncoding sequences interspersed in the coding sequences of genes, and if so, whether the primary RNA transcripts were spliced to remove the noncoding sequences as the mRNA matured. Proof that this was indeed the case came from several experimental procedures. Most direct were the electron-microscope observations of the DNA–RNA heteroduplexes made between functional mRNA molecules and their corresponding chromosomal genes. Single-stranded DNA sections looping out from such heteroduplexes represented gene sequences that had been removed from the mRNA. These have been given the name "introns" (Figure 7-1, page 92). The exact sizes and locations of these introns could be measured more precisely by treating the heteroduplexes with S1 nuclease, which specifically degrades the unpaired DNA bases and leaves genomic DNA fragments bound to the corresponding sequences that persist in the functional mRNA. Such functional gene sequences are now known as exons (because they exit from the nucleus to function in the cytoplasm).

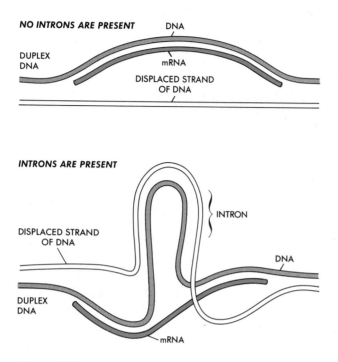

Figure 7-1
DNA–mRNA heteroduplexes, showing the
loops created by the presence of introns in
the DNA.

Introns exist in virtually all mammalian and
vertebrate genes (Figure 7-2), and also in the
genes of eukaryotic microorganisms (such as
yeast), though with much lower frequency. Often
the noncoding introns of a gene contain many
more nucleotides than its coding exons do (Figure
7-3), which accounts, at least in part, for the previously unexplained large sizes of so many primary
RNA transcripts. The number and size of introns
vary widely from one gene to another; genes coding for the long chains of collagen, the connective-
tissue protein, possess nearly 40 introns.

A few genes that do not contain introns have
been found, for example, genes coding for the α
and β forms of interferon. Intronless mammalian
genes can also be generated through recombinant
DNA tricks and tested to see how they function.
For certain genes, the complete removal of introns
has no consequence; they produce fully active
mRNA transcripts. With other genes (such as that
for SV40 T antigen), however, the removal of
their natural introns somehow blocks the exit of
functional mRNA products to the cytoplasm. Per-

haps in these latter cases the newly made transcripts adopt configurations incompatible with
their exit from the nucleus.

Specific Base Sequences Are Found at Exon–Intron Boundaries

By the summer of 1978, just a year after the first
split genes were discovered, the sequences of the
bases at many exon–intron boundaries had been
determined. It was hoped that such sequence data
would provide evidence that intramolecular base
pairing could bring together in a hairpinlike arrangement the upstream and downstream splice
sites in an mRNA precursor. So positioned, specific splicing enzymes could easily carry out the
cutting and joining events that would remove the
introns. However, hairpin loop structures require
a group of upstream bases complementary to a
group of downstream bases, and no such complementary groups of sequences were found at the
splice points. The upstream and downstream splice
sites therefore could not be brought together by
self-complementarity. Yet the base sequences at
the boundaries between exons and introns were
far from random, and after many had been sequenced a pattern emerged. All the sequences at

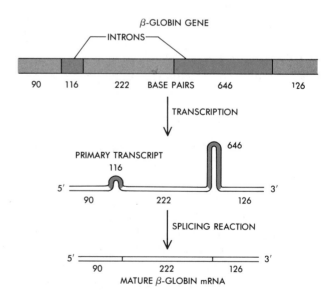

Figure 7-2
Introns are spliced out during maturation of β-
globin mRNA.

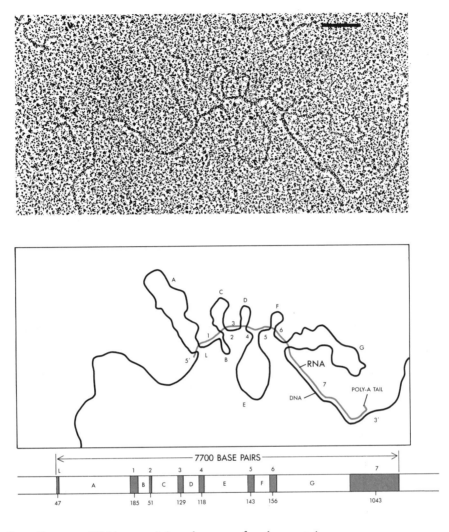

Figure 7-3
The organization of a split gene. DNA containing the gene for the protein
ovalbumin was allowed to hybridize with ovalbumin mRNA. The eight exons
(L, 1–7) of the gene anneal to the complementary regions of RNA, and the seven
introns (A–G) loop out from the hybrid. The 5′ and 3′ ends of the mRNA are
indicated, as is the poly-A tail (Chambon, 1981).

the opposite ends of different introns could be
related to two so-called consensus sequences, with
the first two bases at the 5′ end of each intron
being GT and the last two bases at the 3′ end being
AG (Table 7-1, page 94). These consensus sequences suggested an alternative mechanism that
might bring the opposite ends of introns together
prior to splicing—a mechanism involving small
adapter RNAs that were normally attached to the
splicing enzymes.

A precedent for such positioning RNA was
already known from the discovery that the site-
specific ribonuclease P of *E. coli* contains an essential RNA component. Moreover, large numbers of
functionally obscure small nuclear RNA molecules (snRNA) had long been known to exist in
virtually every type of eukaryotic nucleus. Most
excitingly, when the nucleotide sequence of one
such snRNA—"U1" from rat cells—was examined, a large stretch of bases at its 5′ end was
exactly complementary to the consensus sequences
at the splice sites. Our best guess now is that U1
snRNA forms part of a splicing complex by base
pairing with both ends of an intron to bring them

Table 7–1
Consensus Sequences

Donor Site

	◄—————exon————►				◄—————————intron—————————►				
A	22	38	53	9	0	0	50	61	9	14
G	20	11	12	68	90	0	30	12	77	9
C	25	37	13	4	0	0	4	8	1	11
T	23	4	12	9	0	90	6	9	3	56

Base Frequency

Consensus   AG:GTPuAGT

Acceptor Site

	◄—————————intron—————————►						◄————exon————►		
A	4	6	20	3	85	0	19	11	21
G	7	5	19	1	0	85	40	27	19
C	31	27	25	63	0	0	17	15	26
T	43	47	21	18	0	0	9	32	19

Base Frequency

Consensus $(Py)_nNCAG$:G

90 donor splice junctions and 85 acceptor junctions were
sequenced. The frequency of bases found at various positions
relative to the splice sites were tabulated. Consensus sequences for
splice junctions could be determined.
   Pu = purine
   Py = pyrimidine

into exact register for cutting and splicing (Figure 7-4).

Despite intense efforts, it has proved extremely difficult to prepare extracts of mammalian cells that will splice pre-mRNA reproducibly in the test tube. Yet the splicing of precursors of several yeast tRNAs does occur reproducibly *in vitro*.

Discovery of Autonomous Splicing

Completely unlike all other systems that have been studied so far is the splicing of a 413-base intron found within the precursor of the ribosomal RNA of the ciliate *Tetrahymena*. This intron appears to be removed without the involvement of any enzyme, and thus flies in the face of our past experience that chemical bonds in biological macromolecules are made and broken only in the presence of catalytic enzymes. A decisive experi-

ment was thus required to show that the unspliced precursor did not have traces of a splicing enzyme bound to it. Proof came from an experiment in which the *Tetrahymena* rRNA gene was cloned into a bacterial plasmid constructed in such a way that the rRNA precursor was transcribed in *E. coli.* This artificially made rRNA precursor also showed autonomous splicing, leading to the conclusion that RNA itself must have catalytic

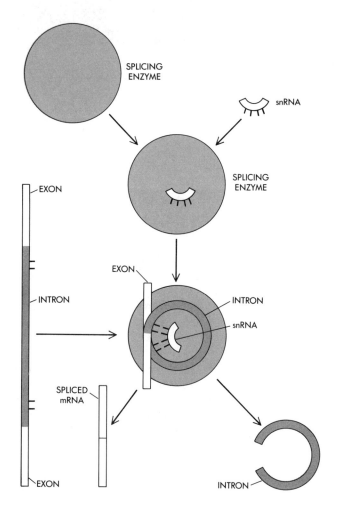

Figure 7-4
A model for mRNA splicing involving small nuclear RNA (snRNA). A few snRNAs have been found to have sequences homologous to the consensus sequences at the ends of introns. This has led to speculation that these snRNAs may be involved in splicing, perhaps by "holding" the ends of the intron in close proximity, where they can be cleaved and spliced by the appropriate enzymes.

capabilities (Figure 7-5). If autonomous splicing is found in other systems, the way we think about RNA will be drastically revised.

Complete Sequencing of the First Mammalian Genes

Given the speed of the new methods, the sequencing of many complete mammalian genes has become a feasible objective, even though the presence of introns means that the sequence of thousands of nucleotides must usually be determined. Several such complete sequences have been established over the past few years, and a considerable number are in the process of being completed. Some of the first examples came from the well-studied globin, immunoglobulin, and ovalbumin gene families.

In the case of the globins, the complete sequences have been established for several different types of genes (for example, α and β chains) from a wide variety of species, including human, rabbit, mouse, and goat. Many polypeptide hormone genes, again from several species, also have been entirely sequenced. These include growth hormone, chorionic somatomammotropin, prolactin, insulin, and preopiomelanocortin genes. Other areas that have received considerable attention are the genes for structural proteins—in particular, actins and tubulins—and the genes coding for the interferons.

Though the time necessary for a highly motivated molecular biologist to sequence an average gene has now been reduced to less than one year, the cost of such procedures is not trivial. Taking into account time and materials, a reasonable estimate might be that each base pair in a gene that is sequenced probably costs 5 to 10 dollars. There is a need for still more automatic and cheaper procedures to further reduce the amount of time involved in establishing a lengthy sequence. Extensive use is already being made of computer programs to facilitate the analysis of DNA sequence data, and in some cases to actually search for the overlaps required when a complete sequence is being constructed out of primary sequence data. Machines now being tested will automatically read the autoradiographs of sequencing

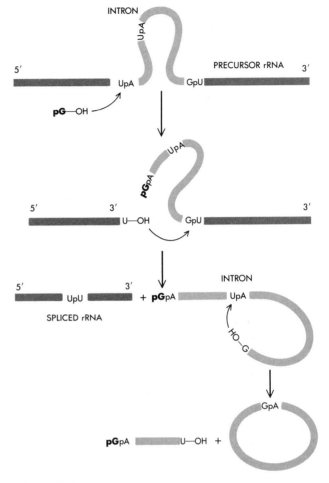

Figure 7-5
Autonomous splicing of the 413-base intron from the rRNA precursor in *Tetrahymena*. The initial cleavage is catalyzed by a GMP molecule that attacks the phosphodiester bond between two specific bases (UpA) at the 5' end of the intron. The GMP is added onto the 5' end of the cleaved DNA and a 3'-hydroxyl is formed at the 3' boundary of the exon. This 3'-hydroxyl attacks the phosphodiester bond between two other bases (GpU) at the 3' end of the intron, and forms a correctly spliced rRNA. The excised intron then undergoes a further cutting-and-joining reaction that generates a circular RNA molecule and a short linear fragment.

gels and feed this information directly into the computer. More such advances are required to enable appreciably larger segments of eukaryotic chromosomes to be studied.

Open Reading Frames in DNA Delineate Protein-Coding Regions

A computer can also be used to analyze a long DNA sequence to determine the location of regions that may code for proteins. The computer is instructed to search for "open reading frames," long stretches of triplet codons that are not interrupted by a translational stop codon. This procedure can be very useful when a cloned DNA fragment is known from, say, some functional assay to contain a certain gene, but when the size of the gene or its location on the fragment is not known. If an open reading frame can be found somewhere in the sequence—especially if the frame has an ATG (the universal translation-initiation codon) near the start—it is very likely that this stretch of sequence is in fact the gene; discovery of an open reading frame does not *prove* the existence of a gene, of course, but it at least delineates an area to home in on. Conversely, the lack of an open reading frame in a stretch of sequence that was thought to contain a gene has been used to determine that some "genes"—chromosomal sequences that hybridize to specific mRNAs—are in fact pseudogenes, nonfunctional relics that arose during the evolution of gene families. Computer searches for open reading frames have even pointed out sequences that code for mRNAs (and probably proteins) that were previously unsuspected. The long terminal repeat (LTR) of mouse mammary tumor virus (Chapter 10) and a stretch of adenovirus DNA, for example, were found to have long open reading frames that have since been found to code for mRNAs. The proteins coded for by these mRNAs have not yet been determined, but no one would have even *looked* for the mRNAs if the open reading frame had not been found.

Leader Sequences at the NH$_2$-Terminal Ends of Secretory Proteins

DNA sequence analysis reveals that many functional proteins first exist in the form of slightly larger precursors containing some 15 to 25 additional amino acids at their NH$_2$-terminal ends. Such "leader" (signal) sequences are diagnostic of proteins that move through cellular membranes to function only after they have been secreted from the cells in which they were made (examples of such proteins are insulin, serum albumin, antibodies, and digestive tract enzymes), or after they have been anchored to the outer surface of a cell membrane (the histocompatibility antigens on the cell surface are an example). A majority of the amino acids found in leaders are hydrophobic, and they somehow function to ensure both the attachment of nascent polypeptide chains to appropriate

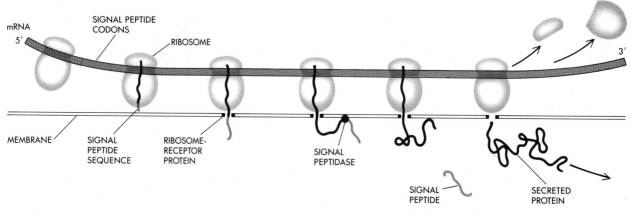

Figure 7-6
Signal sequences. Proteins destined to be secreted from the cell have an N-terminal sequence that is rich in hydrophobic residues. This "signal" sequence binds to the membrane and draws the remainder of the protein through the lipid bilayer. The signal sequence is cleaved off of the protein during this process by an enzyme called signal peptidase.

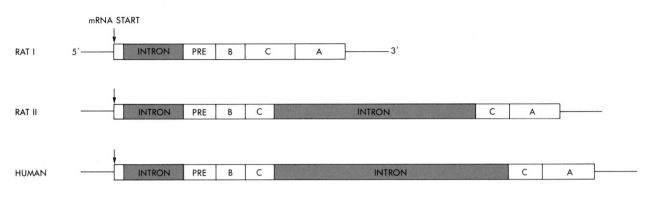

Figure 7-7
A comparison of rat and human insulin genes. Pre, A, B, and C represent the
different peptide domains of the proinsulin molecule.

membranes, and the subsequent passage of the
chains across the lipid bilayers that characterize all
cellular membranes. *In vivo,* leader sequences usually have only a fleeting existence, because they
are cleaved off by specific proteolytic enzymes that
generate the NH_2-terminal amino acids of the
functional secreted products (Figure 7-6).

Introns Sometimes Mark Functional Protein Domains

At first, neither the location nor the number of
introns within a given gene made sense. In rats, for
example, two closely related genes code for insulin—one gene has only one intron and the other
has two. The rat insulin I and rat insulin II genes
have introns of almost identical sizes located immediately downstream from the sequences coding
for the insulin leader. The second intron of the rat
insulin II gene is located within the so-called "C"
segment of the insulin protein precursor that is
digested away to produce the two-chained structure of mature insulin molecules. Humans have
only one insulin gene whose two introns are
located in positions similar to those of the rat insulin II gene (Figure 7-7), thus suggesting the de-
scent of rat and human genes from a common
ancestor. No obvious functional difference marks
the amino acids separated by the second insulin
intron, whose location might be accidental.

In hemoglobin, though, the amino acids constituting the special functional domain surrounding the heme group are clearly delineated by an
intron from the more distal amino acids. As we
describe below, introns within antibody genes are
precisely located between functional domains. For
this reason, much protein evolution may have
been accomplished by genetic recombination
events that brought together domains previously
located on separate genes. It is conceivable that the
long length of many introns helps to ensure that
coding sequences are kept intact during genetic
crossing over.

Alternative Splicing Pathways Generate Different mRNAs from a Single Gene

RNA splicing can also generate different mRNAs
and thus different proteins from one gene, or,
more accurately, from one primary transcriptional
unit. Differential splicing was first seen in the
adenoviruses and then in SV40, in polyoma virus,
and in the mRNAs coding for immunoglobulins.
A recent example involves the mRNA coding for
the hormone calcitonin, a peptide that is normally
produced in large amounts in the thyroid gland.
Although a large amount of calcitonin mRNA is
present in the hypothalamus, very little calcitonin
itself is produced there. Instead, another protein
that is called "calcitonin-gene-related product" or
CGRP, and whose function is still unknown, has
been detected. Both calcitonin and CGRP are produced from the same primary transcript by using
alternative splicing routes. The routes used produce two different mature mRNAs having a common 5' end but different 3' ends: The thyroid

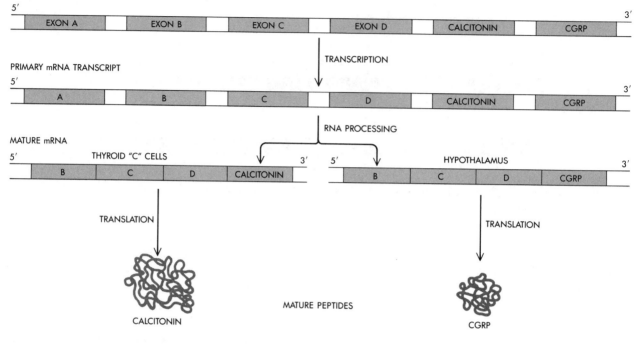

Figure 7-8
The calcitonin gene generates primary mRNA transcript that is spliced to produce two different forms of mature mRNA—that coding for calcitonin, which is produced primarily in the thyroid gland, and that coding for calcitonin-gene-related product (CGRP), which is produced mainly in the hypothalamus.

splicing pathway has the calcitonin sequences at the 3' end, and the hypothalamic pathway has the CGRP sequences (Figure 7-8). As yet we have no idea how cellular differentiation leads to these alternative splicing pathways.

Control Regions Are Found at the 5' and 3' Ends of Genes

As more and more eukaryotic genes have been cloned and sequenced, similarities have been noted in specific sequences near the 5' and 3' ends of different genes. This has led to speculation that these sequences are somehow involved in the control of either transcription or translation.

Virtually all eukaryotic genes have an AT-rich region located about 25 to 30 bases upstream from the site where transcription is initiated (Figure 7-9). Yet this sequence can be altered or deleted entirely with no effect on the rate of transcription. The best guess at this point is that this "TATA box" helps direct RNA polymerase II to the correct initiation site for transcription. Another sequence, roughly CNCAAT, is often found 70 to 80 bases upstream from the start of transcription.

The function of this "CAT box" is unknown; it seems to be present in most eukaryotic genes, yet in the one case where it has been carefully studied (Chapter 8), specific alterations in this region had no effect upon transcription.

At the 3' end of most eukaryotic genes is the sequence AATAAA, which is believed to be a signal for the addition of the poly-A tail to the 3' end of the transcribed mRNA. It is not, however, a transcription-termination signal. Transcription

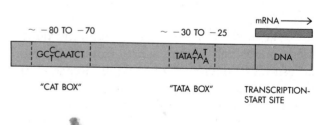

Figure 7-9
Two base sequences, dubbed the "CAT box" and the "TATA box," appear upstream of the transcription-start sites on many eukaryotic genes. These regions are probably control sequences for transcription.

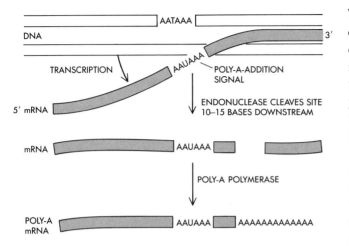

Figure 7-10
The sequence AATAAA appears toward the 3′ end of most eukaryotic genes. It codes for the mRNA sequence AAUAAA, which apparently instructs an endonuclease to cleave the strand at a site 10 to 15 bases downstream; a poly-A tail is then added.

usually continues downstream, past the site of polyadenylation. The sequence AATAAA most likely serves as a signal for a nuclease to clip the nascent RNA chain at a specific site some 10 to 15 bases further downstream. A second enzyme, poly-A polymerase, then adds the 100-to-200-base poly-A tail (Figure 7-10).

There is no obvious consensus sequence at the 5′ end of eukaryotic mRNA coding regions to serve as the ribosome-binding site analogous to the Shine–Dalgarno sequence found on prokaryotic mRNAs. Important differences thus may exist between eukaryotic and prokaryotic mRNAs in how they attach to ribosomes.

Clustered Gene Families May Include Vestigial Evolutionary Relics

With reliable DNA probes for a growing number of vertebrate genes, it has become clear that multiple gene families exist for many proteins. Sometimes multiple copies of a gene have nearly identical sequences, appear in tandem arrangements, and function simultaneously to synthesize certain proteins. The histone genes are one such example. In other instances the members of a gene family, though clustered on the chromosome, are not identical. Such genes specify closely related proteins; embryonic and adult globins are examples.

Within the globin clusters are globinlike sequences that can no longer function because long deletions have removed essential regulatory signals. These segments of DNA could represent vestiges of formerly functional globin genes whose physiological roles have, during evolution, been taken over by other members of the globin gene family (Figure 7-11).

The meaning of other gene clusters is more of a mystery. For example, next to the gene for chicken ovalbumin are two genes (called "X" and "Y") of similar sizes and with identical intron patterns. But the functions of the two corresponding proteins are unknown.

Between closely related genes in a cluster there are often "spacer" segments of DNA that do not code for protein. These are not to be confused with introns, which occur within genes. The spacers frequently contain more DNA than the genes themselves. Whether such spacer DNA is totally useless (junk DNA?) is still unclear. In any case, if most spacers have the same sizes as the spacers between the genes in the hemoglobin family, then as little as 10 percent of the nucleotides along a chromosome actually code for specific amino acids. The number of functional genes within a mammalian cell is thus far lower than would be expected from a cell's total DNA content. Now it is believed that the true number of functional genes will be no greater than 100,000, an estimate that may be decreased as longer stretches of individual vertebrate chromosomes are analyzed.

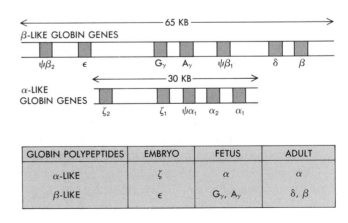

GLOBIN POLYPEPTIDES	EMBRYO	FETUS	ADULT
α-LIKE	ζ	α	α
β-LIKE	ϵ	G_γ, A_γ	δ, β

Figure 7-11
Human globin gene clusters. The ψ genes are the vestigial "pseudogenes."

Gene Families May Be Expanded by Reverse Transcription of mRNA Molecules

As more and more multigene families are thoroughly analyzed in higher vertebrates, occasional examples of intronless chromosomal "genes" are being discovered. When these have been carefully sequenced, they have been found to be perfect copies of the respective mRNAs; that is, all the introns within a given gene have been removed, to the nucleotide. One such intronless mouse α-globin gene exists; interestingly, it is not linked to the globin gene cluster on chromosome 11, but it is found on chromosome 15. Another example is a "pseudogene" that was recently discovered for the human immunoglobulin λ chain. This pseudogene is not on chromosome 22 with its normal counterparts. Its J and C segments (Chapter 9) are not discontinuous, but are precisely joined as they are in normal λ-chain spliced mRNA. Furthermore, at its 3′ end there is a tract of DNA corresponding to the 3′ poly-A tail of mRNA. At their 5′ ends both the mouse α-globin pseudogene and the human immunoglobulin λ-chain pseudogenes have unusual sequences not found in the mature mRNAs of the corresponding normal genes. It is now suspected that such intronless genes arise through the rare reverse transcription of cytoplasmic mRNAs, to create cDNAs that are then somehow reinserted randomly into their respective genomes (Figure 7-12). For such reinserted sequences to be inherited, they must be present in cells that give rise to germ cells. Any such reverse transcription events that occur in somatic cells will of course have no hereditary consequences.

Eukaryotic DNA Contains Interspersed Repetitive Sequences

DNAs from virtually all eukaryotes contain families of repetitive sequences. These sequences, which range in length from 130 to 300 base pairs (bp), depending on the species, may be present in many thousands of copies per genome. For example, in mammals, the most abundant family of repetitive sequences—called the "Alu" family because the sequence contains $\frac{AGCT}{TCGA}$, which is the

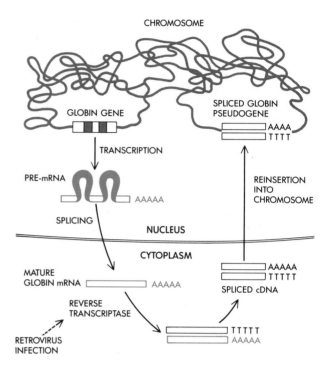

Figure 7-12
Reverse transcription of cytoplasmic mRNAs leads to intronless pseudogenes. The presence of several pseudogenes in the chromosomes of higher eukaryotes has led to speculation that these spliced DNA copies arose through reverse transcription of cytoplasmic mRNA, perhaps during an abortive retrovirus infection. The spliced genes are thought to have then been reinserted into the chromosome by a mechanism that is still unclear.

cleavage site of the Alu restriction enzyme—may be present about 300,000 times per genome (Figure 7-13).

The Alu sequences are found throughout the genome, both within spacer DNA and occasionally within introns. Following the cloning and analysis of these repetitive sequences, it became clear that each repeat is flanked by direct sequence repeats of 7 to 10 bp. At the 3′ ends of the Alu sequences there are traces of a poly-A region reminiscent of the poly-A stretches found at the 3′ ends of most eukaryotic mRNAs. The dispersal of Alu sequences throughout the genome suggests the possibility that new members of the family are generated by the reverse transcription of RNA molecules made from Alu sequences, and the insertion into the genome of the resultant cDNA.

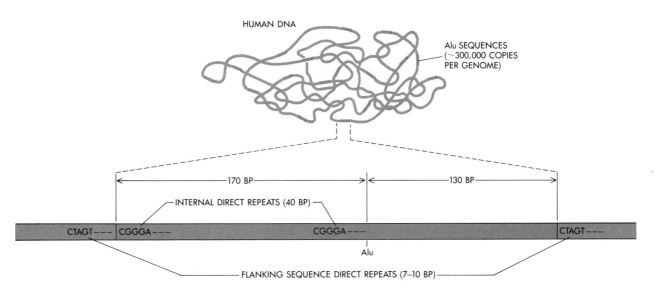

Figure 7-13
The Alu repetitive sequences are dispersed throughout the DNA of humans and other mammals. The human Alu sequence consists of 300 base pairs and typically appears about 300,000 times in a genome. Repeating sequences of about 40 base pairs occur within the Alu sequence, and direct-repeat sequences of 7 to 10 base pairs flank it.

Evolutionarily distinct groups of species have characteristic Alu sequences, and some mechanism that is still to be discovered must ensure that the structures of hundreds of thousands of Alu sequences dispersed throughout the genome are maintained close to the structure of the consensus sequence, in the face of mutations and other spontaneous changes to DNA. Another mystery is the function of the Alu sequences and the other families of dispersed repetitive DNA. Some members may code for several classes of very common cytoplasmic small RNA molecules (approximately 7S in size) whose function remains to be elucidated. But why there are so many copies of these dispersed sequences is a complete mystery at present.

Polypeptide Precursors Give Rise to Protein Hormones

Some years ago it was discovered that several small peptide hormones made in the pituitary are derived by cleavage of a large precursor protein. Corticotropin (ACTH) and β-lipotropin (β-LPH) are derived from a precursor polypeptide with a molecular weight of about 30,000. These two hormones are, in turn, broken into even smaller biologically active peptides such as α-melanotropin (α-MSH), the endorphins, and methionine-enkephalin.

Further progress toward understanding how the common precursor polypeptide was specifically cleaved to yield the active hormones was prevented because it proved impossible to obtain the pure precursor molecule in the amounts necessary for determining its amino acid sequence. All that could be said was that only about one-third to one-half of the precursor molecule ended up in the functional hormones. There was great curiosity about the structure and composition of the half of the precursor about which nothing was known. Perhaps it would be found to give rise to as yet undiscovered hormones.

Recombinant DNA procedures have shed some light on these problems. By early 1979, a cDNA probe of the mRNA for the common precursor of ACTH and β-LPH had been cloned, and the sequence of its 1091 base pairs had been determined. From this base sequence the amino acid sequence of the precursor polypeptide could be deduced. The amino acid sequence revealed the positions of the ACTH and β-LPH sequences in

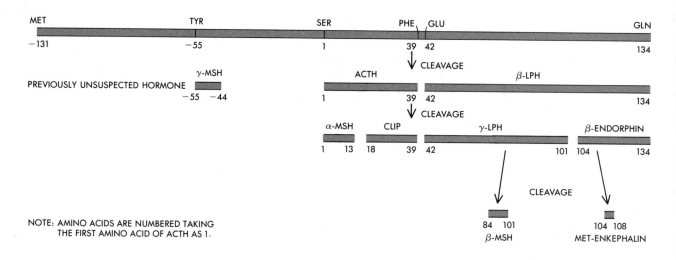

Figure 7-14
Maturation of pituitary hormones. Different pituitary peptide hormones that had been independently identified and studied turned out to be produced by cleavage of the same large polyprotein precursor. (Differential glycosylation of the primary translation product is believed to signal the various cleavages.)

the precursor. Moreover, a gene sequence for a previously unsuspected third hormone named γ-melanotropin (γ-MSH) was discovered (Figure 7-14). So three, not two, related pituitary hormones are produced from the one precursor. In addition, because several nucleotide sequences are repeated in the cDNA, it seems highly likely that during evolution a single ancestral gene underwent triplication to yield a row of three fused genes, each of which has undergone independent mutations to generate three related, but not identical, hormones.

In the future, the most direct way to determine the primary sequences of many proteins will be by cloning and then sequencing their corresponding cDNA probes. If a protein and its mRNA are rare, preparation of corresponding cDNA probes will require patience and ingenuity. It takes about a week to check by *in vitro* translation analysis which protein a single cDNA probe specifies. For rapid progress, enrichment procedures will be needed to achieve higher concentrations of mRNA of the proteins present in small amounts; such proteins represent the vast majority.

READING LIST

Books

Structures of DNA. Cold Spring Harbor Symp. Quant. Biol., vol. 47. Cold Spring Harbor Laboratory, Cold Spring Harbor, N.Y., 1983.

Original Research Papers (Reviews)

DISCOVERY OF RNA SPLICING IN ANIMAL VIRUSES

Berget, S. M., A. J. Berk, T. Harrison, and P. A. Sharp. "Spliced segments at the 5′ termini of adenovirus-2 late mRNA: A role for heterogeneous nuclear RNA in mammalian cells." *Cold Spring Harbor Symp. Quant. Biol.,* 42: 523–529 (1978).

Broker, T. R., L. T. Chow, A. R. Dunn, R. E. Gelinas, J. A. Hassel, D. F. Klessig, J. B. Lewis, R. J. Roberts, and B. S. Zain. "Adenovirus-2 messengers—an example of baroque molecular architecture." *Cold Spring Harbor Symp. Quant. Biol.,* 42: 531–553 (1978).

Westphal, H., and S. P. Lai. "Displacement loops in adenovirus DNA–RNA hybrids." *Cold Spring Harbor Symp. Quant. Biol.,* 42: 555–558 (1978).

SPLIT EUKARYOTIC GENES

Breathnach, R., J. L. Mandell, and P. Chambon. "Ovalbumin gene is split in chicken DNA." *Nature,* 270: 314–319 (1977).

Chambon, P. "Split genes." *Sci. Am.,* 244(5): 60–71 (1981).

Jeffreys, A. J., and R. A. Flavell. "The rabbit β-globin gene contains a large insert in the coding sequence." *Cell,* 12: 1097–1108 (1977).

Tilghman, S. M., D. C. Tiermeier, J. G. Seidman, B. M. Peterlin, M. Sullivan, J. V. Maijel, and P. Leder. "Intervening sequence of DNA identified in the structural portion of a mouse β-globin gene." *Proc. Natl. Acad. Sci. USA,* 75: 725–729 (1978).

Konkel, D., S. Tilghman, and P. Leder. "The sequence of the chromosomal mouse β-globin major gene: Homologies in capping, splicing and poly(A) sites." *Cell,* 15: 1125–1132 (1978).

Lomedico, P., N. Rosenthal, A. Efstratiadis, W. Gilbert, R. Kolodner, and R. Tizard. "The structure and evolution of the two nonallelic rat preproinsulin genes." *Cell,* 18: 545–558 (1979).

Bell, G. I., R. L. Pictet, W. Ruttner, B. Cordell, E. Tischer, and H. M. Goodman. "Sequence of the human insulin gene." *Nature,* 284: 26–32 (1980).

Yamada, Y., V. E. Avvedimento, M. Mudryj, H. Ohkubo, G. Vogeli, M. Irani, I. Pastan, and B. de Crombrugghe. "The collagen gene: Evidence for its evolutionary assembly by amplification of a DNA segment containing an exon of 54 bp." *Cell,* 22: 887–892 (1980).

Wozney, J., D. Hanahan, R. Morimoto, H. Boedtker, and P. Doty. "Fine structural analysis of the chicken pro-α2 collagen gene." *Proc. Natl. Acad. Sci. USA,* 78: 712–716 (1981).

Wozney, J., D. Hanahan., V. Tate, H. Boedtker, and P. Doty. "Structure of the pro α2(I) collagen gene." *Nature,* 294: 129–135 (1981).

EXONS VS. INTRONS

Gilbert, W. "Why genes in pieces?" *Nature,* 271: 501 (1978).

Crick, F. H. C. "Split genes and RNA splicing in evolution of eukaryotic cells." *Science,* 204: 264–271 (1979).

Lewin, R. "On the origin of introns." *Science,* 217: 921–922 (1982). (Review.)

SPLICING MECHANISMS

Breathnach, R., C. Benoist, K. O'Hare, F. Gannon, and P. Chambon. "Ovalbumin gene: Evidence for a leader sequence in mRNA and DNA sequences at the exon–intron boundaries." *Proc. Natl. Acad. Sci. USA,* 75: 4853–4857 (1978).

Breathnach, R., and P. Chambon. "Organization and expression of eucaryotic split genes coding for proteins." *Ann. Rev. Biochem.,* 50: 349–383 (1981).

Abelson, J. "RNA processing and the intervening sequence problem." *Ann. Rev. Biochem.,* 48: 1035–1069 (1979).

Konarska, M., W. Filipowicz, and H. J. Gross. "RNA ligation via 2′-phosphomonoester, 3′, 5′-phosphodiester linkage: Requirement of 2′, 3′-cyclic phosphate termini and involvement of a 5′-hydroxyl polynucleotide kinase." *Proc. Natl. Acad. Sci. USA,* 79: 1474–1478 (1982).

Murray, V., and R. Holliday. "Mechanism for RNA splicing of gene transcripts." *FEBS Letters,* 106: 5–7 (1979).

Rogers, J., and R. Wall. "A mechanism for RNA-splicing." *Proc. Natl. Acad. Sci. USA,* 77: 1877–1878 (1980).

Lerner, M. R., J. A. Boyle, S. M. Mount, S. L. Wolin, and J. A. Steitz. "Are snRNPs involved in splicing?" *Nature,* 283: 220–224 (1980).

Yang, V. W., M. R. Lerner, J. A. Steitz, and S. J. Flint. "A small nuclear ribonucleoprotein is required for

splicing of adenoviral early RNA sequences." *Proc. Natl. Acad. Sci. USA,* 78: 1371–1375 (1981).

Cech, T. R., A. J. Zaug, and P. J. Grabowski. "In vitro splicing of the ribosomal RNA precursor of Tetrahymena: Involvement of a guanosine nucleotide in the excision of the intervening sequence." *Cell,* 27: 487–496 (1981).

Grabowski, P. J., A. J. Zaug, and T. R. Cech. "The intervening sequence of the ribosomal RNA precursor is converted to a circular RNA in isolated nuclei of Tetrahymena." *Cell,* 23: 467–476 (1981).

Kruger, K., P. J. Grabowski, A. J. Zaug, J. Sands, D. E. Gottschling, and T. R. Cech. "Self-splicing RNA: Autoexcision and autocyclization of the ribosomal RNA intervening sequence of Tetrahymena." *Cell,* 31: 147–157 (1982).

ALTERNATIVE SPLICING

Anava, S. G., V. Jonas, M. G. Rosenfeld, E. S. Ong, and R. M. Evans. "Alternative RNA processing calcitonin gene expression." *Nature,* 298: 240–244 (1982).

Crabtree, G. R., and J. A. Kant. "Organization of the rat γ-fibrinogen gene: Alternative mRNA splice patterns produce the γA and γB (γ') chains of fibrinogen." *Cell,* 31: 159–166 (1982).

UPSTREAM (5') AND DOWNSTREAM (3') CONTROL SEQUENCES

Benoist, C., K. O'Hare, R. Breathnach, and P. Chambon. "The ovalbumin gene sequence of putative control regions." *Nuc. Acids Res.,* 8: 127–143 (1979).

Benoist, C., and P. Chambon. "In vivo sequence requirements of the SV40 early promoter region." *Nature,* 290: 304–310 (1981).

Ghosh, P. K., P. Lebowitz, R. J. Frisque, and Y. Gluzman. "Identification of a promoter component involved in positioning the 5' termini of simian virus 40 early mRNAs." *Proc. Natl. Acad. Sci. USA,* 78: 100–104 (1981).

Hen, R., P. Sassone-Corsi, J. Corden, M. P. Gaub, and P. Chambon. "Sequences upstream from the T-A-T-A box are required *in vivo* and *in vitro* for efficient transcription from the adenovirus serotype 2 major late promoter." *Proc. Natl. Acad. Sci. USA,* 79: 7132–7136 (1982).

Contreras, R., D. Gheysen, J. Knowland, A. van de Voorde, and W. Fiers. "Evidence for the direct involvement of DNA replication origin in synthesis of late SV40 RNA." *Nature,* 300: 500–505 (1982).

Lai, C.-J., R. Dhar, and G. Khoury. "Mapping the spliced and unspliced late lytic SV40 RNAs." *Cell,* 14: 971–982 (1978).

Nevins, J. R., and J. E. Darnell. "Steps in the processing of Ad2 mRNA: Poly(A)+ nuclear sequences are conserved and poly(A) addition precedes splicing." *Cell,* 15: 1477–1493 (1978).

Fitzgerald, M., and T. Shenk. "The sequence 5'-AAUAAA-3' forms part of the recognition site for polyadenylation of late SV40 mRNAs." *Cell,* 24: 251–260 (1981).

CLUSTERED GENE FAMILIES

Fritsch, E., R. Lawn, and T. Maniatis. "Molecular cloning and characterization of the human β-like globin gene cluster." *Cell,* 19: 959–972 (1980).

Slauer, J., C. Shen, and T. Maniatis. "The chromosomal arrangement of human α-like globin genes: Sequence homology and α-globin gene deletions." *Cell,* 20: 119–130 (1980).

Hentschel, C. C., and C. C. Birnstiel. "The organization and expression of histone gene families." *Cell,* 25: 301–313 (1981). (Review.)

PSEUDOGENES

Proudfoot, N. J., and T. Maniatis. "The structure of a human β-globin pseudogene and its relationship to β-globin gene duplication." *Cell,* 21: 537–544 (1980).

Leder, P., N. Hansen, D. Konkel, A. Leder, Y. Nishioka, and C. Talkington. "Mouse globin system: A functional and evolutionary analysis." *Science,* 209: 1336–1342 (1980). (Review.)

Little, P. F. R. "Globin pseudogenes." *Cell,* 28: 683–684 (1982). (Review.)

Hollis, G. F., P. A. Hieter, O. W. McBride, D. Swan, and P. Leder. "Processed genes: A dispersed human immunoglobulin gene bearing evidence of RNA-type processing." *Nature,* 296: 321–325 (1982).

Reilly, J. G., R. Ogden, and J. J. Rossi. "Isolation of a mouse pseudo tRNA gene encoding CCA—a possible example of reverse flow of genetic information." *Nature,* 300: 287–289 (1982).

Lemischka, I., and P. A. Sharp. "The sequences of an expressed rat α-tubulin gene and a pseudogene with an inserted repetitive element." *Nature,* 300: 330–335 (1982).

INTERSPERSED REPETITIVE SEQUENCES

Britten, R. J., and D. E. Kohne. "Repeated sequences in DNA." *Science,* 161: 529–540 (1968).

Jelinek, W. R., T. P. Toomey, L. Leinwand, C. H. Duncan, P. A. Biro, P. V. Choudary, S. M. Weissman, C. M. Rubin, C. M. Houck, P. L. Deininger, and C. W. Schmid. "Ubiquitous interspersed repeated DNA sequences in mammalian genomes." *Proc. Natl. Acad. Sci. USA,* 77: 1398–1402 (1980).

Schmid, C. W., and W. R. Jelenik. "The Alu family of

dispersed repetitive sequences." *Science,* 216: 1065–1070 (1982). (Review.)

Singer, M. F. "SINEs and LINEs: Highly repeated short and long interspersed sequences in mammalian genomes." *Cell,* 28: 433–434 (1982). (Review.)

Duncan, C. H., P. Jagadeeswaran, R. R. C. Wang, and S. M. Weissman. "Structural analysis of templates and RNA polymerase III transcripts of Alu family sequences interspersed among the human β-like globin genes." *Gene,* 13: 185–196 (1981).

Jagadeeswaran, P., B. G. Forget, and S. M. Weissman. "Short interspersed repetitive DNA elements in eucaryotes: Transposable DNA elements generated by reverse transcription of RNA PolIII transcripts." *Cell,* 26: 141–142 (1981). (Review.)

Weiner, A. "An abundant cytoplasmic 7S RNA is partially complementary to the dominant interspersed middle repetitive DNA sequence family in the human genome." *Cell,* 22: 209–218 (1980).

Van Arsdell, S. W., R. A. Denison, L. B. Bernstein, A. M. Weiner, T. Manser, and R. F. Gesteland. "Direct repeats flank three small nuclear RNA pseudogenes in the human genome." *Cell,* 26: 11–17 (1982).

POLYPEPTIDE PRECURSORS OF PROTEIN HORMONES

Herbert, E., and M. Uhler. "Biosynthesis of polyprotein precursors to regulatory peptides." *Cell,* 30: 1–2 (1982). (Review.)

Nakanishi, S., A. Inoue, T. Kita, M. Nakamura, A. Chang, S. Cohen, and S. Numa. "Nucleotide sequence of cloned cDNA for bovine corticotropin-β-lipotropin precursor." *Nature,* 278: 423–427 (1979).

Comb, M., P. H. Seeburg, J. Adelman, L. Eiden, and E. Herbert. "Primary structure of the human Met- and Leu-enkephalin precursor and its mRNA." *Nature,* 295: 663–666 (1982).

Gubler, U., P. Seeburg, B. J. Hoffman, L. P. Gage, and S. Udenfriend. "Molecular cloning establishes proenkephalin as precursor of enkephalin-containing peptides." *Nature,* 295: 206–208 (1982).

Noda, M., Y. Furutani, H. Takahashi, M. Toyosato, T. Hirose, S. Inayama, S. Nakanishi, and S. Numa. "Cloning and sequence analysis of cDNA for bovine adrenal preproenkephalin." *Nature,* 295: 202–206 (1982).

Land, H., G. Schütz, H. Schmale, and D. Richter. "Nucleotide sequence of cloned cDNA encoding bovine arginine vasopressin-neurophysin II precursor." *Nature,* 295: 299–303 (1982).

Scheller, R. H., J. F. Jackson, L. B. McAllister, J. H. Schwartz, E. R. Kandell, and R. Axell. "A family of genes that codes for ELH, a neuropeptide eliciting a stereotyped pattern of behavior in Aplysia." *Cell,* 28: 707–719 (1982).

8

In Vitro Mutagenesis

With the advent of recombinant DNA technology and ancillary techniques such as DNA sequencing, we can now examine, at the molecular level, the DNA sequences that are involved in the control of gene expression. The classical approach to genetics is to create *in vivo* mutations randomly throughout the genome, and then isolate those that display a particular phenotype. These mutants are then analyzed to determine which gene or genes have been altered. The precise nature of the mutation itself can be determined by DNA sequencing. An almost converse method (in fact, the method has been called "reverse genetics") is to create specific mutations in a DNA segment *in vitro,* and to analyze the effects of these changes on the organism *in vivo* following reintroduction of the mutant gene. It is now possible to create mutations—either deletions, insertions, or specific base changes—at predetermined sites in a DNA molecule.

Deletions

The simplest *in vitro* mutation that can be constructed is the deletion of a DNA segment between two restriction enzyme sites. Circular SV40 DNA, for instance, can be partially digested with a given restriction enzyme that would normally produce several fragments. The conditions of digestion can be maintained so that each molecule gets, on the average, two or three cuts. The resultant molecules are recircularized using DNA ligase. The population will then consist of molecules that have had a specific restriction enzyme fragment excised. Because deletions made in this way tend to be rather large, this approach is typically used in preliminary analysis to determine the functions of relatively large areas of a cloned DNA molecule. For instance, deletion of the SV40 *Hin*dIII B or C fragment destroys transforming activity, whereas deletion of the smallest *Hin*dIII fragment does not, but does prevent packaging of SV40 into virus. Once these relatively large areas of DNA have been associated with given functions, finer mutations can be made to determine more precisely the functional units of DNA.

Smaller deletions can be produced in a circular DNA molecule by cleaving it with a restriction enzyme that linearizes it (cleaves it once). The linear molecules are then treated with an enzyme called exonuclease III (exo III), which starts from each 3' end of the DNA and chews away single strands in a 3'-to-5' direction, creating a population of DNA molecules with single-stranded tails at each end. The tails can then be degraded with S1 nuclease, which specifically attacks single-stranded DNA, resulting in duplex DNA molecules with deletions. More recently, the enzyme Bal 31 exonuclease has been used; this enzyme chews away *both* strands from the ends of linear DNA molecules. These nucleolytic reactions can be controlled by varying the time of incubation, the temperature, and the enzyme concentration to make deletions that can range from 20 to 2000 bases of DNA. The deleted molecules are then

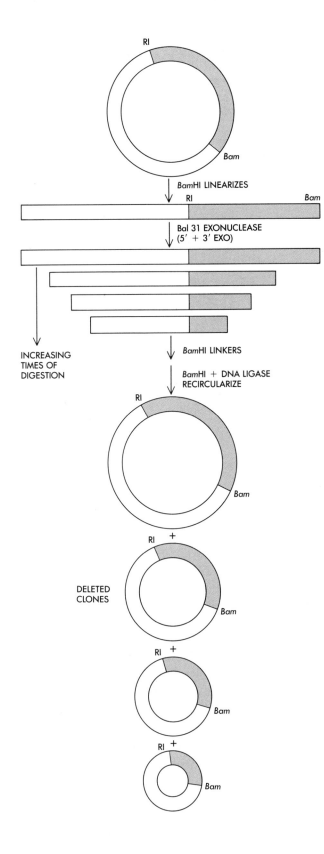

RI

BamHI LINEARIZES

RI   Bam

Bal 31 EXONUCLEASE
(5' + 3' EXO)

INCREASING
TIMES OF
DIGESTION

BamHI LINKERS

BamHI + DNA LIGASE
RECIRCULARIZE

RI

Bam

RI   +

DELETED
CLONES

Bam

RI   +

Bam

RI   +

Bam

recircularized with DNA ligase (usually after addition of a synthetic oligonucleotide linker) and used to transform *E. coli* (Figure 8-1).

A variation of this technique was used to study the DNA sequences that are important in the regulation of transcription of the herpes simplex virus thymidine kinase gene (HSV tk). Two libraries of deletion mutants were constructed with the exo III and S1 nucleases: One set of mutations started from a site well beyond the 5' end and proceeded towards the gene; the other set started at a point in the gene and proceeded in the opposite direction through the 5' end of the gene.

DNA sequencing was used to define the precise endpoints of the two sets of deletions. Forty-three different mutants in which the deletions came from the 5' direction and 42 mutants in which the deletions came from the 3' direction were found to have terminated in a small 140-nucleotide stretch surrounding the 5' end of the tk gene. (The ends of all the deletions contained a synthetic *Bam* HI linker). Next, a search was made for deletion pairs that came from opposite directions and that terminated exactly 10 bases apart. When two such deletions were recombined through the *Bam* HI linker, the 10 bases of the linker replaced the 10 nucleotides in the normal sequence. This was called a "clustered point mutant" because most, though generally not all, of the 10 bases in the linker were different from the wild-type sequence (Figure 8-2, page 108). When two deletion mutants coming from opposite directions had their termini 15 or 20 bases apart, then recombination of these termini through the *Bam* HI linker resulted in a 5-or-10-base-pair deletion of DNA, as well as the base changes caused by the *Bam* HI linker. In this way, the spacing between control regions was systematically and precisely varied.

Figure 8-1
Small deletions are made in a circular DNA molecule by cleaving it once with *Bam* HI and treating it with Bal 31 exonuclease, which digests the ends of both strands. The size of the deletion can be controlled by varying incubation time, temperature, and enzyme concentration. The strand is then recircularized by adding *Bam* HI linkers and DNA ligase.

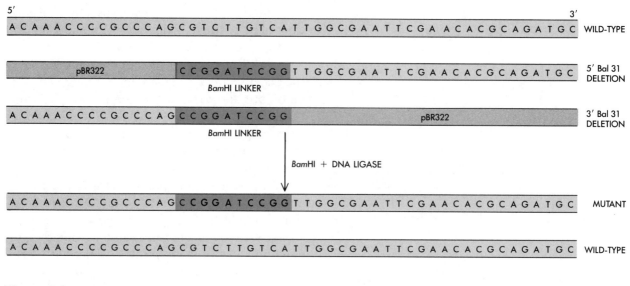

Figure 8-2
Creation of a clustered point mutation resulting in an eight-base substitution. Pairs of mutants in which the members terminate exactly ten bases from one another are selected. Identical *Bam*HI linkers are added to the ends; these ends are cleaved with *Bam*HI and joined with DNA ligase. A DNA molecule that differs from the wild-type sequence only in the ten bases that constituted the linkers is thus formed. Two of these bases happen to match the bases in the wild-type sequence; the other eight (boldface) represent the clustered point mutation.

A set of clustered point mutations or small deletions was constructed in this way throughout the promoter region of the HSV tk gene. When such mutants were tested in an *in vitro* transcription assay, it was found that three distinct regions are required for efficient expression of tk mRNA. One is the "TATA box" (Chapter 7, page 98). This AT-rich stretch of nucleotides about 25 to 30 bases upstream from the cap site is believed to direct initiation at the correct nucleotide. Two other regions near the tk gene were also found to be critical for transcription: GC-rich stretches at about -50 and -90. Base substitutions in either of these regions greatly reduce transcriptional efficiency. The spacing between these three regions was also found to be important: Deletions that bring the -90 region closer to the -50 reduce transcription, as do deletions that bring the -50 region closer to the TATA box.

Insertions

Insertion of a synthetic oligonucleotide linker at a given site in a cloned DNA molecule can be used either to interrupt the normal DNA sequence or to generate a new restriction site that can serve as a starting point for further mutagenesis. Pancreatic DNase in the presence of manganese ions introduces random double-stranded breaks in a DNA molecule. Normally this enzyme would completely degrade a DNA molecule into small pieces; however, if cloned DNA is treated with a very small amount of pancreatic DNase for a very short time, each DNA molecule will be cut, on the average, only once. These linear molecules can be isolated from a gel. Synthetic oligonucleotide linkers can then be added to them so that they may be recircularized with DNA ligase. This will result in a population of DNA molecules with the linker sequence randomly inserted (Figure 8-3). Alternatively, cloned DNA molecules can be cleaved with a restriction enzyme that would normally cut each DNA molecule several times, but a very small amount is used so that each molecule is cut only once. Insertion of linkers into this population will result in a less random set of insertion mutants than that generated using pancreatic DNase, since each

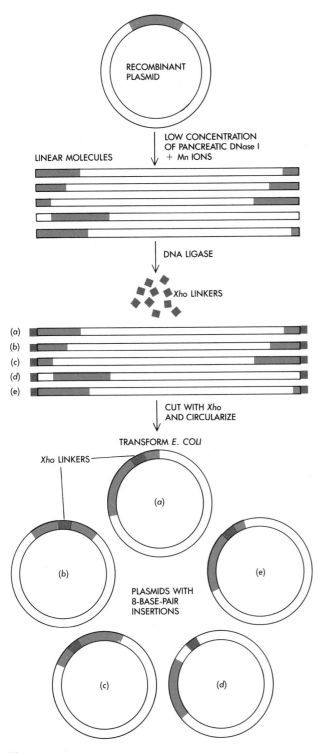

Figure 8-3
Creation of insertion mutants via synthetic oligonucleotide linkers. Recombinant plasmids are
cleaved only once at random locations by treatment
with a tiny amount of pancreatic DNase I. *Xho*
linkers added to the resulting linear molecules are
cut open with *Xho*. The molecules are
recircularized with DNA ligase, producing plasmids
that contain eight-base-pair inserts.

insertion will be at a known restriction enzyme
site.

This technique was used to define DNA sequences important in mating-type regulation in
yeast (Chapter 11). The yeast genome contains
three copies ("cassettes") of genes that control
mating type: one expressed copy (the "MAT"
copy) and two silent copies (called "HMR" and
"HML"). The silent genes are kept silent by the
action of another gene set called "SIR." Insertion
mutations around the HMR locus showed that the
negative regulation of this gene by SIR is dependent upon two regions of DNA, one lying on each
side of the silent cassette. Mutations in one of the
regions (the "E," or "essential," region) resulted
in complete loss of control by SIR, whereas mutations in the other (the "I," or "important," region) caused only a partial loss of regulation.
Thus, a region of negative transcriptional control
extending about 3000 bases was defined.

Substitutions: Deamination of Cytosine

Specific single base changes (point mutations) can
also be generated in a DNA molecule *in vitro.* One
method is based on the fact that cytosine residues
in single-stranded, but not double-stranded, DNA
can be chemically deaminated to produce uracil by
the action of bisulfite ions. Thus, if small single-
stranded regions can be created in a cloned DNA
molecule, the C residues in this region can be
specifically changed to U (which is read as T when
the molecule is replicated).

There are two methods of creating single-
stranded regions of DNA around a restriction enzyme site. One method is to use the enzyme exonuclease III to chew away single strands of DNA
in the 3'-to-5' direction. The result will be single-
stranded tails, the lengths of which will depend on
the extent of the reaction with exo III. The other
method, which is more controllable, is based on
the fact that many DNA polymerases will, in the
absence of dNTPs, degrade single strands of duplex DNA in the 3'-to-5' direction; if only *one*
dNTP is present in the reaction, the polymerase
will degrade single strands of duplex DNA until
it reaches a point at which the base on the other
strand is complementary to the available nucleotide. The exonucleolytic activity will then stop.
Both phage T4 DNA polymerase and the so-called

"Klenow" or large fragment of *E. coli* DNA polymerase I have this property. For example, if a DNA molecule is cut with the restriction enzyme *Sma* I, which recognizes the sequence CCCGGG, and this cleaved molecule is treated with Klenow polymerase in the presence of only dATP, the enzyme starts at the *Sma* I site and chews away each 3′ end of the duplex DNA until it reaches an A residue. This is usually only a few bases away from the enzyme site; thus small single-stranded regions are generated on each strand. The molecule is then treated with bisulfite, which changes the C residues to U. Next Klenow polymerase is added in the presence of *all four* dNTPs and the single-stranded regions are repaired; the uracil residues created by the bisulfite pair with A. GC base pairs are thus mutated to AT pairs (Figure 8-4).

Using this method, a number of single and double base changes were generated in a cloned proline tRNA molecule. The mutations were GC → AT changes clustered around a *Sma* I site inside the tRNA molecule. When tested in *in vitro* transcription assays, some of these mutants were found to be very poorly transcribed. This indicated that the promoter for the tRNA genes (which are transcribed by RNA polymerase III) is actually *inside* the gene, rather than off the 5′ end. This "intragenic promoter" has been found for all genes transcribed by RNA polymerase III, in contrast to genes transcribed by RNA polymerase I or II or by the prokaryotic RNA polymerases.

Substitutions: Incorporation of Nucleotide Analogs

Another method of producing single base changes near a restriction site is to create small single-stranded regions in a cloned DNA molecule. The molecule is treated with a restriction enzyme in the presence of ethidium bromide. This drug causes the restriction enzyme to "nick" the DNA at the restriction site (that is, to cut only one of the DNA strands rather than both). This nick can be extended into a small gap by using a low concentration of DNA polymerase isolated from a bacterium called *M. luteus*. In the absence of triphosphates, *M. luteus* polymerase will degrade only five or six bases of DNA in the 5′-to-3′ direction, starting from the nick. The result is a DNA molecule with a five-or-six-base gap in it. The gap is then repaired with Klenow polymerase in the presence of dATP, dCTP, dGTP, and *N*-4-hydroxycytosine instead of dTTP. This nucleotide analog is incorporated in place of T, and, because the 6-keto–enol ratio ($C_6 = O / C_6 OH$) of this compound is almost 1, it can pair equally well with

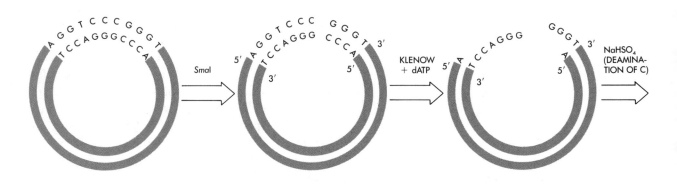

Figure 8-4
Creation of a substitution mutant through deamination of cytosine. A *Sma* I cleavage site is treated with Klenow polymerase in the presence of dATP. The enzyme chews away the single strands in the 3′-to-5′ direction, until it reaches an A residue on each strand. The addition of bisulfite changes the exposed C residues to U. Klenow polymerase is added again, this time with all four dNTPs; the result is a repaired molecule in which two of the GC base pairs have been mutated to AU (or AT).

either A or G. Incorporation of a G residue in the complementary strand will result in an AT → GC mutation (Figure 8-5, page 112).

Substitutions: Misincorporation of Nucleotides

Finally, single base changes near a restriction site in DNA can be generated by misincorporation of a nucleotide during enzymatic repair of a gap, created as just described. The gapped molecules are treated with *E. coli* or T4 DNA polymerase in the presence of only three of the four dNTPs. Normally, the polymerase stops when it reaches a point at which it has no complementary nucleotide for the base in the opposite strand; a small percentage of the time, however, the polymerase will misincorporate one of the other three nucleotides in place of the one that is missing, and then continue to repair the gap completely. This results in circular plasmids with single base mismatches. When the plasmids are reintroduced into *E. coli,* the mismatches will be repaired during the next round of DNA replication. Half of the replicating strands will recreate the wild-type sequences, and the other half will create mutant sequences with a single base-pair change. Since the base change was made at the restriction site at which the molecule

was originally nicked, the site has been altered and the mutant molecule will not be cleaved by that enzyme (Figure 8-6, page 113). The daughter plasmid molecules can be quickly screened for the presence or absence of the enzyme site to determine whether they are mutant or wild-type. One potential problem with this technique is the fact that most polymerases have a "proofreading" function, in the form of a nuclease that excises a mismatched nucleotide as soon as it is incorporated. This can be circumvented by using α-thiophosphate nucleotides during the gap repair; these molecules cannot be excised by the polymerase's proofreading exonuclease.

Mutants May Be Constructed by Using Oligonucleotides with Defined Sequences

All of the techniques we have described for generating specific base changes in a DNA molecule are dependent upon the presence of a restriction enzyme site near the region of interest. Of course, the location of such a site is a matter of random chance. Another method for making specific point mutants in a cloned DNA molecule does not depend at all on the presence of a convenient restriction site. The method utilizes synthetic oligonucleotides of defined sequence (Chapter 5, page 63). When the sequence of a cloned DNA mole-

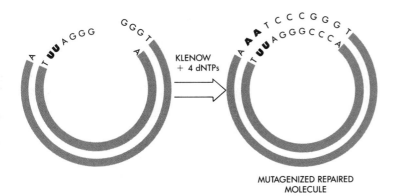

KLENOW
+ 4 dNTPs

MUTAGENIZED REPAIRED
MOLECULE

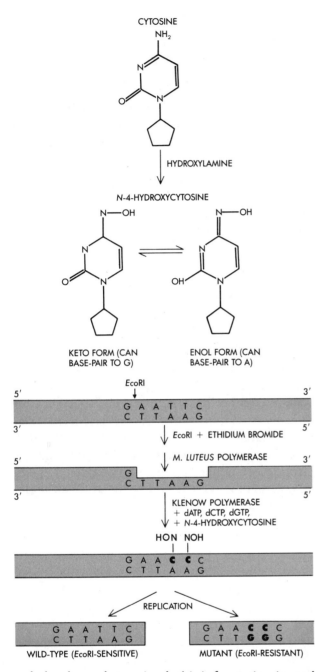

CYTOSINE

HYDROXYLAMINE

N-4-HYDROXYCYTOSINE

KETO FORM (CAN
BASE-PAIR TO G)

ENOL FORM (CAN
BASE-PAIR TO A)

EcoRI

5'   3'
G A A T T C
C T T A A G
3'   5'

EcoRI + ETHIDIUM BROMIDE

M. LUTEUS POLYMERASE

5'   3'
G
C T T A A G
3'   5'

KLENOW POLYMERASE
+ dATP, dCTP, dGTP,
+ N-4-HYDROXYCYTOSINE

HON NOH
| |
G A A C C C
C T T A A G

REPLICATION

G A A T T C   G A A C C C
C T T A A G   C T T G G G
WILD-TYPE (EcoRI-SENSITIVE)   MUTANT (EcoRI-RESISTANT)

Figure 8-5

N-4-Hydroxycytosine can be obtained from cytosine by treatment with hydroxylamine. *N*-4-Hydroxycytosine can pair equally well with A or G. A DNA strand is nicked at an *Eco* RI site by cleaving it with this restriction enzyme in the presence of ethidium bromide. The nick can be extended into a small gap by using *M. luteus* polymerase. The gap is then repaired by treating it with Klenow polymerase in the presence of dATP, dCTP, dGTP, and *N*-4-hydroxycytosine. The latter replaces T in the resulting double strand, since in its enol form it can pair with A. During replication, the *N*-4-hydroxycytosine residues are interpreted as C residues, so the complementary strand is formed with G residues opposite the inserts, instead of the original A residues. The result is a mutant that resists *Eco* RI cleavage because the recognition site for that enzyme has been altered.

cule has been determined, this information is used to make an oligonucleotide that is 12 to 15 bases long and that is complementary to the region to be mutated, but with one or two mismatches. This oligonucleotide is mixed with a single-stranded clone of the complementary strand of the original molecule, carried in an M13 phage vector. Although the oligonucleotide is not a perfect match, it will anneal to the single-stranded clone if the hybridization conditions are not very stringent (the annealing must be done at a low temperature and in the presence of high salt), and if the mis-

matches are in the middle of the oligonucleotide rather than at an end. The mismatched oligonucleotide serves as a primer for the action of DNA polymerase to synthesize the remainder of the complementary strand. The double-stranded molecule, now containing one or two mismatches, is introduced into *E. coli,* where the mismatches will be repaired, to recreate either the wild-type sequence or the mutant one (Figure 8-7).

Theoretically, 50 percent of the daughter molecules will be wild-type and 50 percent mutant; in practice, however, the percentage of mutant molecules is much lower (usually 10 to 15 percent). [This is probably for one or both of the following reasons: (1) The DNA polymerase, having synthesized the entire complementary strand starting from the primer, sometimes continues polymerizing after going around the circle, so that it actually displaces the primer and recreates the original wild-type sequence. (2) The original wild-type DNA strand is "marked" somehow, perhaps by methylation, in such a way that when the mismatched duplex is introduced into *E. coli,* it is repaired to the wild-type sequence more often than to the mutant sequence.] Mutant molecules can be distinguished from wild-type ones in two ways. If the base change either created or destroyed a restriction enzyme site, then several M13 clones can be quickly assayed for the presence or absence of that site. Alternatively, the oligonucleotide that was originally used to make the mutation can be used to distinguish mutant from wild-type

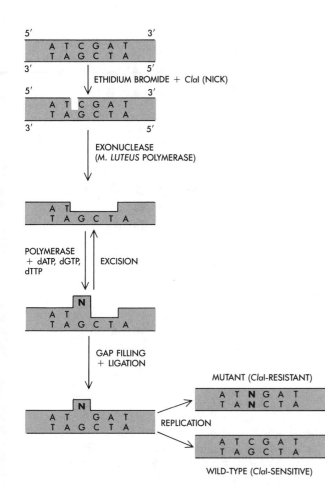

Figure 8-6
Creation of a substitution mutant through misincorporation of a nucleotide. The restriction enzyme *Cla*I is used in the presence of ethidium bromide to nick a cloned DNA molecule. The nick is extended into a small gap by using *M. luteus* polymerase. The gap is then repaired by treating it with *E. coli* or T4 DNA polymerase and three of the four dNTPs—dCTP is omitted. Occasionally the enzyme will try to substitute A, T, or G opposite the G residue in the intact chain. If the mismatched nucleotide is not excised, the result will eventually be a set of daughter plasmid molecules that contain a mutant base pair.

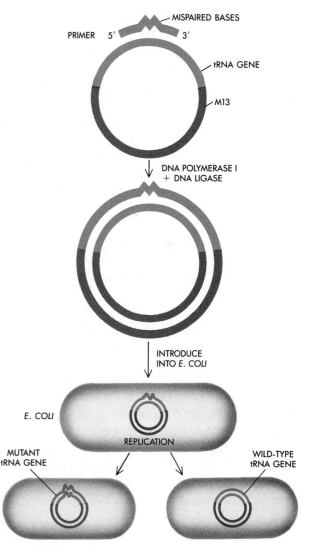

Figure 8-7
Creation of a substitution mutant by use of a synthetic oligonucleotide. The 12-to-15-base oligonucleotide is constructed so that it is complementary to a region of a DNA strand, but with one or two mismatches. When mixed with a clone of the complementary strand, the oligonucleotide will anneal to it even though the match is not exact, as long as the hybridization conditions are not stringent and the mismatches are in the middle of the oligonucleotide segment. The segment then serves as a primer for DNA polymerase I, which synthesizes the remainder of the complementary strand. When the resulting double-stranded molecule is introduced into *E. coli,* the molecule replicates to recreate either the original wild-type sequence or the mutant sequence.

molecules: As was described above, an oligonucleotide will hybridize to a complementary
sequence with one or two mismatches if the
stringency of hybridization is kept low; if the temperature of the hybridization reaction is raised,
however, the oligonucleotide will form a stable
duplex only with a sequence to which it is perfectly
complementary. The oligonucleotide that was
used to make the mutation can be labeled with ^{32}P
and used as a probe to screen bacterial colonies on
nitrocellulose filters, as described earlier (Figure
6-7). If the temperature of the hybridization is
raised in 5°C increments, a point can usually be
reached at which the labeled oligonucleotide will
hybridize only to the mutant molecules (to which
it is perfectly complementary) and not to the wild-
type molecules (Figure 8-8).

A synthetic oligonucleotide was used to create
a synthetic suppressor tRNA. A 12-base oligonucleotide was synthesized to be complementary to
the anticodon loop of a cloned lysine tRNA molecule, except that the bases complementary to the
lysine anticodon AAA were changed to be complementary to the termination codon UAG. This
mismatched oligonucleotide was used to generate
a mutant lysine tRNA molecule. The mutant
tRNA was injected into frog oocytes with a cloned
mutant β-globin gene obtained from a thalassemia
patient (Chapter 17). This globin gene had the
termination codon UAG at amino acid position
17, instead of the codon AAG which codes for
lysine. Normally, the synthesis of β-globin polypeptide in oocytes would stop at this UAG codon.
When the mutant tRNA gene is also microinjected, however, lysine was incorporated at this
position, and synthesis of the β-globin protein proceeded normally.

Oligonucleotide-directed mutagenesis is also
proving valuable for the study of protein structure
and function. If the DNA sequence of the protein-
coding region of a gene is known, then a synthetic
oligonucleotide can be used to specifically change
one amino acid codon to another. When this mutant gene is reintroduced into cells by means of
techniques that we will describe later (Chapter
14), it will produce a protein with exactly one
amino acid change. The effects of such an alteration on the protein's function can then be assessed.

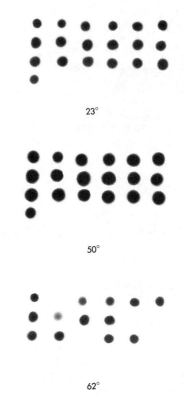

23°

50°

62°

Figure 8-8
Distinguishing mutants created by defined-
sequence oligonucleotides. M13 clones, some
of which contain a specific base change, were
obtained as described in Figure 8-7. DNA
from these clones was spotted onto a
nitrocellulose filter. The oligonucleotide that
had been used to create the mutation was
labeled with ^{32}P and hybridized to the filter.
The filter was then washed at successively
higher temperatures until the M13 clones
containing the mutant sequence could be
distinguished from those harboring the wild
type. (Courtesy of Mark Zoller.)

This technique is currently being used to study the
amino acids that are responsible for the correct
insertion and anchorage of membrane proteins,
and also for the study of the structure of certain
tumor virus proteins that may be involved in oncogenic transformation.

Given the recent advances in DNA sequencing and in oligonucleotide synthesis, oligonucleotide-directed mutagenesis could prove to be by far
the most versatile method of creating specific point
mutations in a DNA molecule.

READING LIST

Original Research Papers (Reviews)

DELETIONS

Lai, C. J., and D. Nathans. "Deletion mutants of SV40 generated by enzymatic excision of DNA segments from the viral genome." *J. Mol. Biol.,* 89: 179–193 (1974).

Mertz, J. E., J. Carbon, M. Herzberg, R. W. Davis, and P. Berg. "Isolation and characterization of individual clones of simian virus 40 mutants containing deletions, duplications and insertions in their DNA." *Cold Spring Harbor Symp. Quant. Biol.,* 39: 69–84 (1974).

Beier, D. R., and E. T. Young. "Characterization of a regulatory region upstream of the ADR 2 locus of S. cerevisiae." *Nature,* 300: 724–728 (1982).

Grummt, I. "Nucleotide sequence requirements for specific initiation of transcription by RNA polymerase I." *Proc. Natl. Acad. Sci. USA,* 79: 6908–6911 (1982).

McKnight, S. L., and R. Kingsbury. "Transcriptional control signals of a eukaryotic protein-coding gene." *Science,* 217: 316–324 (1982).

Mirault, M. E., R. Southgate, and E. Delwart. "Regulation of heat shock genes: A DNA sequence upstream of Drosophila hsp genes is essential for induction." *EMBO J.,* 1: 1279–1282 (1982).

Struhl, K. "The yeast *his3* promoter contains at least two distinct elements." *Proc. Natl. Acad. Sci. USA,* 79: 7385–7389 (1982).

INSERTIONS

Heffron, F., M. So, and B. J. McCarthy. "In vitro mutagenesis of a circular DNA molecule using synthetic restriction sites." *Proc. Natl. Acad. Sci. USA,* 75: 6012–6016 (1978).

Abraham, J., J. Feldman, K. A. Nasmyth, J. N. Strathern, A. J. S. Klar, J. R. Broach, and J. B. Hicks. "Sites required for position-effect regulation of mating type information in yeast." *Cold Spring Harbor Symp. Quant. Biol.,* 47: 989–998 (1983).

BASE CHANGES

Chu, C. T., D. S. Parris, R. A. F. Dixon, F. E. Farber, and P. A. Schaffer. "Hydroxylamine mutagenesis of HSV DNA and DNA fragments: Introduction of mutations into selected regions of the viral genome." *Virology,* 98: 168–181 (1979).

Shortle, D., and D. Nathans. "Regulatory mutants of simian virus 40: Constructed mutants with base substitutions at the origin of viral DNA replication." *J. Mol. Biol.,* 131: 801–817 (1979).

Shortle, D., K. Koshland, G. M. Weinstock, and D. Botstein. "Segment-directed mutagenesis: Construction in vitro of point mutations limited to a small predetermined region of a circular DNA molecule." *Proc. Natl. Acad. Sci. USA,* 77: 5375–5379 (1980).

Hayatsu, H. "Bisulfite modification of nucleic acids and their constituents." *Prog. Nuc. Acid Res. Mol. Biol.,* 16: 75–124 (1976).

Shortle, D., and D. Botstein. "Single-stranded gaps as localized targets for in vitro mutagenesis." In J. F. Lemont and W. M. Generosa, eds., *Molecular and Cellular Mechanisms of Mutagenesis.* Plenum, New York, 1982, pp. 147–156. (Review.)

Weiher, H., and H. Schaller. "Segment-specific mutagenesis: Extensive mutagenesis of a *lac* promoter/operator element." *Proc. Natl. Acad. Sci. USA,* 79: 1408–1412 (1982).

Ciampi, M. S., D. A. Melton, and R. Cortese. "Site-directed mutagenesis of a tRNA gene: Base alterations in the coding region affect transcription." *Proc. Natl. Acad. Sci. USA,* 79: 1388–1392 (1982).

Sakonju, S., D. Bogenhagen, and D. D. Brown. "A control region in the center of the 5S gene directs specific initiation of transcription, I: The 5' border of the region." *Cell,* 19: 13–26 (1980).

Bogenhagen, D., S. Sakonju, and D. D. Brown. "A control region in the center of the 5S gene directs specific initiation of transcription, II: The 3' border of the region." *Cell,* 19: 27–36 (1980).

Shortle, D., J. Pipas, S. Lazarowitz, D. DiMaio, and D. Nathans. "Constructed mutants of simian virus 40." In J. K. Setlow and A. Hollaender, eds., *Genetic Engineering, Principles and Methods,* vol. 1. Plenum, New York, 1979, pp. 73–92. (Review.)

Hochstadt, J., H. L. Ozer, and C. Shopsis. "Genetic alteration in animal cells in culture." *Curr. Top. Microbiol. Immunol.,* 94/95: 244–308 (1981). (Review.)

Everett, R. D., and P. Chambon. "A rapid and efficient method for region- and strand-specific mutagenesis of cloned DNA." *EMBO J.,* 1: 433–437 (1982).

Hirose, S., K. Takeuchi, and Y. Suzuki. "*In vitro* characterization of the fibroin gene promoter by the use of single-base substitution mutants." *Proc. Natl. Acad. Sci. USA,* 79: 7258–7262 (1982).

Peden, K. W., and D. Nathans. "Local mutagenesis within deletion loops of DNA heteroduplexes." *Proc. Natl. Acad. Sci. USA,* 79: 7214–7217 (1982).

Zakour, R. A., and L. A. Loeb. "Site-specific mutagenesis by error-directed DNA synthesis." *Nature*, 295: 708–710 (1982).

MUTANT CONSTRUCTION BY USING
OLIGONUCLEOTIDES WITH DEFINED SEQUENCES

Gillam, S., C. R. Astell, and M. Smith. "Site specific mutagenesis using oligodeoxyribonucleotides: Isolation of a phenotypically silent ϕX174 mutant, with a specific nucleotide deletion, at very high frequency." *Gene*, 12: 129–137 (1980).

Smith, M., and S. Gillam. "Constructed mutants using synthetic oligodeoxyribonucleotides as site-specific mutagens." In J. K. Setlow and A. Hollaender, eds., *Genetic Engineering, Principles and Methods,* vol. 3. Plenum, New York, 1981, pp. 1–32. (Review.)

Shortle, D., D. DiMaio, and D. Nathans. "Directed mutagenesis." *Ann. Rev. Genetics,* 15: 265–294 (1981).

Wallace, R. B., M. Schold, M. J. Johnson, P. Dembek, and K. Itakura. "Oligonucleotide directed mutagenesis of the human β-globin gene: A general method for producing specific point mutations in cloned DNA." *Nuc. Acids Res.,* 9: 3642–3656 (1981).

Dalhadie-McFarland, G., L. W. Cohen, A. D. Riggs, C. Morin, K. Itakura, and J. H. Richards. "Oligonucleotide-directed mutagenesis as a general and powerful method for studies of protein function." *Proc. Natl. Acad. Sci. USA,* 79: 6409–6413 (1982).

Laski, F. A., R. Belagaje, U. L. RajBhandary, and P. Sharp. "An amber suppressor tRNA derived by site-specific mutagenesis." *Proc. Natl. Acad. Sci. USA,* 79: 5813–5817 (1982).

Temple, G. F., A. M. Dozy, K. L. Roy, and Y. W. Kan. "Construction of a functional human suppressor tRNA gene." *Nature,* 296: 537–540 (1982).

Winter, G., A. R. Fersht, A. J. Wilkinson, M. Zoller, and M. Smith. "Redesigning enzyme structure by site-directed mutagenesis." *Nature,* 756–758 (1982).

9

Rearranging Germ-Line DNA Segments to Form Antibody Genes

For several decades an enormous number (perhaps in the millions) of different antibody (immunoglobulin) molecules have been known to exist, each characterized by a unique site that can bind to specific molecular determinants (antigens). Many immunologists thought initially that all antibodies were made of the same polypeptide chains and that their uniqueness arose from the way their newly synthesized identical polypeptide chains folded around the respective antigens. This theory was proved wrong. Each antibody has its own amino acid sequence, and each antibody-producing cell (plasma cell) makes only one antibody. At first, this was a disturbing discovery because it seemed to imply that a separate gene would have to exist for each separate antibody. If so, perhaps a large fraction, if not the majority, of the vertebrate DNAs would have to be devoted to coding antibody molecules. But such speculations could not be tested until protein chemists established the basic structure of the antibody molecule.

The Basic Structure of Antibody Molecules Is Established

The first insights began to emerge in the early 1960s, when it was realized that the fundamental antibody unit consists of two identical light (L) chains of molecular weight 17,000 and two identical heavy (H) chains of molecular weight 35,000, held together by disulfide bonds (Figure 9-1). (The terms "light" and "heavy" refer to the differences in the molecular weight of the chains.) Each such four-chain unit contains two identical binding sites for antigens, with a site being formed partly

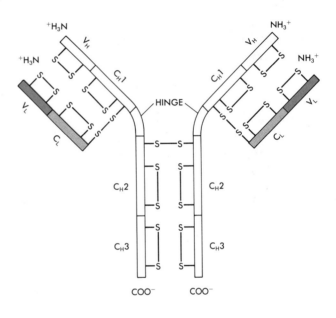

Figure 9-1
The structure of an antibody protein. Two light chains (color) and two heavy chains (white) are held together by disulfide bonds. The light chains and the heavy chains each contain one variable unit (V_L or V_H) at their amino-terminal ends. The light chains also contain one constant unit (C_L); the heavy-chain constant portion has four domains ($C_H 1$, $C_H 2$, $C_H 3$, and the hinge region).

by specific amino acids of the light chain and partly by specific heavy-chain amino acids. Once the basic antibody layout had been established, the amino acid sequences of the component light and heavy chains were determined by using the homogeneous antibodies made by specific myeloma cells. Myelomas are cancerous antibody-producing

(plasma) cells, and in any one animal all the cells of a myeloma tumor are the descendants of one original cancer cell. This explains why all the antibody molecules from any one myeloma have the same amino acid sequence.

Both light- and heavy-chain sequences vary from one type of antibody to another, but in a way that no one would have predicted initially. Although each chain has unique sequences, almost all of this specificity is restricted to about 100 amino acids at the amino-terminal ends (the variable, or V, regions). Half of each light chain and three-quarters of each heavy chain have almost identical sequences (the constant, or C, regions) (Figure 9-1, page 117).

Separate Genes for V and C Segments Are Proposed

To account for the constant and variable portions of the chains, William Dreyer and Claude Bennett at the California Institute of Technology put forward a bold hypothesis in 1965. They proposed that the V and C regions are coded by separate genes, and that in the germ line the C-region segments (C_L for the light chain, C_H for the heavy chain) are each coded for by only one gene, but that the V regions (V_H and V_L) are coded for by many thousands of genes. Dreyer and Bennett further proposed that a functional antibody is formed when genetic recombination in the precursor to the plasma cell brings one of the V genes next to its respective C gene to yield $V_L C_L$ and $V_H C_H$ genes. This perceptive hypothesis won few early converts because it flew in the face of the general belief that the arrangement of DNA within a given chromosome was effectively immutable, except at meiosis during the formation of the sex cells.

Messenger RNA Probes Are Used to Obtain Support for the Joining of V and C Genes

It was impossible to test this V–C joining hypothesis until there were direct molecular probes that might identify the putative V and C genes. In the first experiments, which were done between 1974 and 1976, mRNA probes that were more than 50 percent pure were made from myeloma cells. These mRNA probes were mixed, under conditions favoring hybridization, with unfractionated homologous myeloma DNA to see if it was possible to count the number of genes for the single antibody type that was present in a given myeloma cell. The answer for at least one light-chain mRNA probe was that there were a very low number of genes, and perhaps only one.

Such experiments, however, could not distinguish V from C sequences, nor could they indicate differences in the relative locations of V and C sequences in embryonic cells compared with myeloma cells. To do that required a way to cut up the total myeloma and embryonic cell DNAs into reproducible pieces, a procedure that became possible only with the ready availability of restriction enzymes. Using them, Susumu Tonegawa at the Basel Institute of Immunology in Switzerland observed in the spring of 1976 that V and C sequences that were linked together on the same DNA restriction fragment from an antibody-producing mouse myeloma cell were not similarly linked together in embryonic DNA (Figure 9-2). This classic experiment was done with necessarily impure mRNA probes because of the still-effective prohibitions against cloning cDNA molecules. But as soon as the first suitable cDNA vector was approved, the appropriate cDNA probes for V and C regions were made and their specificity directly determined by DNA sequencing. These regions were then used for selecting genomic segments that carried the respective antibody genes. Possession of such reagents has revolutionized our understanding of the molecular basis of antibody diversity.

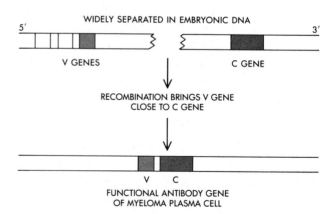

Figure 9-2
Functional antibody genes are produced by genetic recombination. The V and C genes in embryonic myeloma DNA were discovered to be widely separated, whereas they were found close together in mature myeloma antibody-producing cells.

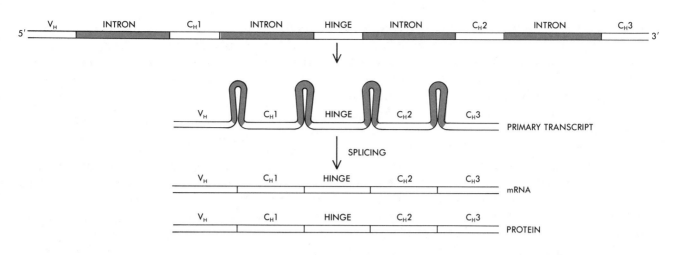

Figure 9-3
The protein domains of immunoglobulin heavy chains are separated by introns.

Functional Antibody Genes Are Isolated from Myeloma Cells

The cDNA probes that established the nature and number of antibody genes were made from specific myelomas whose antibody products had already been sequenced. Direct comparisons were thus possible between the nucleotide sequences of functional antibody genes and the amino acid sequences of the antibodies they specified. Many introns were found immediately, and, most importantly, most were located at junctions between functional domains. In the light chain, an intron separates almost all of the amino-terminal leader sequences from the V segment, and a second intron divides the V from the C sequences. Within heavy-chain genes, introns separate functionally related domains (in other words, they separate exons coding for domains) more extensively. Each of the three domains of the C_H protein is clearly delineated by introns, as is the so-called hinge region lying between the first and second C_H domains (Figure 9-3). All of these observations have supported the hypothesis that proteins have evolved by the rearrangement of exons.

Embryonic Cells Are Sources of Unjoined V and C Genes

The structures of the V and C segments before they are joined to create functional antibody genes were revealed by cloning the appropriate genomic DNA segments from embryonic cells and hybridizing them with probes specific for the V and C regions. C-Region probes invariably were very specific, whereas V-region probes often hybridized to many different V genes. Such cross-hybridization reflects the fact that the V regions of different antibodies often differ by only a few amino acid substitutions. Now we have evidence for at least 200 V_L genes and an equal number of V_H genes. Note that because the specificity of an antibody is determined by both its V_L and V_H components, the number of potentially different antibodies is at least $V_L \times V_H$, or $200 \times 200 = 40,000$.

In contrast to the multiplicity of V_L and V_H genes, there exist only two C_L genes (C_λ and C_κ), located on different chromosomes, and a single cluster of linked C_H genes. In the mouse, this C_H cluster is located on chromosome 12 and consists of eight genes that are spaced out over some 200 kilobases (kb) of DNA. The separate C_H genes reflect different functional roles for their gene products, with, for example the C_μ gene coding for the early-appearing immunoglobulin M, and the several C_γ genes coding for immunoglobulins that have generally higher specificity and that predominate in the later stages of an immune response.

Multiple J (Joining) Segments Are Attached to Genomic C (Constant) Segments

The number of potential light chains is greatly increased by the presence of a cluster of related, but not identical, J (for joining) segments that reside upstream of each C_L gene. The number of

potential heavy chains is likewise increased by a J-segment cluster located upstream from each C_H cluster. The linkage of a V segment to a C segment may occur next to any of these J segments, and, depending on which J segment is used, a different group of amino acids will be found inserted between the amino acids encoded by the V segment and those encoded by the C segment. RNA splicing thus occurs in such a way that only the J segment used in the V–C joining event is retained (Figure 9-4). How splicing can be so regulated remains totally mysterious, as does the nature of the events that join the V and C genes together.

The joining event itself is slightly variable, which creates additional diversity at the V–JC combining site. The site occurs, probably not by chance, within the nucleotides coding for amino acids that help form the cavity into which antigens bind. One model for DNA joining postulates that the respective V and C segments, initially located far apart on the same DNA molecule, are brought together by a recombination process that eliminates the intervening sequences. The sequences at the ends of the V and JC segments are comple-

mentary and could allow formation of hydrogen-bonded hairpin loops that would align the appropriate bases ready for cutting and rejoining. There is some experimental support for this proposal.

Three Discontinuous Regions of DNA Code for Heavy-Chain Amino Acids

At first it was believed that heavy-chain formation followed the same pattern of events as light-chain formation. But when the appropriate germ-line V_H and $J_H C_H$ regions were cloned, sequenced, and compared to the sequences found in the respective myeloma heavy chains, it became apparent that a second group of internal amino acids were not coded by either the V_H or $J_H C_H$ segments but had to arise from a third DNA segment. This third segment, given the name D (for diversity), is sited between the V_H and $J_H C_H$ segments. It consists of a tandem multigene family of related sequences. The existence of multiple D segments, any one of which has the possibility of being inserted into the functional heavy-chain gene, still further increases the potential number of heavy-chain specificities $(V_H \times D_H \times J_H)$ (Figure 9-5). Like the J-encoded amino acids, the D-encoded amino acids help to form the antigen-binding sites, so the diversity they create is likely to be biologically useful.

A DNA Elimination Event Allows a V_H Gene to Be Attached to Two Different C_H Genes

Recombinant DNA procedures have also swiftly solved the puzzle of how a given heavy-chain variable gene (V_H) can be attached first to a constant segment (C_μ) that is characteristic of the immunoglobulin M class, and then, during the later immune response, can transfer its linkage to a constant segment (C_γ) characteristic of the immunoglobulin G class. No change of immunological specificity occurs during this transfer, because the heavy-chain types differ only in their "constant" components, all of which are coded by a group of genes clustered together on the same chromosome. How this switch happens became crystal-clear as soon as the appropriate C_μ and C_γ genes were cloned and their sequences were compared with those of a functional gene coding for a known γ-class heavy chain (MOPC 141).

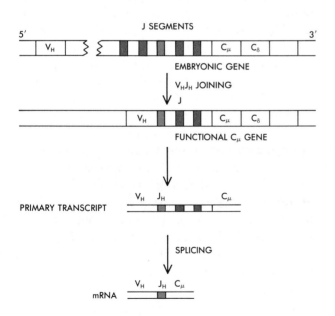

Figure 9-4
A V_H gene is linked to a C_H gene by means of a J (joining) segment that is located in a cluster of such segments upstream of the C_H genes. After the initial recombination event, RNA splicing removes all of the other J segments to produce the mature mRNA.

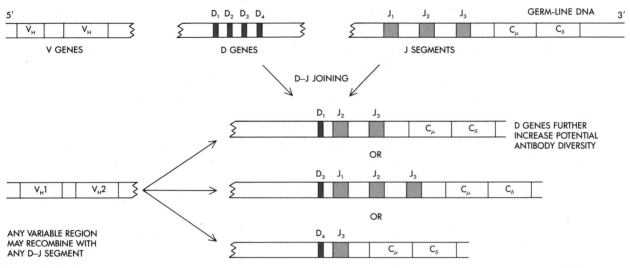

Figure 9-5
D (diversity) segments join to J segments to create additional potential diversity in antibody molecules.

The key observation was the finding of J-segment bases in the gene coding for the heavy chain of MOPC 141, despite the absence of any J segment near the corresponding embryonic C_γ gene. In contrast, several J segments were found at the beginning of the embryonic C_μ gene, one of which exactly corresponded to that observed in the functional C_γ gene of MOPC 141. Two recombination events are therefore necessary to generate a functional gene for γ-class heavy chains. The first joining event attaches a V_H gene to the C_μ gene at one of its flanking J segments, thereby allowing synthesis of a μ-class heavy chain. This synthesis continues until a second recombination event removes most of the C_μ gene sequences and links the previously joined $V_H J_H$ segment to the intron sequences flanking a C_γ gene. This leads to the synthesis of a γ-class heavy chain whose J-coded sequences bear witness to the prior $V_H J_H C_H$ arrangement (Figure 9-6).

Alternative Splicing Allows Single Cells to Make Both μ- and δ-Class Heavy Chains with Identical V_H Segments

Recombinant DNA procedures have also clarified another previously puzzling observation—namely, that one cell can simultaneously make heavy chains that are of two different types but that contain the same variable region. The initial explanation pro-

posed for this situation was that one type of heavy chain might be translated from a very stable long-lived mRNA that persisted in the cytoplasm long after the corresponding gene had been eliminated. Analysis of the structure of the genes involved with recombinant DNA techniques, however, suggested an alternative. The variable region (V), a joining region (J), and the two constant regions C_μ and C_δ were found to be contiguous. The whole complex might thus be translated into a

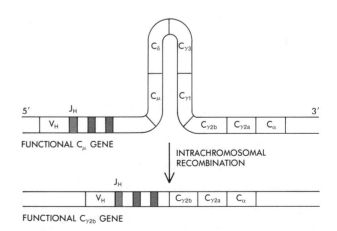

Figure 9-6
Genetic recombination changes a functional μ-class gene to a functional γ-class gene. This recombination event occurs after the formation of the initial V_H–J_H link.

large precursor RNA molecule from which, by differential splicing, either one or the other of the constant regions is eliminated (Figure 9-7). This mechanism was confirmed with cDNA probes prepared from the appropriate mRNA. Differential splicing of a common precursor RNA generates two distinct heavy-chain mRNAs. A dual-splicing potential also determines whether certain classes of antibodies are bound to the plasma membranes of the cells in which they are made or are secreted to the outside.

Somatic Mutations Provide a Further Source of Immunoglobulin Diversity

Before the existence of recombinant DNA, there was seemingly endless debate as to whether the specificity for antibody genes lay largely in germ-line DNA segments or whether most diversity was created by somatic mutations that occurred during the multiplication of antibody-producing cells and their precursors. With the discovery of V and

(D) JC joining it became unambiguously clear that much antibody diversity was carried in germ-line DNA segments. Virtually simultaneously, however, it was found that the exact amino acid sequence of many antibodies does not precisely correspond with that predicted by their respective germ-line sequences. The first direct evidence for somatic diversification came from analysis of the mouse light-chain variable regions. Of 19 λ proteins examined, 12 had the germ-line sequence while 7 others had one to three amino acid differences. Since then, somatic diversification has also been found in V_H segments, and significant proportions of antibody variability must now be ascribed to somatic mutational events. Although these mutations occur both in the DNA sequences that control the specificity of antigen binding (the V regions) and in those that code for the so-called "framework" (C) regions, most mutations are observed to affect antigen binding. Equally important "mutant" antibodies are largely restricted to the later stages of the immunological response, in

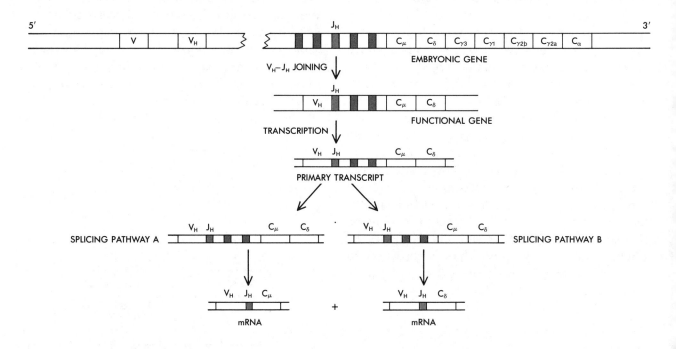

Figure 9-7
Alternative splicing can generate antibody molecules with different heavy-chain determinants. Here a large precursor RNA molecule is differentially spliced so as to produce two different kinds of RNA: that coding for a μ-class heavy chain and that coding for a δ-class heavy chain.

which C_γ- (or C_α-) chain synthesis predominates. It has thus been asked whether the antibodies produced by somatic mutations are more efficient (whether they bind antigen better) than the early-appearing antibodies whose specificity comes entirely from germ-line sequences. Though immunological theory predicts that this should occur, preliminary experiments have failed to support this idea.

Establishing the Genes of the Major Histocompatibility Complex (MHC) Proteins and Their Protein Antibodies Through Gene Cloning

Some 30 years ago, the rejection of foreign skin grafts (transplantations) in mice was found to depend on a group of proteins (H2) that recognized the skin cells as "nonself." It was determined that these proteins were coded by closely linked genes that were mapped to a region of chromosome 17. With time, the number of these so-called "histocompatibility proteins" was found to be far greater than was first perceived, and the "major histocompatibility complex" (MHC) is now known to code for three very different classes of proteins. Class I consists of the genes for the highly polymorphic H2 transplantation antigens (in humans, the HLA antigens), which are expressed on virtually all cell surfaces, as well as the genes for other surface molecules restricted to specific types of differentiated cells (for example, blood-forming cells). The class II (immune-response) genes encode a specific group of cell surface molecules that are present only on lymphocytes and that control the extent of specific immunological responses. Several components of the complement system that links an immune response to the desired destruction of an unwanted foreign cell by lysis constitute the class III genes.

Our understanding of the exact chemistry of these MHC proteins has lagged far behind our understanding of the chemistry of the immunoglobulins themselves, and only with the advent of cloning procedures is a comprehensive picture of the MHC structures beginning to emerge. In general, the class I H2 and H2-like proteins contain a 45,000-dalton transmembrane protein attached

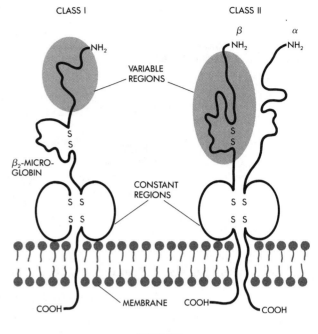

Figure 9-8
Major histocompatibility complex proteins and their protein antibodies. Class I and class II MHC proteins have variable and constant regions in their cytoplasmic domains. The constant regions show some homology with the constant regions of immunoglobulin molecules. Class I MHC proteins are associated with another protein called β_2-microglobulin.

noncovalently to a 12,000-dalton protein, β_2-microglobulin, that is separately encoded on chromosome 2. Each 45,000-dalton chain contains three external domains comprising approximately 90 amino acids, a transmembrane region of approximately 40 amino acids, and a short cytoplasmic region of about 30 amino acids (Figure 9-8). This structure reflects the exon–intron arrangement, with separate exons encoding the signal leader peptide, each of the exterior domains, the transmembrane region, and the three regions of the cytoplasmic component.

The organization of the class I genes along chromosome 17 has been revealed through cloning 40-kb fragments of mouse DNA into cosmids and looking for overlaps between clones contain-

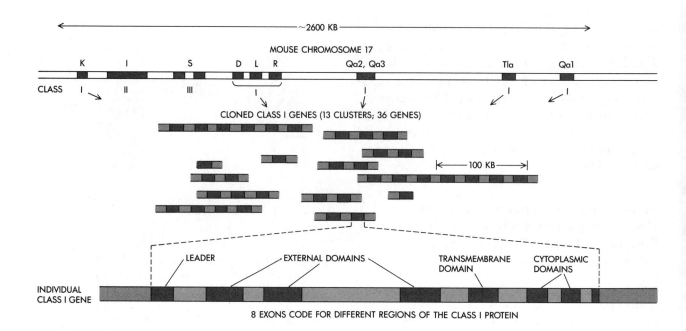

Figure 9-9
The genes coding for MHC class I, II, and III proteins are linked on mouse chromosome 17. The 36 class I genes have been cloned on cosmids and were found to occur in 13 clusters containing varying numbers of genes.

ing class I genes (Figure 9-9). So far, 36 class I genes encompassing some 837 kb of DNA have been found. Now the class II and class III genes are in the process of being mapped, and through chromosome walking the complete molecular map of the mouse MHC complex could be obtained over the next several years.

Recombinant DNA analysis has thus already profoundly affected immunological research. The separate origins of immunological diversity are in the process of being sorted out. Equally important, by being able to go directly to the MHC genes, which control the extended nature of immunological responses, we are much closer to explaining the phenomenological descriptions that have constituted cell immunology than we would have been if we had been limited to the biological approaches of the cellular immunologist.

READING LIST

Books

Hood, L. E., I. L. Weissman, and W. B. Wood. *Immunology.* Benjamin-Cummings, Menlo Park, Cal., 1978.

Original Research Papers (Reviews)

AMINO ACID SEQUENCE EVIDENCE FOR THE JOINING OF SEPARATE V AND C GENES

Dreyer, W. J., and J. D. Bennett. "The molecular basis of antibody formation: A paradox." *Proc. Natl. Acad. Sci. USA,* 54: 864–869 (1965).

Hood, L. E. "Two genes, one polypeptide chain—fact or fiction?" *Fed. Proc.,* 31: 179–187 (1972).

Weigert, M., and R. Riblet. "Genetic control of antibody variable regions." *Cold Spring Harbor Symp. Quant. Biol.,* 41: 837–846 (1977).

Weigert, M., L. Gatmaitan, E. Loh, J. Schilling, and L. E. Hood. "Rearrangement of genetic information may produce immunoglobulin diversity." *Nature,* 276: 785–790 (1978).

LIGHT-CHAIN FORMATION FROM SEPARATE V AND C GENES

Hozumi, N., and S. Tonegawa. "Evidence for somatic rearrangement of immunoglobulin genes coding for variable and constant regions." *Proc. Natl. Acad. Sci. USA,* 73: 3628–3632 (1976).

Tonegawa, S., C. Brack, N. Hozumi, and V. Pirrotta. "Organization of immunoglobulin genes." *Cold Spring Harbor Symp. Quant. Biol.,* 42: 921–931 (1978).

HEAVY-CHAIN JOINING EVENTS

Sakano, H., J. H. Rogers, K. Huppi, C. Brack, A. Traumecker, R. Maki, R. Wall, and S. Tonegawa. "Domains and the hinge region of an immunoglobulin heavy chain are encoded in separate DNA segments." *Nature,* 277: 627–633 (1979).

Davis, M. M., K. Calame, P. W. Early, D. L. Livant, R. Joho, I. L. Weissman, and L. E. Hood. "An immunoglobulin heavy-chain gene is formed by at least two recombinational events." *Nature,* 283: 733–739 (1980).

Sakano, H., R. Maki, Y. Kurosawa, W. Roeder, and S. Tonegawa. "Two types of somatic recombination are necessary for the generation of complete immunoglobulin heavy chain genes." *Nature,* 286: 676–683 (1980).

Sakano, H., Y. Kurosawa, M. Weigert, and S. Tonegawa. "Identification and nucleotide sequence of a diversity DNA segment (D) of immunoglobulin heavy chain genes." *Nature,* 290: 562–565 (1981).

Siebenlist, U., J. V. Ravetch, S. Korsmeyer, T. Waldmann, and P. Leder. "Human immunoglobulin D segments encoded in tandem multigenic families." *Nature,* 294: 631–635 (1981).

Ravetch, J. V., U. Siebenlist, S. Korsmeyer, T. Waldmann, and P. Leder. "Structure of the human immunoglobulin locus: Characterization of embryonic and rearranged J and D genes." *Cell,* 27: 583–591 (1981).

ALTERNATIVE RNA SPLICING EVENTS

Early, P., J. Rogers, M. Davis, K. Calame, M. Bond, R. Wall, and L. E. Hood. "Two mRNAs can be produced from a single immunoglobulin μ gene by alternative RNA processing pathways." *Cell,* 20: 313–319 (1980).

Moore, K. W., J. Rogers, T. Hunkapiller, P. Early, C. Nottenburg, I. Weissman, H. Bazin, R. Wall, and L. E. Hood. "Expression of IgD may use both DNA rearrangement and RNA splicing mechanism." *Proc. Natl. Acad. Sci. USA,* 78: 1800–1804 (1981).

HEAVY-CHAIN CLASS SWITCHES

Sledge, C., D. S. Fair, B. Black, R. G. Krueger, and L. E. Hood. "Antibody differentiation: Apparent sequence identity between variable regions shared by IgA and IgG immunoglobulins." *Proc. Natl. Acad. Sci. USA,* 73: 923–927 (1976).

Kataoka, T., T. Kawakami, N. Takahashi, and T. Honjo. "Rearrangement of immunoglobulin γ1-chain gene and mechanisms for heavy-chain class switch." *Proc. Natl. Acad. Sci. USA,* 77: 919–923 (1980).

Marcu, K. B. "Immunoglobulin heavy-chain constant-region genes." *Cell,* 29: 719–721 (1982). (Review.)

CONTRIBUTION OF SOMATIC MUTATIONS TO IMMUNOLOGICAL DIVERSITY

Baltimore, D. "Somatic mutation gains its place among the generators of diversity." *Cell,* 26: 295–296 (1981). (Review.)

Gearhart, P. J., N. D. Johnson, R. Douglas, and L. E. Hood. "IgG antibodies to phosphorylcholine exhibit more diversity than their IgM counterparts." *Nature,* 291: 29–34 (1981).

Selsing, E., and U. Storb. "Somatic mutation of immunoglobulin light-chain variable-region genes." *Cell,* 25: 47–58 (1981).

Crews, S., J. Griffin, H. Huang, K. Calame, and L. E. Hood. "A single V_H gene segment encodes the immune response to phosphorylcholine: Somatic mutation is correlated with the class of the antibody." *Cell,* 25: 59–66 (1981).

Bothwell, A. L. M., M. Paskind, M. Reth, T. Imanishi-Kari, K. Rajewski, and D. Baltimore. "Heavy chain variable region contribution to the NP[b] family of antibodies: Somatic mutation evident in a γ2a variable region." *Cell,* 24: 625–637 (1981).

THE MAJOR HISTOCOMPATIBILITY COMPLEX

Hood, L. E., M. Steinmetz, and R. Goodenow. "Genes of the major histocompatibility complex." *Cell,* 28: 685–687 (1982). (Review.)

Steinmetz, M., K. W. Moore, J. G. Frelinger, B. T. Sher, F.-W. Shen, E. A. Boyse, and L. E. Hood. "A pseudogene homologous to mouse transplantation antigens: Transplantation antigens are encoded by eight exons that correlate with protein domains." *Cell,* 25: 683–692 (1981).

Steinmetz, M., A. Winoto, K. Minard, and L. E. Hood. "Clusters of genes encoding mouse transplantation antigens." *Cell,* 28: 489–498 (1982).

Moore, K. W., B. T. Sher, Y. H. Sun, K. A. Eakle, and L. E. Hood. "DNA sequence of a gene encoding a BALB/c mouse L[d] transplantation antigen." *Science,* 215: 679–682 (1982).

Malissen, M., B. Malissen, and B. R. Jordan. "Exon/intron organization and complete nucleotide sequence of an HLA gene." *Proc. Natl. Acad. Sci. USA,* 79: 893–897 (1982).

Margulies, D. H., G. A. Evans, L. Flaherty, and J. G. Seidman. "H-2 like genes in the *Tla* region of mouse chromosome." *Nature,* 295: 168–170 (1982).

Steinmetz, M., K. Minard, S. Horvath, J. McNicholas, J. Srelinger, C. Wake, E. Long, B. Mach, and L. Hood. "A molecular map of the immune response region from the major histocompatibility complex of the mouse." *Nature,* 300: 35–42 (1982).

10

Tumor Viruses

It is by now virtually impossible to conceive of working on tumor viruses without using recombinant DNA methods. Given the ability to fragment precisely and then clone viral DNA and to systematically insert or delete prescribed DNA sequences in viral genomes, the essence of how tumor viruses act is now within our grasp. However, if restriction enzymes had not been discovered, the massive entry of molecular biologists into the "war on cancer" might have had no greater impact than past misadventures of biochemists when they sought to diagnose the metabolic defects that lead to cancer.

The discrete relatively small amounts of DNA found within the smaller DNA tumor viruses, like SV40 (5243 bp), polyoma virus (5292 bp), and the adenoviruses (about 36,000 bp), allowed preliminary molecular analysis as soon as the first of the various restriction enzymes became available in the early 1970s. The restriction fragments produced by enzymes like *Eco* RI were few in number, and generally they could be routinely separated from each other on agarose gels. The SV40 and polyoma virus nucleotide sequences could thus be worked out and distinctions made between the genes that function early versus those that function late in the viral cycles (Figure 10-1), without the need for recombinant DNA procedures. Cloning of the separate fragments of the DNA—forbidden under the Asilomar guidelines—would only have hastened this first-stage molecular analysis.

The first such gene cloning was done in 1978 in France and England, where the requirements regarding physical containment for recombinant DNA experiments with tumor viral genomes were

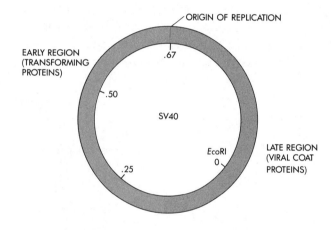

Figure 10-1
The SV40 chromosome. The early-region primary transcript codes for the transforming (large T and small t) proteins. The late primary transcript codes for the viral coat proteins (VP1, VP2, and VP3).

decided on a case-by-case basis, and were less exacting than the rigid, prohibitively stringent American requirements. By January 1979, however, it became possible to clone tumor virus DNA in the United States. As soon as it was allowed, specific restriction enzyme fragments from the DNA of tumor viruses were cloned to produce specific probes for isolating the viral DNA integrated into transformed cells.

At that point, detailed analysis of the way the retroviruses (RNA tumor viruses containing reverse transcriptase) transform cells also became dependent upon recombinant DNA techniques. Gene cloning procedures have made it simpler to study DNA than to study RNA. With these meth-

ods both cDNA copies of retrovirus RNA and proviral DNA (the integrated copy of the retrovirus genome) with its flanking host chromosome sequences have been isolated and analyzed in great detail. The remarkable results obtained have quickly led to a highly credible unified hypothesis of retroviral oncogenesis (oncology is the study of tumors) which has already profoundly changed the way we study cancer.

Cloning Integrated Forms of DNA Tumor Viruses

Most, if not all, cancerous cells transformed by DNA viruses like SV40 and adenovirus 2 have viral DNA inserted into their chromosomes, and the expression of the DNA is responsible for the cancerous phenotypes. Beginning in 1972, molecular biologists began to examine the amounts of integrated tumor virus DNA, using as probes DNA fragments obtained by dissecting viral genomes with restriction enzymes. However, meaningful analysis of the arrangement of the integrated viral DNAs and of their flanking host chromosome sequences became possible only when recombinant DNA procedures became available.

Now it looks as if the integration processes are virtually haphazard, occurring neither at specific sites nor within any specific sequence in either the viral or host chromosomal DNA. Some integrated segments reveal almost bizarre rearrangements of viral sequences, and we suspect that the recombination events that mediate such integrations often involve either replicating DNA or DNA being transcribed into RNA. Both replication and transcription necessarily produce limited regions of single-stranded DNA; these could form the temporary base-paired bridges that would be neces-

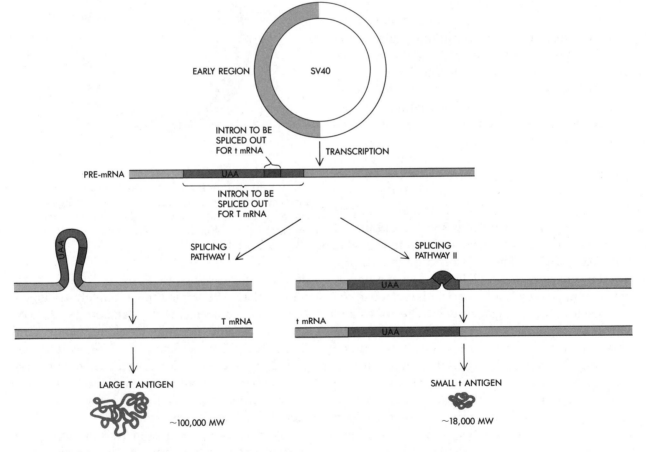

Figure 10-2
Differential splicing of the same primary early transcript leads to the large T and small t proteins. A translational stop codon (UAA) is spliced out to make large T mRNA, but the stop codon remains in the small t mRNA. Therefore the small t mRNA, though much longer than the large T mRNA, makes a smaller protein.

sary to hold the molecules together as they participate in nonhomologous (illegitimate) recombination events.

Tumor Proteins of SV40 and Polyoma

Initially both the SV40 and polyoma genomes were thought to contain a single "early" gene coding for the key tumor (T) protein whose presence made some cells cancerous. The first genetic analyses using viral mutants suggested that this protein, which functions early in the viral life cycle to initiate viral DNA replication, also plays a vital role in the maintenance of the cancerous state. How this single protein with a molecular weight of approximately 100,000 could carry out this dual role was not at all obvious. There was much relief when the initial discovery of RNA splicing during adenovirus replication was followed quickly by the finding that the primary RNA transcript made from the early SV40 region is spliced in two different ways. One splicing choice leads to the "large T" antigen with the MW of about 100,000; this is located in the nucleus, where it binds to the origin of SV40 DNA replication. The second splicing choice produces a cytoplasmically located "small t" protein that has a MW of about 18,000 (Figure 10-2). An even more complicated splicing pattern occurs with the transcript from the polyoma virus early region, which is spliced to produce a "middle T" protein as well as large T and small t molecules.

The exact molecular functions of the different T proteins of SV40 and polyoma viruses remain unknown, but they are being investigated using *in vitro* mutagenesis to make specific deletions that lead to specifically altered T proteins (Figure 10-3). It seems that the fully cancerous phenotype created by SV40, polyoma, or adenoviral DNA requires the functioning of more than one type of T protein. One T protein may act by effectively immortalizing the transformed cell, giving it the capacity for unlimited multiplication—perhaps by causing the synthesis of an essential growth factor. A second class of T protein appears to bind to the inner surface of the outer plasma membrane, which somehow leads to a cascade of changes similar to those produced by the "sarc" protein coded by Rous sarcoma virus (discussed later in this chapter).

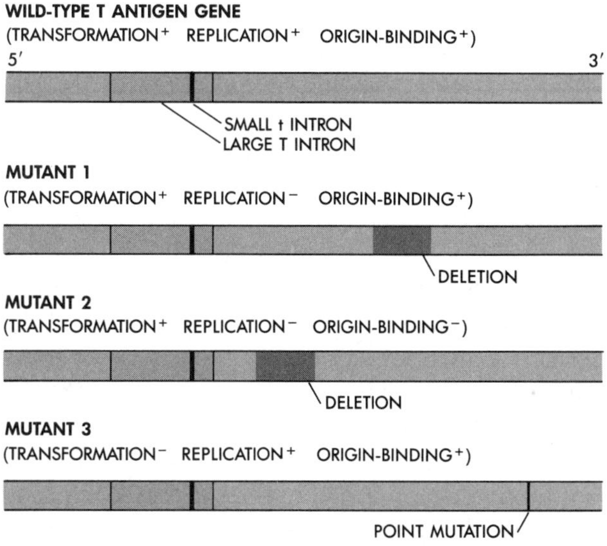

Figure 10-3
Functional domains of the large T protein were determined by assessing the activity of several mutant T proteins, obtained either as spontaneous SV40 mutants *in vivo*, or by *in vitro* mutagenesis of a cloned T antigen gene.

Overlapping Genes Code for the Structural Proteins That Surround the SV40 and Polyoma Chromatin

The DNA in each SV40 (and polyoma) particle is complexed with cell histones to form a circular, nucleosome-containing minichromosome that folds up into a compact structure that is surrounded by an outer protein shell (capsid). Originally it was thought that only one protein (VP1), present in multiple copies, was used to make the capsid. Later, however, two additional proteins (VP2 and VP3) were found in lesser amounts. All three are coded within the late region of the SV40 (and polyoma) genome, and through DNA sequencing, as well as through elucidating some key polypeptide sequences, the precise organization of their respective genes has been revealed. VP1 is translated off a 16S mRNA molecule. Though VP2 and VP3 share the same carboxyl-terminal amino acids, one is not the precursor of the other, and they are translated from different viral mRNA molecules of slightly larger sizes (19S and 18S, respectively).

The 16S, 18S, and 19S mRNAs all arise through the processing of the same primary late

transcript. The more prevalent splicing route produces the 16S (VP1) mRNA, with less common splicing routes yielding the 19S (VP2) or 18S (VP3) mRNAs. Quite unexpected was the realization that the genes for VP1 and VP2 (VP3) partially overlap. Moreover, the reading frame used for the translation of VP1 is different from that employed for VP2 or VP3 (Figure 10-4). This finding was not the first occasion in which overlapping genes were observed in a viral system. DNA sequence analyses had earlier shown the overlapping of two genes of the bacteriophage ØX174. Until now, gene overlapping has been observed only with viral genomes that must be packaged within outer capsids of fixed size. In the absence of such size limitations, overlapping genes seem a most inefficient way to evolve, since most base-pair substitutions would lead simultaneously to amino acid changes in both polypeptide products.

Different Control Signals for the Initiation of Early and Late SV40 and Polyoma mRNA Synthesis

Though both early and late SV40 (and polyoma) RNA chains are synthesized by RNA polymerase II (the polymerase type used to make host cell mRNA chains), the signals that determine the timing of their syntheses differ. Early viral mRNA initiation appears similar to that found for most host cell mRNA, in that the start sites are preceded by TATA- and CAT-like sequences. As mentioned earlier, such sequences are not now believed to control the frequency of initiation, since these elements can be deleted using recombinant DNA procedures without grossly affecting the transcription rates of the genes they precede. Instead, the TATA and CAT boxes are believed to function in defining the exact nucleotides where the initiation of early chains occurs. The frequency of early-chain initiation itself is now believed to be controlled by further upstream sequences called enhancers (Chapter 15, page 194).

Synthesis of late mRNA chains begins only after replication of their SV40 (or polyoma) tem-

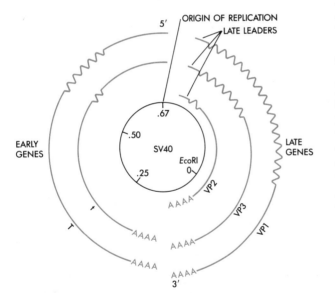

Figure 10-4
(a) A schematic drawing of SV40 DNA, with its processed early and late RNA transcripts and their protein products. The solid colored lines are processed RNA chains; spliced-out regions are indicated by zigzag sections. The numbers in the circle refer to the locations in fractional genome length relative to the single *Eco*RI cleavage site. (b) A section of the nucleotide sequence of SV40 DNA, showing parts of the 110-bp overlap between the VP1 amino-terminal end and the VP2 (VP3) carboxyl end. Different reading frames are used for the VP1 and VP2 (VP3) genes, so the overlap region codes for two completely different sets of amino acids.

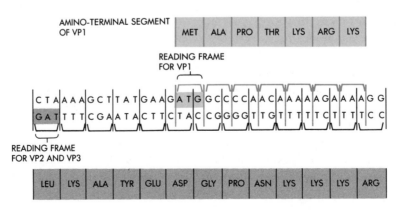

plates has commenced. Late mRNA chains show much more variability in their starting sites (these may occur within a 300-nucleotide range) than do early mRNA chains, which almost always start within a set of 10 contiguous nucleotides. This uncertainty in where the late transcript begins may reflect the absence of TATA- or CAT-like sequences upstream of its starting points.

The first step in late-chain initiation may be the binding of RNA polymerase II molecules to specific sites on the single-stranded DNA regions found in the immediate vicinity of DNA replication forks. Binding to exposed single-stranded DNA could explain why the so-called enhancer region appears to play no role in controlling the frequency of late, as opposed to early, SV40 (or polyoma) mRNA synthesis. Enhancer elements are now thought to create limited nucleosome-free regions to which RNA polymerase preferentially binds. The similar nucleosome-free regions that fleetingly exist at DNA replicating forks might also be preferential RNA-polymerase-binding sites.

A Region of Approximately 100 Base Pairs Encompasses the SV40 (and Polyoma) Origin of DNA Replication

The nucleotides involved in the initiation of SV40 and polyoma DNA replication overlap those sequences just mentioned that control the synthesis of the early and late primary mRNA transcripts (Figure 10-5). "Origins" of DNA replica-

tion were first well defined through electron-microscope observation of replicating SV40 and polyoma DNA and recently further delineated through the study of precisely located deletion mutants that retain the ability to replicate. The approximately 100-bp essential region contains a long stretch of contiguous AT base pairs, as well as several inverted repeats capable of forming hairpin loops when denatured into their component single strands. They also provide the three sites to which the large T protein tightly binds. Recombinant DNA procedures have allowed us to insert such origin sequences into other sites within SV40 DNA to see if such SV40 DNA molecules could initiate replication at these new sites. Such artificially inserted origins do, in fact, work, providing further proof that all the control signals involved in initiation of DNA synthesis are encompassed within this 100-bp "origin" segment.

The Complex Organization of RNA Tumor Viruses (Retroviruses)

That the replication of RNA tumor viruses involves the synthesis of a DNA intermediate was merely the first of many surprising facts that have been learned about these viruses. To start with, the genomes of RNA tumor viruses are dimers of two identical RNA chains, usually containing some 8 to 10 kilobases (kb) each, although defective viruses with as little as 3.2 kb of RNA per chain have been found. Why the genome is dimeric is still to be revealed; we generally talk about these

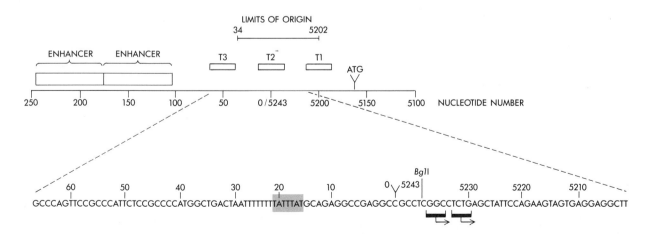

Figure 10-5
The nucleotide sequence of SV40 around its origin of replication, showing the presence of the TATA box governing the starting points for early mRNA synthesis. T1, T2, and T3 are the three sets of nucleotides to which the large T antigen tightly binds. To the left of the origin sequences are two tandem repeated 72-bp enhancer sequences. ⊢▬▬⊣ represents starting points for early mRNA synthesis.

viruses as if they had only a single RNA molecule. The genomic RNA of the RNA tumor viruses has two characteristics of eukaryotic mRNA molecules: a capped 5′ end and a poly-A tail at the 3′ end. Typically they contain nucleotide sequences coding for three viral proteins: a structural protein (GAG) which associates with the RNA in the core of the virus particle; the reverse transcriptase (POL) which makes the DNA complement; and a glycoprotein (ENV) which resides in the lipoprotein envelope of the particles, where it is needed for binding the virus to the surface of host cells on infection (Figure 10-6). These three viral proteins are all found in the virus particles.

The particles of all the RNA tumor viruses have basically similar organizations. The genomic RNA is associated in a central core with the GAG protein and small amounts of reverse transcriptase. The core is surrounded by a membrane envelope that contains the ENV glycoprotein; this membrane is essentially similar in construction to the plasma membrane of cells.

Highly Oncogenic Retroviruses Contain Specific Oncogenic Sequences

The highly oncogenic viruses transform cells in culture and induce tumors rapidly in host animals. Many such viruses have a fourth gene, the oncogene, which specifies a protein responsible for the transformation of the host cell to the cancerous state. The weakly oncogenic viruses, which do not transform cells in culture and which induce tumors only very slowly, lack an oncogene. The RNAs of several viruses of both classes have now been completely sequenced. This was done originally by directly sequencing the RNA. Nowadays gene cloning methods are used to make cDNA that is sequenced by the rapid DNA-sequencing methods.

Viruses with RNA genomes that replicate

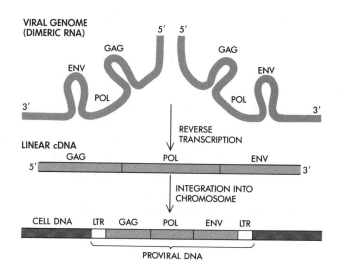

Figure 10-7
The retroviral genome is reverse-transcribed into cDNA. The cDNA then integrates into the chromosome, where it is called a provirus.

using proviruses as DNA intermediates are now called retroviruses. To this group belong not only the highly and weakly oncogenic RNA viruses, but also other morphologically similar viruses capable of causing a variety of other diseases, as well as infectious agents not known to cause disease.

Proviruses Maintain the Same Gene Order as RNA Genomes

When an RNA tumor virus infects a susceptible cell, the reverse transcriptase present in the viral core synthesizes double-stranded DNA copies of the viral RNA. The viral DNA moves to the cell nucleus, where it can be found in a linear form and in several circular forms. Within the nucleus, viral DNA molecules become inserted into host chromosomes to form a provirus by a recombination event between viral and cellular DNA (Figure 10-7). The integrated provirus has exactly the same sequence of genes as the viral DNA molecule; the viral DNA that is inserted is probably one of the circular forms. It is still not known how the cellular DNA is cut to give the two free ends between which the viral DNA slips. But it has been established that during the synthesis and integration of the viral DNA, sequences that distinguish the 5′ end from the 3′ end of viral RNA become repeated at both ends of the DNA. As a result, the integrated provirus has at both ends identical long terminal repeats (LTRs) of several hundred base pairs (Figure 10-7). Moreover, the

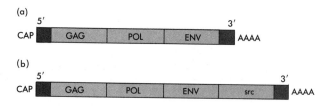

Figure 10-6
The genome organization of retroviruses.
(a) The arrangement of the three essential genes. (b) The genome of Rous sarcoma virus, which contains the oncogene src.

4 to 6 base pairs of host DNA immediately adjacent to the two viral LTRs are tandem duplicates; such a situation is the signature of transposable genetic elements, or transposons (Chapter 11, page 140). It would not, therefore, be surprising if the mechanism of integration of retroviral DNA had much in common with the mechanism by which transposons move about the genome.

The Promoter of RNA Synthesis Is Located Within an LTR

The DNA provirus integrated in a cell chromosome serves as a template for the synthesis of RNA chains identical to those found in the virus particle. The promoter for this transcription is located in the left-hand (5') LTR sequence, and positioned in such a way that the primary transcripts begin exactly with the 5' nucleotide of the genomic RNA molecules (Figure 10-8). Some of the primary transcripts are spliced to yield mRNA for the ENV protein. Other viral RNA molecules are spliced and translated to yield the GAG protein, while the original (unspliced?) transcript is translated into a GAG–reverse-transcriptase polyprotein (GAG–POL). This is cleaved proteolytically to liberate reverse transcriptase (which is made in at least ten-fold smaller amounts than the other viral proteins).

Instead of being spliced and serving as

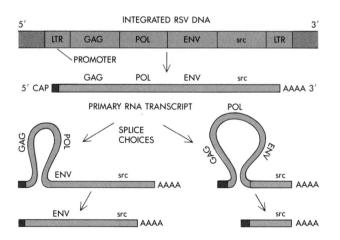

Figure 10-8
The only known promoter for viral RNA synthesis is contained in the left-hand long terminal repeat (LTR). The mRNAs for the different viral proteins arise through differential splicing of the same primary transcript. The primary transcript itself can be translated to make the GAG–POL fusion protein.

mRNA, some of the primary transcripts of the provirus become genomic RNA molecules to be packaged into virus particles that bud from the surface of infected cells. Within an infected cell, the transcripts that are destined to become genomic RNA are now believed to be somehow kept separate from those that function as messengers.

Since the left-hand LTR acts as a promoter for the transcription of the viral DNA, it might be expected that the right-hand LTR would promote the transcription of cellular DNA downstream of the provirus. In some cases this "downstream promotion" has been detected.

Retroviral Oncogenes Often Code for Protein Kinases

Rous sarcoma virus (RSV), which was discovered in 1911, is the prototype of the class of RNA tumor viruses that have oncogenes that rapidly transform cells to the cancerous state. The oncogene (v-src) of RSV specifies a sarc protein with an MW of about 60,000. Sarc has protein kinase activity, and it can phosphorylate tyrosine residues in proteins. Once this was established, several of the other highly oncogenic RNA tumor viruses were studied to see if they, too, have oncogenes in addition to the GAG, POL, and ENV genes. It was found that some do; however, others of these viruses, instead of having the normal three viral genes plus an oncogene, have an oncogene inserted into either the GAG, POL, or ENV gene. This finding explained why these particular viruses are "defectives" unable to replicate without a coinfecting normal ("helper") virus (Figure 10-9, page 134). The insertion of the oncogene destroys the function of one of the three genes whose products are essential for viral replication. As expected, cells infected with this latter group of highly oncogenic viruses were found to contain a fusion protein consisting of part of either GAG, POL, or ENV.

A number of these oncogenic fusion proteins also have protein kinase activity, leading to the obvious hypothesis that the transformations they bring about result from the phosphorylation of key cellular regulatory proteins; this phosphorylation presumably somehow changes their function. Unfortunately, we still do not know the names, much less the function, of most of the cellular proteins that are phosphorylated. Even less clear is which, if any, of these proteins have a central role in regulating cell multiplication. It has been learned,

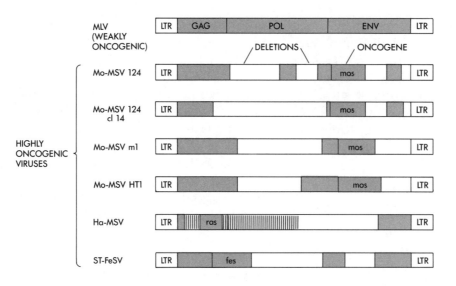

Figure 10-9
The genomic organization of some oncogenic mammalian retroviruses. "Mos,"
"ras," and "fes" are distinct cellular oncogenes that different retroviruses have
assimilated. The insertion of an oncogene into the retroviral genome almost always
destroys a gene that is necessary for viral growth, and often leads to deletions of
other regions of viral DNA or to insertions of cellular DNA. These highly
oncogenic viruses are therefore "defectives" and can be grown only in the presence
of helper viruses. The cross-hatched region in Ha-MSV represents inserted rat DNA.

however, that most of the v-src protein in cells
transformed by RSV is not in the nucleus, but
rather is close to the interface of the plasma membrane, where it may influence the cell's response
to external signals regulating its multiplication or
physical movement.

Normal Cellular Genes Are the Progenitors of Retroviral Oncogenes

Another question that has been investigated is
where the v-src gene of RSV (and the other oncogenes of the highly oncogenic viruses) comes
from. Is v-src a cellular gene that has somehow
become incorporated into the virus? To test this
hypothesis, the DNA of normal cells was screened
(initially with a DNA copy of the v-src gene; more
recently using cloned viral DNAs as probes) for a
cellular counterpart of v-src. Such a gene (c-src)
was found not only in normal chicken cell DNA
but in the DNA of mammals, fish, and even *Drosophila*. Subsequent gene cloning experiments
proved that the c-src gene of normal vertebrate
cells is, except for its possession of introns, very
similar (though not identical) to that of RSV.

That retroviral oncogenes arise from assimila-
tion of normal cellular genes is further supported
by an experiment in which a strain of RSV that had

lost part of its v-src gene was propagated in
chicken cells. Among the progeny were viruses
with fully reconstituted v-src genes. The missing
portion of the viral copy of the gene had been
supplied by the cellular gene.

Following these discoveries with RSV, a number of other RNA tumor viruses from a variety of
host species—chicken, turkey, mouse, rat, and
cat—were shown to have oncogenes that are
closely related to normal cellular genes. Alto-
gether some fifteen genes present both in normal
cells and in fast-transforming tumor viruses have
been identified so far.

Hybridization studies show that the cellular
counterparts of many of these viral oncogenes are
expressed at a low level in normal cells. This, and
the fact that several of these genes are known to
have undergone no major changes during the
whole of vertebrate evolution from fishes to humans, argues that they code for key regulatory
proteins.

Deciding Whether Retroviral Oncogenesis Results from the Overexpression or the Misexpression of Normal Cellular Genes

Within retroviral genomes, the oncogenes are
under the control of the strong viral LTR promot-

ers. This positioning leads to the synthesis of abnormally large amounts of the putative regulatory proteins. It is possible that the excessive amounts of a key control protein that is present only in small amounts in healthy cells upsets a vital regulatory pathway and changes the cell's phenotype to that of a malignant cancer cell. An alternative hypothesis for retroviral oncogenesis is that the small differences in nucleotide sequence that exist between the viral oncogenes and their cellular counterparts lead to significant changes in the functioning of their protein products; these changes somehow lead in turn to the creation of specific cancer cells. Fortunately, now that both viral and cellular genes can be isolated, cloned, and manipulated *in vitro,* the merits of each of these two hypotheses can be rigorously explored.

Cancer Induction by Weakly Oncogenic Retroviruses

Many retroviruses have only the three viral genes GAG, POL, and ENV, and no oncogenes. Although these viruses do not transform cells in cul-

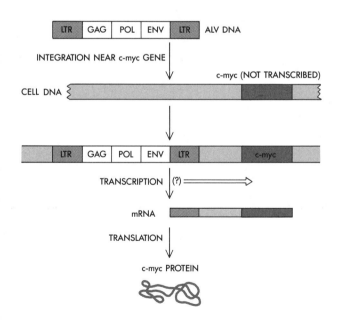

Figure 10-10
Activation of a silent oncogene through nearby insertion of avian leukemia virus (ALV) proviral DNA. In virtually all tumors caused by ALV (which is weakly oncogenic), the proviral DNA has integrated very close to (usually within 3 or 4 kb of) a cellular oncogene called "c-myc." By a mechanism that is still unclear, this insertion activates the c-myc gene.

ture, a large number of them are able to induce tumors in animals after a long latent period (of many months, as compared to the few weeks needed by highly oncogenic viruses). The key to how the weakly oncogenic viruses cause tumors came from the isolation, cloning, and sequencing of their proviruses from tumor cells. In tumors induced by chicken lymphoma viruses, which have no oncogenes, the provirus is almost always inserted in the immediate vicinity of a cellular oncogene progenitor, the c-myc gene, related to the v-myc gene present in the highly oncogenic MC29 virus (Figure 10-10). This location of the lymphoma provirus close to the c-myc gene may result in its overexpression. In most cases, transcription begins in a proviral LTR region and proceeds into the c-myc gene downstream, so that the amount of c-myc protein can be greatly enhanced. The net result is the same as infection with a strongly oncogenic virus—namely, a great increase in the expression of a cellular regulatory gene. By studying the cellular genes in the vicinity of proviruses in tumors caused by the retroviruses without oncogenes, we may be able to discover regulatory genes that have not yet been identified by other means.

A Generalized Theory for the Induction of Cancer

We now suspect that many, if not most, cancers arise through the overexpression (misexpression) of key cellular regulatory genes. According to this theory, malignant cells are created through either: (1) the introduction of additional (modified) copies of a key regulatory gene under the control of a strong promoter, as occurs in infection by a highly oncogenic retrovirus; (2) the introduction of a promoter that enhances the expression of a resident cellular regulatory gene, as occurs in infection by a weakly oncogenic virus; or (3) a mutation that arises within a cellular regulatory gene and alters its expression, as can occur as a result of exposure to a chemical carcinogen, to ionizing radiation, or to UV light. As we shall discuss later (Chapter 14), several of the cancer genes that have recently been isolated from cultured cancer cells derived from spontaneous human tumors have turned out to be very similar to the previously identified oncogenes in some highly oncogenic viruses.

Without recombinant DNA, we would probably not have been able to make—and certainly not so quickly—these profound advances in our understanding of how retroviruses cause cancer. In the future, gene cloning and *in vitro* mutagenesis will allow dissection of viral oncogenes to reveal which regions of the gene are crucial for transformation. It will become possible to define precisely any small differences between the viral genes and their cellular counterparts—differences that may be crucial for the regulation of the genes' expression or the range of their functions. Furthermore, we now seem on the threshold of being able to delineate a broad class of regulatory genes that have been conserved during evolution, and that affect cell differentiation and multiplication.

READING LIST

Books

Tooze, J., ed. *DNA Tumor Viruses: Molecular Biology of Tumor Viruses,* 2nd ed., revised. Cold Spring Harbor Laboratory, Cold Spring Harbor, N.Y., 1981.

Weiss, R., N. Teich, H. Varmus, and J. Coffin, eds. *RNA Tumor Viruses: Molecular Biology of Tumor Viruses,* 2nd ed. Cold Spring Harbor Laboratory, Cold Spring Harbor, N.Y., 1982.

Varmus, H., and A. J. Levine, eds. *Readings in Tumor Virology.* Cold Spring Harbor Laboratory, Cold Spring Harbor, N.Y., 1983.

Viral Oncogenes. Cold Spring Harbor Symp. Quant. Biol., vol. 44. Cold Spring Harbor Laboratory, Cold Spring Harbor, N.Y., 1980.

Original Research Papers (Reviews)

DISCOVERY THAT SPLICING PRODUCES TWO FORMS OF SV40 T ANTIGEN

Crawford, L. V., C. N. Cole, A. E. Smith, E. Paucha, P. Tegtmeyer, K. Rundell, and P. Berg. "The organization and expression of simian virus 40's early genes." *Proc. Natl. Acad. Sci. USA,* 75: 117–121 (1978).

INTEGRATED FORMS OF DNA TUMOR VIRUSES

Sambrook, J., R. Green, J. Stringer, T. Mitchison, S.-L. Hu, and M. Botchan. "Analysis of the sites of integration of viral DNA sequences in rat cells transformed by adenovirus 2 or SV40." *Cold Spring Harbor Symp. Quant. Biol.,* 44: 569–584 (1980).

Stringer, J. R. "DNA sequence homology and chromosomal deletion at a site of SV40 DNA integration." *Nature,* 296: 363–366 (1982).

THE ROLE OF SV40 AND POLYOMA T(TUMOR) PROTEINS IN TRANSFORMATION

Hassell, J. A., W. C. Topp, D. B. Rifkin, and P. Moreau. "Transformation of rat embryo fibroblasts by cloned polyoma virus DNA fragments containing only part of the early region." *Proc. Natl. Acad. Sci. USA,* 77: 3978–3982 (1980).

Ito, Y., and N. Spurr. "Polyoma virus T antigens expressed in transformed cells: Significance of middle T antigen in transformation." *Cold Spring Harbor Symp. Quant. Biol.,* 44: 149–157 (1980).

Lania, L., D. Gandini-Attardi, M. Griffiths, B. Cooke, D. deCicco, and M. Fried. "The polyoma virus 100K large T antigen is not required for the maintenance of transformation." *Virology,* 101: 217–232 (1980).

Novak, Y., S. M. Dilworth, and B. E. Griffin. "Coding capacity of a 35% fragment of the polyoma virus genome is sufficient to initiate and maintain cellular transformation." *Proc. Natl. Acad. Sci. USA,* 77: 3278–3282 (1980).

Rassoulzadegan, M., A. Cowie, A. Carr, N. Glaichenhaus, R. Kamen, and F. Cuzin. "The roles of individual polyoma virus early proteins in oncogenic transformation." *Nature,* 300: 713–718 (1982).

Smith, A. E., R. Smith, and E. Paucha. "Characterization of different tumor antigens present in cells transformed by simian virus 40." *Cell,* 18: 335–346 (1979).

Smith, A. E., R. Smith, B. Griffin, and M. Fried. "Polyoma virus middle-T has associated protein kinase." *Cell,* 18: 915–924 (1979).

OVERLAPPING GENES

Contreras, R., R. Rogiers, A. Van de Voorde, and W. Fiers. "Overlapping of the VP_2-VP_3 gene and the VP_1 gene in the SV40 genome." *Cell,* 12: 529–538 (1977).

Soeda, E., J. R. Arrand, and B. E. Griffin. "Polyoma virus DNA. The complete nucleotide sequence of the gene that codes for the polyoma virus capsid protein VP1 and overlaps the VP2/VP3 gene." *J. Virol.,* 33: 619–630 (1980).

START SIGNALS FOR SV40 (AND POLYOMA) mRNA SYNTHESIS

Benoist, C., and P. Chambon. "*In vivo* sequence requirements of the SV40 early promoter region." *Nature,* 290: 304–310 (1981).

Ghosh, P. K., P. Lebowitz, R. J. Frisque, and Y. Gluzman. "Identification of a promoter component involved in positioning the 5′ termini of simian virus 40 early mRNA." *Proc. Natl. Acad. Sci. USA,* 78: 100–104 (1981).

Mathis, D. J., and P. Chambon. "The SV40 early region TATA box is required for accurate *in vitro* initiation of transcription." *Nature,* 290: 310–315 (1981).

Contreras, R., D. Gheysen, J. Knowland, A. van de Voorde, and W. Fiers. "Evidence for the direct involvement of DNA replication origin in synthesis of late SV40 RNA." *Nature,* 300: 500–505 (1982).

ORIGINS OF SV40 DNA REPLICATION

Cole, C. N., T. Landers, S. P. Goff, S. Manteiul-Brutlag, and P. Berg. "Physical and genetic characterization of deletion mutants of simian virus 40 constructed *in vitro*." *J. Virol.,* 24: 277–294 (1977).

Subramanian, K. N., and T. Shenk. "Definition of the boundaries of the origin of DNA replication in simian virus 40." *Nuc. Acids Res.,* 5: 3635–3642 (1978).

Soeda, E., J. R. Arrand, N. Smolar, and B. E. Griffin. "Sequence from early region of polyoma virus DNA containing the viral replication origin and encoding small, middle and (part of) large T antigens." *Cell,* 17: 357–370 (1979).

Tjian, R. "Protein-DNA interactions at the origin of simian virus 40 DNA replication." *Cold Spring Harbor Symp. Quant. Biol.,* 43: 655–662 (1979).

RETROVIRAL GENOMES AND THEIR INTEGRATED DNA PROVIRAL FORMS

Hughes, S. "Synthesis, integration, and transcription of the retrovirus provirus." *Curr. Top. Microbiol. Immunol.,* 103 (1983) (in press). (Review.)

Temin, H. M. "Function of the retrovirus long terminal repeat." *Cell,* 28: 3–5 (1982). (Review.)

Varmus, H. E. "Form and function of retroviral proviruses." *Science,* 216: 812–820 (1982). (Review.)

RETROVIRAL TRANSFORMING GENES AND THEIR CELLULAR HOMOLOGUES

Martin, G. S. "Rous sarcoma virus: A function required for the maintenance of the transformed state." *Nature,* 227: 1021–1023 (1970).

Duesberg, P. H., and P. K. Vogt. "Differences between the ribonucleic acids of transforming and nontransforming avian tumor viruses. "*Proc. Natl. Acad. Sci. USA,* 67: 1673–1680 (1970).

Spector, D. H., H. E. Varmus, and J. M. Bishop. "Nucleotide sequences related to the transforming gene of avian sarcoma virus are present in the DNA of uninfected vertebrates." *Proc. Natl. Acad. Sci. USA,* 75: 4102–4106 (1978).

Bister, K., and P. H. Duesberg. "Structure and specific sequences of avian erythroblastosis virus RNA: Evidence for multiple classes of transforming genes among avian tumor viruses." *Proc. Natl. Acad. Sci. USA,* 76: 5023–5027 (1979).

Roussel, M., S. Saule, C. Lagrou, C. Rommens, H. Beug, T. Graf, and D. Stehelin. "Three new types of viral oncogene of cellular origin specific for haematopoietic cell transformation." *Nature,* 281: 452–455 (1979).

Hanafusa, T., L.-H. Wang, S. M. Anderson, R. E. Karess, W. S. Hayward, and H. Hanafus. "Characterization of the transforming gene of Fujinami sarcoma virus." *Proc. Natl. Acad. Sci. USA,* 77: 3009–3013 (1980).

Bishop, J. M., S. A. Courtneidge, A. D. Levinson, H. Oppermann, N. Quintrell, D. K. Sheiness, S. R. Weiss, and H. E. Varmus. "Origin and function of avian retrovirus transforming genes." *Cold Spring Harbor Symp. Quant. Biol.,* 44: 919–930 (1980).

Karess, R. E., and H. Hanafusa. "Viral and cellular *src* genes contribute to the structure of recovered avian sarcoma virus transforming protein." *Cell,* 24: 155–164 (1981).

DeFeo, D., M. A. Gonda, H. A. Young, E. H. Chang, D. R. Lowy, E. M. Scolnick, and R. W. Ellis. "Analysis of two divergent rat genomic clones homologous to the transforming gene of Harvey murine sarcoma virus." *Proc. Natl. Acad. Sci. USA,* 78: 3328–3332 (1981).

Bishop, J. M. "Oncogenes." *Sci. Am.,* 246(3): 80–92 (1982).

THE STRUCTURE AND FUNCTION OF ONCOGENIC PROTEINS

Purchio, A. F., E. Erikson, J. S. Brugge, and R. L. Erikson. "Identification of a polypeptide encoded by the avian sarcoma virus *src* gene." *Proc. Natl. Acad. Sci. USA,* 75: 1567–1571 (1978).

Collet, M. S., and R. L. Erikson. "Protein kinase activity associated with the avian sarcoma virus *src* gene product." *Proc. Natl. Acad. Sci. USA,* 75: 2021–2024 (1978).

Levinson, A. D., H. Oppermann, L. Levintow, H. E. Varmus, and J. M. Bishop. "Evidence that the transforming gene of avian sarcoma virus encodes a protein kinase associated with a phosphoprotein." *Cell,* 15: 561–572 (1978).

Hunter, T., and B. M. Sefton. "Transforming gene product of Rous sarcoma virus phosphorylates tyrosine." *Proc. Natl. Acad. Sci. USA,* 77: 1311–1315 (1980).

Levinson, A. D., S. A. Courtneidge, and J. M. Bishop. "Structural and functional domains of the Rous sarcoma virus transforming protein (pp60^{src})." *Proc. Natl. Acad. Sci. USA,* 78: 1624–1628 (1981).

Rohrschneider, L. R. "Immunofluorescence of avian sarcoma virus-transformed cells: Localization of the *src* gene product." *Cell,* 16: 11–24 (1979).

Willingham, M. C., G. Jay, and I. Pastan. "Localization of the ASV *src* gene product to the plasma membrane of transformed cells by electron microscopic immunocytochemistry." *Cell,* 18: 125–134 (1979).

Neil, J. C., J. F. Delamarter, and P. K. Vogt. "Evidence for three classes of avian sarcoma viruses: Comparison of the transformation-specific proteins of PRCII, Y73, and Fujinami viruses." *Proc. Natl. Acad. Sci. USA,* 78: 1906–1910 (1981).

Witte, O. N., N. Rosenberg, and D. Baltimore. "A normal cell protein cross-reactive to the major Abelson murine leukaemia virus gene product." *Nature,* 281: 396–398 (1979).

Langbeheim, H., T. Y. Shih, and E. M. Scolnick. "Identification of a normal vertebrate cell protein related to the p21 *src* of Harvey murine sarcoma virus." *Virology,* 106: 292–300 (1980).

Shih, T. Y., A. G. Papageorge, P. E. Stokes, M. O. Weeks, and E. M. Scolnick. "Guanine nucleotide-binding and autophosphorylating activities associated with the p21^{src} protein of Harvey murine sarcoma virus." *Nature,* 287: 686–691 (1980).

Oskarsson, M., W. L. McClements, D. G. Blair, J. V. Maizel, and G. F. Vande Woude. "Properties of a normal mouse cell DNA sequence (sarc) homologous to the *src* sequence of Moloney sarcoma virus." *Science,* 207: 1222–1224 (1980).

ONCOGENESIS BY SLOWLY TRANSFORMING RETROVIRUSES

Hayward, W. S., B. G. Neel, and S. M. Astrin. "Activation of a cellular *onc* gene by promoter insertion in ALV-induced lymphoid leukosis." *Nature,* 290: 475–480 (1981).

11

Movable Genes

In the early 1950s, Barbara McClintock, working at the Cold Spring Harbor Laboratory on the maize plant, began to call attention to the novel behavior of genetic elements that she called "controlling elements." These genetic elements, which were first noticed because they inhibited the expression of other maize genes with which they came into close contact, did not have fixed chromosomal locations. Instead, they seemed to move about the maize genome. The controlling elements could be inserted and excised; after excision, the function of a previously dormant gene often returned. The genes that associated with the controlling elements were rendered unstable and had high mutation rates because of the instability of the controlling element.

For many years, the corn plant provided the only genetic system in which such movable elements were observed. Then there arose some preliminary evidence that several of the highly mutable genetic loci in *Drosophila* might be associated with movable control elements. But most geneticists paid little attention to such loci until the discovery, in the late 1960s, that certain highly pleiotropic mutations (mutations affecting several functions) in *E. coli* resulted from the insertion of large segments of DNA called insertion sequences (ISs) (Figure 11-1). Significantly, many independent insertion events involved exactly the same DNA sequences. Four main blocks of insertion sequences were found and labeled: IS1, IS2, IS3, and IS4. Multiple copies of the IS1 and IS3 elements are scattered throughout the *E. coli* chromosome. Not only are the individual IS elements capable of movement, but when present in closely spaced pairs they can move as a unit, carrying along the genes lying between them. These complex units are called transposons (Figure 11-1).

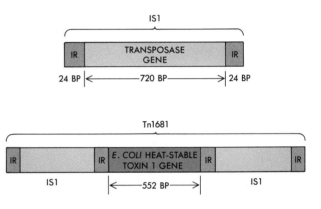

Figure 11-1
An insertion sequence and a transposon it generates. IS1 contains a gene for a "transposase" enzyme that is responsible for its movement. This gene is flanked by 24-bp inverted repeats (IRs). Below, a transposon (Tn) consists of two of these IS elements flanking another gene (in this case, the *E. coli* gene for heat-stable toxin 1, which causes diarrhea).

With the arrival of recombinant DNA, it quickly became possible to clone each of the IS elements along with its particular flanking sequences. Through comparisons of the sequences at the junctions between IS elements and chromosomal DNA, it became obvious that these elements can insert themselves into many locations in the *E. coli* chromosome. Equally important, the adjoining host chromosomal sequence at the left-hand junction point was always found to be identical to the host chromosomal sequence at the right-hand junction. This arrangement indicates that the actual insertion event probably involves a staggered cut within the target host chromosome; this cut

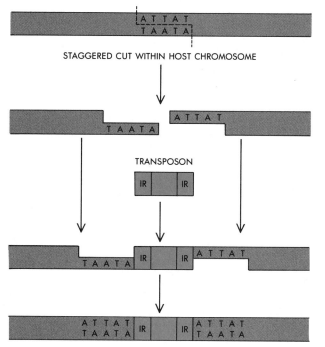

Figure 11-2
A model for transposon insertion. The host
chromosome sequences flanking the transposon
are usually found to be identical, implying that a
staggered cut is involved in the insertion of the
transposable element.

creates single-stranded tails into which the donor
transposon is placed (Figure 11-2).

Transposon Movement Involves the Creation of a New Daughter Transposon

It now appears that transposons do not actually
move, since they do not disappear from their initial site when they appear at a new location. Instead, the parental transposon gives rise to a new
copy that becomes inserted elsewhere. How this
happens is still not completely known, but there is
general agreement that the process involves staggered cuts made in both the donor transposon and
the target host molecules, followed by the joining
of donor to target chains to produce a chi-shaped
(X-shaped) intermediate in which the transposon
sequence is flanked by DNA replicating forks
(Figure 11-3). Replication of the transposable elements creates two transposons, one at the original
site and the other inserted within the target DNA.
Transposition terminates with a site-specific crossover within the transposons, to regenerate the
exact original movable element and daughter element flanked by newly created short direct repeats
(Figure 11-3).

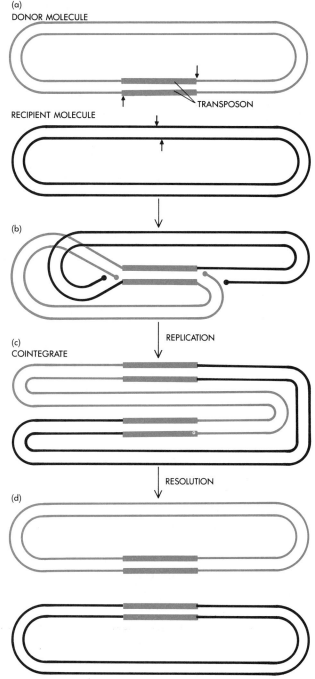

Figure 11-3
A model for transposition. (a) Single-stranded cuts
are made at the ends of the transposon on the donor
molecule and staggered cuts 5 to 7 bases
apart in the target DNA (recipient molecule).
(b) Each transposon end is attached to a protruding
target DNA end. Replication from the two forks
copies the transposon to produce structure (c).
(c) The resulting "cointegrate" consists of two
copies of the transposon in a direct-repeat
orientation. (d) A site-specific crossover at the
internal resolution or "res" site produces the starting donor molecule and the recipient molecule,
now containing a copy of the transposon.

Transposition thus requires the existence of several specific nuclease activities. One creates the staggered cuts at the opposite ends of the movable elements, while another makes the staggered breaks at the target site. Both these enzymatic activities appear to be carried out by single transposase molecules, proteins coded by genes located on the transposable elements themselves; transposase has been shown by mutational analysis to play a vital role in transposition. Synthesis of transposase mRNA is itself regulated by a repressor protein, also coded by the transposon (Figure 11-4). By binding to the appropriate operator segment on the transposon DNA, this repressor blocks the production of transposase so that only rarely does a movable element generate a new copy of itself. Interestingly, the repressor also catalyzes the site-specific crossover step ("resolution") that terminates transposition. Thus, given the appropriate signal, the repressor can act as enzymatic mediator of site-specific recombination. How the protein performs these two quite distinct functions as repressor and resolvase is not known.

Movable Genetic Elements May Be Common Features of All Organisms

We now suspect that transposable elements are a major feature of all DNA. Their existence has already been broadly documented at the molecular level in yeast and in *Drosophila* through cloning and subsequent sequence analysis. In yeast (*Saccharomyces cerevisiae*), there are approximately 35 copies of the most abundant transposon, Ty1 (5600 bp), per cell. Ty1 contains near its ends two direct repeats called delta sequences (340 bp), in addition to terminal direct repeats of its 5-bp chromosome-integration site. Several comparable movable elements exist in *Drosophila,* the most common being copia (5000 bp). It too occurs many times per genome, and likewise has at its ends long direct repeats of a common sequence (300 bp) and short terminal direct repeats of its 5-bp chromosome-integration site (Figure 11-5). Copia elements differ greatly in location and number between different *Drosophila* strains; the same is true for Ty1 elements in yeast strains. This sug-

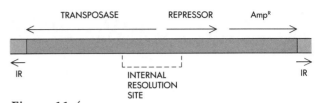

Figure 11-4
The structure of Tn3, a prokaryotic transposon. Tn3 consists of a transposase gene, a repressor gene coding for a protein that represses the transcription of the transposase gene, and the *E. coli* ampicillin-resistance gene. The repressor protein is also believed to function as a "resolvase." The intermediate in the transposition reaction (see Figure 11-3) is believed to be a DNA molecule with two complete copies of the transposon in a direct-repeat orientation. Through recognizing a specific DNA sequence (the res site) in each of the transposon copies, the resolvase is involved in the site-specific recombination reaction that resolves the intermediate into the donor molecule and the recipient molecule, now containing the inserted transposon.

gests that on appropriate occasions these elements are actively transposed.

The fact that Ty1 in yeast and copia in *Drosophila* apparently integrate randomly at so many chromosomal target sites argues against their having a major role in development *per se.* Instead, they may function to create genetic diversity, either by changing the expression of target genes through the insertion or deletion of promoterlike elements, or by transposing donor genes to new genomic sites, bringing them under the control of new regulatory elements.

Using Mobile Elements to Genetically Engineer *Drosophila* Embryos

In most *Drosophila* strains, there exists another type of dispersed movable element called "P"

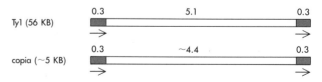

Figure 11-5
The structures of two eukaryotic transposons, Ty1 and copia.

which is quite distinct from *copia*. P is approximately 3 kilobases in length with 31-bp inverted repeats at its ends. P elements contain three internal long open reading frames that most likely code for three proteins, one of which is a transposase; these give P the capacity to move from one chromosomal location to another in a controlled fashion. In its basic structure, a P element thus closely resembles bacterial transposons like Tn3.

Within the P⁺ strains, however, P does not appear to move and generate copies at new locations. It has been found that when P⁺ sperm fertilize eggs borne by P⁻ females, the resulting hybrid progeny fail to produce many viable germ cells (a phenomenon called "hybrid dysgenesis"). The rare germ cells that do generate progeny often lead to genetically changed individuals in the second generation. Recently it has been shown that hybrid dysgenesis results from the mobilization of P elements after their insertion into P⁻ eggs (Figure 11-6). For a reason still to be worked out (most likely the absence of a repressor substance within the recipient P⁻ eggs), the P elements begin to transpose, and many of the daughter P elements lethally insert themselves into vital genes. Such transpositions occur at noticeable rates only in germ cells, as opposed to somatic cells. So the progeny of P⁺ × P⁻ crosses are normal except for their failure to produce viable (nonmutated) sperm or eggs.

The realization that the P element is mobile in embryos arising from P⁻ eggs has provided a means of genetically engineering *Drosophila* embryos. In this procedure, desired genes (for example, that for the easily observed eye-color marker "rosy") are first inserted within central regions of P elements. Such modified P elements lose their ability to be mobilized because the insertion process inactivates the transposase gene. They can, however, be mobilized in the presence of "helper" normal P elements, which provide the necessary P transposition enzymes. Successful high-level insertion of defective P elements containing foreign genes into the chromosomes of early *Drosophila* embryos occurs after the plasmids are simultaneously microinjected with wild-type P elements. In some of the resulting nuclei only modified P elements will be inserted, and since

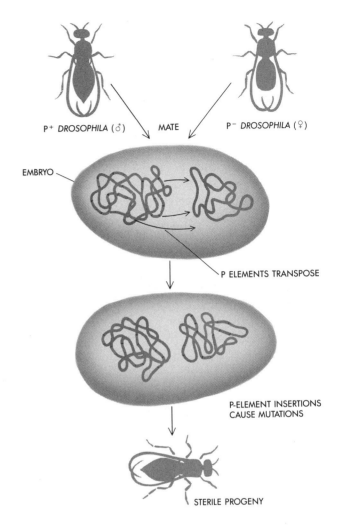

Figure 11-6
The phenomenon known as hybrid dysgenesis results from the mobilization of DNA sequences called P elements in *Drosophila* embryos. When a sperm from a P-carrying strain fertilizes an egg from a non-P strain, the P elements transpose throughout the genome, usually disrupting vital genes.

they cannot move, they give rise to viable genetically engineered progeny (Figure 11-7).

In these experiments the P elements were inserted into the plasmid pBR322 and propagated
first in *E. coli.* After their injection into P⁻ embryos, only DNA from the P element itself, as
opposed to pBR322 DNA, became inserted into
recipient P⁻ chromosomes. Moreover, complete P
elements, as opposed to incomplete P fragments,
were most frequently observed. Integration thus
occurs by transposition of a transposonlike element, rather than by an unspecific illegitimate
crossoverlike event that would have resulted in
insertion of both P element and pBR322 sequences.

Transposon-mediated gene transfers have the
great advantage that they occur at high frequency
and do not result in detectable rearrangement of
the integrated DNA. In the rosy gene transfer
experiments, 8 percent of the injected embryos
developed into adult flies. Thirty-nine percent of
these flies yielded progeny possessing the rosy-
containing P transposons. Equally advantageous
are the relatively large well-delineated DNA segments that can be so transferred. Engineered P
elements of 9 kb and 12 kb transpose at high rates
even though they are several times longer than
normal 3-kb P elements. Conceivably, the upper
practical size limit of an engineered P element will
be 40 kb, the largest DNA molecule that can be
effectively propagated as a cosmid in *E. coli.*

Isolation of the Ds Transposable Element from Maize

Though movable genetic elements were first described in corn (maize), only recently have the first
such elements been isolated and cloned. In her
early seminal experiments, Barbara McClintock
discovered two types of movable elements. One,
Ds, remains stationary within its chromosomal site
(for example, the bronze gene) unless a second
element, Ac, is also present. In the process of moving, the Ds elements can induce a wide spectrum
of chromosomal abnormalities, including insertions, rearrangements, breakages, and deletions.
Movement of Ds into the adh (alcohol dehydrogenase) gene inactivates it, leading to cells that cannot grow in the absence of oxygen. Because
alcohol dehydrogenase is a relatively abundant

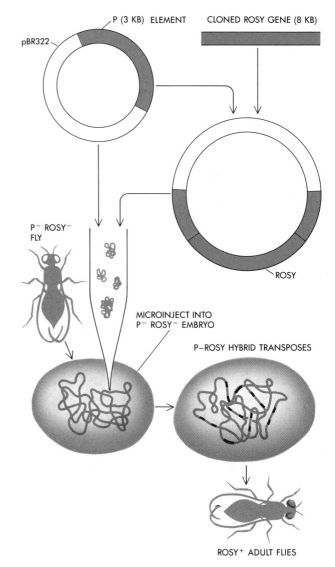

Figure 11-7
Genetic engineering of *Drosophila* using P
elements. A cloned copy of the rosy gene
(coding for xanthine dehydrogenase, an
eye-color marker) is inserted into a cloned copy
of a P element. This insertion disrupts the
putative transposase gene, so the P–rosy hybrids
are not mobile; however, if such hybrids are
inserted into P⁻ rosy⁻ embryos along with
wild-type P elements, they can transpose and be
inserted, and the rosy gene can function
normally.

enzyme, the cloning of its gene was a fairly straightforward task, and allowed comparisons to be made between the normal adh gene and the adh gene modified by the insertion of a Ds element. It was found that a mutant adh gene created by inserting 402 bp of Ds DNA was inactivated. At the ends of the Ds insert were inverted repeats of 11 bp. Flanking the Ds inserts were 8-bp direct repeats of the segment of the adh gene into which the Ds element had inserted itself.

Most likely, Ds represents a deleted form of the larger complete Ac element, which is now believed to be a complete transposon coding for its own transposase, and to have the same basic genomic structure as the P element of *Drosophila.* Proof (or disproof) for this belief will come from the cloning of Ac itself, a task that should soon be accomplished.

Have RNA Tumor Virus Genomes Evolved from Movable Genetic Elements?

Like the DNA tumor viruses, the RNA tumor viruses (retroviruses) transform normal cells into their cancerous equivalents by integrating themselves into host cell chromosomes. But as we discussed earlier (Chapter 10), they do so in a radically different fashion. After the infecting viral RNA chromosome enters a cell, it is transcribed into a cDNA complement by the enzyme reverse transcriptase, which is coded by the viral nucleic acid. The proviral cDNA then becomes inserted into a host chromosome.

Sequence analyses of the ends of several integrated retroviral genomes aroused much excitement when the adjacent cellular DNA sequences on both sides of the integrated viral DNA were found to be the same. The five bases of cellular DNA on either side of the provirus were invariably identical. In addition, at the two ends of the integrated proviruses there were identical blocks (LTRs) of viral DNA that were several hundred base pairs long, and that were similar in size and direct orientation to the δ-like sequences of copia (but not P) elements of *Drosophila.*

These structural similarities suggested that the copia and Ty1 elements might be the proviral forms of yet-to-be-discovered retroviruses. Firm support for this idea now comes from analysis of retroviruslike particles found within *Drosophila*

cells growing in culture. The RNA chains from such particles have the 5-kb size expected for retroviruses. In addition, reverse transcriptase is found in the isolated viruslike particles, and the 5-kb RNA chains code for their structural proteins. Even more importantly, the 5-kb RNA forms DNA–RNA hybrids with copia DNA, indicating that the similar chain lengths of the copia DNA and "retroviral" RNA is not an accident, but that the copia DNA *is* the proviral form of the retroviral RNA.

It is thus conceivable that retroviruses are movable genetic elements that have acquired the ability to move from cell to cell by possessing genes for proteins that aggregate around their RNA transcripts to form infectious virus particles. Alternatively, some classes of movable genes could represent descendants of functional retroviruses that have lost their ability to form viruslike particles.

Are There Functionally Two Different Classes of Transposons?

Now it seems likely that all movements of the copia- or Ty1-like elements involve RNA intermediates that, in turn, are converted by reverse transcriptase back into cytoplasmically located DNA proviral forms prior to their eventual insertion into new chromosomal sites. In contrast, the movement of bacterial transposons like Tn3, the P elements of *Drosophila,* and the Ac (or Ds) elements of corn is thought to occur through the temporary formation of a fused chromosomal DNA segment whose subsequent replication generates the daughter transposon. Mediating these fusion events are the transposases, which are DNA-cutting-and-joining proteins that are coded by the transposable elements themselves.

At the same time, we should note an important structural similarity between the retrovirus-like transposons and the P-element-like transposons. Both have inverted terminal repeats at their ends. Such sequences are thought to be recognized by the respective transposase prior to the cutting events that initiate the transposition. This raises the question of whether the inverted sequences of LTRs might have evolved to bind to the enzymes that mediate the integration of proviral cDNA into host cell chromosomal DNA.

Whether all movable elements have a common DNA ancestor thus cannot now be answered. When we understand in much greater detail the way the known transposons move, we may have a more comprehensive picture.

Sex Changes in Yeast by Gene Replacement: The Cassette Model

Haploid yeast cells have two sexual forms, *a* and α, which, when mixed, fuse into diploid cells. The sex of a haploid cell is determined by the mating-type gene, which can be either the *a* or the α form. In some strains of yeast the mating type is a stable property. In others it changes rapidly; the mating type can be converted from one form to the other at nearly every cell generation. Such rapid changes of mating type are not explained by any conventional genetic models. This has prompted yeast geneticists to explore models in which the molecular organization of the mating-type gene could change from one state to another. It was found that the sex of a yeast cell was dependent on the expression of either the *a* or the α gene at a specific chromosomal site called the "mating-type locus."

Important experiments revealed that *a* and α are also present at two other sites near the mating-type locus. These findings led to the cassette model of sex switching. This model postulated that the two additional genes are "silent" copies of mating-type genes, one an unexpressed *a* type, the other an unexpressed α type. In this scheme, sex switching occurs when a copy of one of the silent genes is inserted into the mating-type locus and the previous tenant is displaced. Each time this occurs, the yeast mating type changes (Figure 11-8). At each cell division, α cells convert to *a* cells and *a* cells convert to α cells with a 90 percent efficiency. In 5 percent of the cells there is no change, and in the other 5 percent the resident gene is replaced by a new copy of the same type. Proof of the cassette model was obtained in 1979 when the DNA of the functional mating-type locus was cloned and used to select the predicted silent genes. The base sequences of the *a* and α silent genes closely resembled the base sequences of the corresponding functional genes in the mating-type locus.

The mechanism for moving copies of silent genes and inserting them into the mating-type locus has clear parallels with the movement of transposons. In both processes, a copy of the parental genetic element is inserted at the new site in the chromosome while the parental element remains intact and in place. An enzyme that makes a very specific cut at the beginning of the active

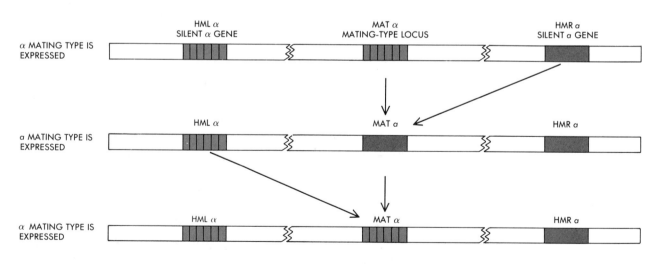

Figure 11-8
The cassette model for mating-type interconversion in yeast. The mating type of yeast (*a* or α) is determined by the expression of either the *a* or the α gene at the mating-type locus. However, silent copies ("cassettes") of the mating-type genes exist at two other loci (HML and HMR) on the same yeast chromosome. Copies of these silent genes can be inserted with high frequency into the mating-type locus, so that the sex of the yeast cell can change virtually continuously. The mechanism is believed to be similar to transposon movement.

mating-type locus has already been found. But how the insertion of the copy of the previously silent gene leads to the elimination of the existing tenant of the mating-type locus remains totally obscure.

Antigenic Changes in Trypanosomes Through Gene Switching

Trypanosomes, unicellular parasites of humans and other mammals, cause some of the most pernicious tropical diseases (Figure 11-9). In Africa they are responsible for sleeping sickness in people and related diseases in horses, camels, and cattle. If not treated, trypanosome infections are usually fatal; the costs in human suffering and economic loss to domestic herds are enormous. African wild mammals, in contrast to domestic animals and people, usually survive with a chronic infection that serves as a reservoir of the parasite, which cycles back and forth between the mammals and the tsetse fly. In the tsetse fly's gut, trypanosomes migrate to the salivary gland, from which as many as 10,000 of them may be discharged into the bloodstream of a mammal when the fly feeds.

Within the bloodstream the trypanosomes multiply exclusively by binary fission, one dividing into two and so on, without any sexual stage. (In this respect they are unlike the malarial parasites, which operate inside host cells.) One of the most intriguing features of trypanosome infections of mammals is a cycling parasitemia; about every 7 to 10 days a wave of new trypanosomes appears in the bloodstream and the parasites in each successive wave have a different antigen on their cell surface. A single trypanosome can specify more than 100 different surface antigens, one after the other, and this ability to change surface antigens means that the trypanosomes are ideally adapted to escaping destruction by the immune system of the host. No sooner has the host mounted an immune response by making antibody to the trypanosome's surface antigen than the trypanosome switches that surface antigen. The antibody being made is useless against the new antigen, and the parasite keeps a step ahead of its host.

The antigen that elicits the immune response is a membrane glycoprotein, a protein with sugar residues attached as a side chain. Each switch in

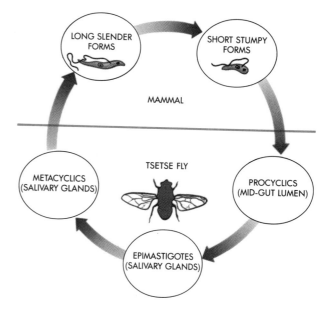

Figure 11-9
The life cycle of a trypanosome.

antigenicity of the trypanosome results from the complete replacement of one variant surface glycoprotein (VSG) by another, from the repertoire of over 100 different glycoproteins each parasite is capable of specifying. How is this switching, which must involve sequential, selective expression of a battery of genes, achieved?

Through the use of cloned cDNA molecules copied from mRNAs for different VSGs, it was quickly established that each trypanosome has a copy of each VSG gene, regardless of whether or not that gene has been or is being expressed. In other words, individual surface glycoprotein genes are not created by recombining gene segments in the way that antibody genes are created. Instead, each surface glycoprotein gene exists in its entirety all the time, and the complete set accounts for as much as 10 percent of the trypanosome's DNA. Secondly, although the VSG genes are distinct, they fall into several families of comparatively closely related genes, and in any one family the 5' end of the coding region is more highly conserved than the 3' region.

In most cases, when a particular surface glycoprotein gene is expressed, that gene is duplicated. The extra copy, called the expression-linked copy, persists only while it is being expressed, and it is found at a different site in the DNA than that

containing the master copy of the gene (Figure 11-10). This means that expression of most of these genes is accompanied by a gene duplication event and a transposition of the duplicate copy to the site at which it is expressed. This mechanism, which is reminiscent of mating-type switching in yeast, allows the expression of the set's individual genes in an indefinite sequence. Moreover, the switching on and off is reversible because the master copy is always retained.

Interest now focuses on the mechanism of transposition to the expression site of the duplicate copy of a gene, and on the mechanisms that regulate the expression at that site. The evidence to date suggests that the recombination event that introduces a gene into the expression site certainly involves pairing of sequences that are located at the 3' end of each glycoprotein gene and that are homologous to sequences at the expression locus. When mRNAs obtained from different populations of trypanosomes expressing the same surface glycoprotein were compared, small differences in the sequences at the 3' (untranslated) ends of the mRNAs were found. This was interpreted to mean that each time a glycoprotein gene is inserted into the expression site, recombination takes place in a homologous region of the DNA, but not necessarily at precisely the same base pair. So far, however,

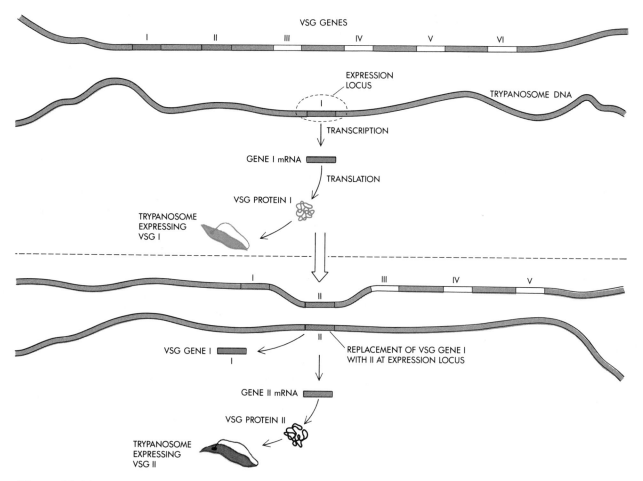

Figure 11-10
Expression of different trypanosome variant surface glycoproteins (VSGs) is due to genome rearrangement. The genome of a trypanosome contains hundreds of genes that code for different VSGs. Only one VSG gene, that located at a specific chromosomal site called the expression locus, is transcribed at any given time. However, another VSG gene can displace the first one at the expression locus, leading to the expression of a different VSG on the surface of the trypanosome. This process can continue with high frequency apparently indefinitely, and the organism can thus escape immune surveillance.

no such extensive homology has been found between the 5' end of glycoprotein genes and the DNA at the expression site. From the evidence available, it also seems that insertion of the gene at the expression site does not break some negative control that keeps the master gene silent, but rather places the duplicate gene copy under the control of a promoter that allows it to be transcribed into RNA.

There is clearly still much to be learned about the fine details of gene switching in trypanosomes, but recombinant DNA technology has already provided a satisfying explanation of the molecular events that underlie the trypanosome's ability to avoid destruction by its host's immune system. The mating-type switching in yeast and the more elaborate antigenic variation shown by trypanosomes are examples of the cassette mechanism, one solution to the problem of achieving alternating and potentially reversible expression of two or more functionally related genes.

Antigenic Changes in *Neisseria gonorrhoeae* Through Gene Rearrangement

Genomic rearrangement leading to the expression of altered surface proteins has also been found in *Neisseria gonorrhoeae,* the bacterium responsible for the venereal disease. The ability of *N. gonorrhoeae* to attach to and invade mucosal cells depends, at least in part, on the presence of structures called "pili" on the outer membrane of the organism. These pili are made up of hundreds of identical protein subunits, each with a molecular weight of 18,000. Different clinical isolates of *N. gonorrhoeae,* however, have antigenically distinct pili; that is, antibodies raised against one type of pilus from one isolate will not react with the pili of another isolate. The antigenic variation among pili is due to variability at the C-termini of the pilus proteins. *N. gonorrhoeae* expressing one type of pilus protein can switch to a pilus$^-$ state, and then switch back to a pilus$^+$ state in which it is expressing a different pilus protein. These switches are accompanied by genomic rearrangements of the pilus protein genes.

It appears that the versatility of *N. gonorrhoeae* is analogous to that of trypanosomes, which we have just discussed. *N. gonorrhoeae* has hundreds of pilus protein genes, only one of which is expressed at any one time (perhaps by being at an "expression locus"). That copy can be deleted from the locus leading to the pilus$^-$ state and replaced by another pilus protein gene. This may be important in immune evasion by the *N. gonorrhoeae* organism.

READING LIST

Books

Bukhari, A. I., J. A. Shapiro, and S. L. Adya, eds. *DNA Insertion Elements, Plasmids, and Episomes.* Cold Spring Harbor Laboratory, Cold Spring Harbor, N.Y., 1977.

Shapiro, J., ed. *Mobile Genetic Elements.* Academic, New York, 1983.

Movable Genetic Elements. Cold Spring Harbor Sympo. Quant. Biol., vol. 45. Cold Spring Harbor Laboratory, Cold Spring Harbor, N.Y., 1981.

Original Research Papers (Reviews)

MAIZE (CLASSIC PAPERS)

McClintock, B. "Chromosome organization and genic expression." *Cold Spring Harbor Symp. Quant. Biol.,* 16: 13–47 (1951).

McClintock, B. "Controlling elements and the gene." *Cold Spring Harbor Symp. Quant. Biol.,* 21: 197–216 (1957).

McClintock, B. "The control of gene action in maize." *Brookhaven Symp. in Biol.,* 18: 162–184 (1965).

BACTERIA

Cohen, S. N., and J. S. Shapiro. "Transposable genetic elements." *Sci. Am.,* 242(2): 40–49 (1980).

Calos, M. P., and J. H. Miller. "Transposable elements." *Cell,* 20: 579–595 (1980). (Review.)

Shapiro, J. A. "Molecular model for the transposition and replication of bacteriophage Mu and other transposable elements." *Proc. Natl. Acad. Sci. USA,* 76: 1933–1937 (1979).

Grindley, N. D. F., and D. J. Sherratt. "Sequence analysis at IS1 insertion sites: Models for transposition." *Cold Spring Harbor Symp. Quant. Biol.,* 43: 1257–1261 (1979).

Gill, R. E., F. Heffron, and S. Falkow. "Identification of the protein encoded by the transposable element Tn3 which is required for its transposition." *Nature,* 282: 797–801 (1979).

Reed, R. R. "Resolution of cointegrates between transposons γδ and Tn3 defines the recombination site." *Proc. Natl. Acad. Sci. USA,* 78: 3428–3432 (1981).

Reed, R. R. "Transposon-mediated site-specific recombination: A defined in vitro system." *Cell,* 25: 713–719 (1981).

Reed, R. R., and N. D. F. Grindley. "Transposon-mediated site-specific recombination in vitro: DNA cleavage and protein-DNA linkage at the recombination site." *Cell,* 25: 721–728 (1981).

Reed, R. R., G. I. Shibuya, and J. A. Steitz. "Nucleotide sequence of γδ resolvase gene and demonstration that its gene product acts as a repressor of transcription." *Nature,* 300: 381–383 (1982).

DROSOPHILA

Rubin, G. M., D. J. Finnegan, and D. S. Hogness. "The chromosomal arrangement of coding sequences in a family of repeated genes." *Prog. Nuc. Acid Res. Mol. Biol.,* 19: 221–226 (1976).

Dunsmuir, P., W. J. Brorein Jr., M. A. Simon, and G. M. Rubin. "Insertion of the *Drosophila* transposable element *copia* generated a 5 base pair duplication." *Cell,* 21: 575–579 (1980).

Levis, R., P. Dunsmuir, and G. M. Rubin. "Terminal repeats of *Drosophila* transposable element *copia:* Nucleotide sequence and genomic organization." *Cell,* 21: 581–588 (1980).

Engels, W. R., and C. R. Preston. "Identifying P factors in *Drosophila* by means of chromosome breakage hotspots." *Cell,* 26: 421–428 (1981).

Periquet, G., and D. Anxolabehere. "Elements causing hybrid dysgenesis on the second chromosome of *Drosophila melanogaster.*" *Mol. Gen. Genet.,* 186: 309–314 (1982).

Spradling, A. C., and G. M. Rubin. "Transposition of cloned P elements into *Drosophila* germ-lined chromosomes." *Science,* 218: 341–347 (1982).

Rubin, G. M., and A. C. Spradling. "Genetic transformation of *Drosophila* with transposable element vectors." *Science,* 218: 348–353 (1982).

RETROVIRUSES

Dhar, R., W. L. McClements, L. W. Enquist, and G. F. Vande Woude. "Nucleotide sequences of integrated Moloney sarcoma provirus long terminal repeats and their host and viral junctions." *Proc. Natl. Acad. Sci. USA,* 77: 3937–3941 (1980).

Shoemaker, C., S. Goff, E. Gilboa, M. Paskind, S. W. Mitra, and D. Baltimore. "Structure of a cloned circular Moloney murine leukemia virus DNA molecule containing an inverted segment: Implications for retrovirus integration." *Proc. Natl. Acad. Sci. USA,* 77: 3932–3936 (1980).

Shimotohno, K., S. Mizutani, and H. M. Temin. "Sequence of retrovirus provirus resembles that of bacterial transposable elements." *Nature,* 285: 550–554 (1980).

Flavell, A. J., and D. Ish-Horowicz. "Extrachromosomal circular copies of the eukaryotic transposable element *copia* in cultured *Drosophila* cells." *Nature,* 292: 591–595 (1981).

Shiba, T., and K. Saigo. "Retrovirus-like particles containing RNA homologous to the transposable element *copia* in *Drosophila melanogaster*." *Nature*, 302: 119–124 (1983).

YEAST

Cameron, J. R., E. Y. Loh, and R.W. Davis. "Evidence for transposition of dispersed repetitive DNA families in yeast." *Cell*, 16: 739–751 (1979).

Roeder, G. S., and G. R. Fink. "Transposable elements in yeast." In J. Shapiro, ed., *Mobile Genetic Elements*. Academic, New York, 1983, pp. 300–328. (Review.)

MAIZE (CONTEMPORARY WORK)

Schwartz, D., and J. Osterman. "A pollen selection system for alcohol-dehydrogenase-negative mutants in plants." *Genetics*, 83: 63–65 (1976).

McClintock, B. "Mechanisms that rapidly reorganize the genome." *Stadler Genet. Symp.*, 10: 25–48 (1978).

Freeling, M., and D. S. K. Cheng. "Radiation-induced alcohol dehydrogenase mutants in maize following allyl alcohol treatment selection of pollen." *Genet. Res.*, 31: 107–130 (1978).

Freeling, M., and J. A. Birchler. "Mutants and variants of the alcohol dehydrogenase-1 gene in maize." In J. K. Setlow and A. Hollaender, eds., *Genetic Engineering—Principles and Methods*, vol. 4. Plenum, New York, 1981, pp. 223–264.

Burr, B., and F. A. Burr. "*Ds* controlling elements of maize at the shrunken locus are large dissimilar insertions." *Cell*, 29: 977–986 (1982).

Gerlach, W. L., A. J. Pryor, E. S. Dennis, R. J. Ferl, M. M. Sachs, and W. J. Peacock. "cDNA cloning and induction of the alcohol dehydrogenase gene (*Adh1*) of maize." *Proc. Natl. Acad. Sci. USA*, 79: 2981–2985 (1982).

McCormick, S., J. Mauvais, and N. Fedoroff. "Evidence that the two sucrose synthetase genes in maize are related." *Mol. Gen. Genet.*, 197: 494–500 (1982).

TRANSPOSABLE MATING-TYPE GENES

Hicks, J. B., J. N. Strathern, and I. Herskowitz. "The cassette model of mating-type interconversion." In A. I. Bukhari, J. A. Shapiro, and S. L. Adhya, eds., *DNA Insertion Elements, Plasmids, and Episomes*. Cold Spring Harbor Laboratory, Cold Spring Harbor, N.Y., 1977, pp. 457–462.

Hicks, J., J. N. Strathern, and A. J. S. Klar. "Transposable mating-type genes in *Saccharomyces cerevisiae*." *Nature*, 282: 478–483 (1979).

Strathern, J. N., E. Spatola, C. McGill, and J. B. Hicks. "Structure and organization of transposable mating-type cassettes in *Saccharomyces* yeast." *Proc. Natl. Acad. Sci. USA*, 77: 2839–2843 (1980).

Nasmyth, K. A., K. Tatchell, B. D. Hall, C. Astell, and M. Smith. "A position effect in the control of transcription at yeast mating-type loci." *Nature*, 289: 244–250 (1981).

GENE REARRANGEMENTS TO PRODUCE ALTERED SURFACE PROTEIN

Borst, P., and G. A. M. Cross. "Molecular basis for trypanosome antigenic variation." *Cell*, 29: 291–303 (1982). (Review.)

Meyer, T. F., N. Mlawer, and M. So. "Pilus expression in Neisseria gonorrhoeae involves chromosomal rearrangement." *Cell*, 30: 45–52 (1982).

12

The Experimentally Controlled Introduction of DNA into Yeast Cells

The baker's yeast *Saccharomyces cerevisiae* is one of the most useful eukaryotic organisms for the study of the regulation of gene expression. Because of its small genome (only four times the size of *E. coli*'s) and short generation time (a few hours), yeast can be experimentally manipulated as easily as most prokaryotes. At the same time, it can be used to study some extremely complex phenomena specific to eukaryotes, including chromosome structure, mitotic and meiotic cell division, RNA splicing, and so on. The genetics of yeast has been well worked out, and hundreds of mutations affecting nutritional requirements, mating, cell division, and radiation sensitivity have been isolated and mapped by conventional methods. Moreover, yeast is a particularly attractive organism for the geneticist because it can be maintained in either the haploid or the diploid state. Complementation between genetic markers can be easily tested by mating pairs of haploid strains that each carry one of the markers. The resulting diploid will reveal whether there is complementation. The diploid can then be induced to undergo meiosis to yield four haploid cells, the products of meiotic division. Recessive markers can be followed easily in the haploids, which greatly simplifies analysis of linkage and recombination. With the advent of recombinant DNA technology and the development of techniques to introduce DNA into yeast cells, this organism has recently been used to make significant advances in our understanding of the molecular biology of eukaryotes.

For example, classical yeast genetics suggested the existence of positive and negative regulatory functions acting to control gene expression. The cloning of these regulatory genes is opening the molecular details of their functions to inspection and manipulation. Various structural properties of the yeast genome have been deduced from the behavior of recombinant DNA when it is transformed into yeast. As we shall describe, these investigations have identified particular chromosomal origins of replication, centromeric sequences (attachment points to the mitotic spindle), and functional telomeres (chromosomal ends). In addition, these studies have provided insight into the genetic properties of tandem reiterated sequences, like the ribosomal RNA genes, and dispersed repeated sequences, like the transposable elements.

Further, the availability of cloned genes provided an opportunity to study the structure of the chromatin at particular genes and to investigate the role of chromatin proteins in the regulation of gene expression. Finally, using yeast it is possible to mutagenize cloned genes (Chapter 8, page 109), reintroduce the mutant genes into their proper positions in the genome, and then assess the effects of the mutations on the genes' function. This "reverse genetics" will expand significantly the number of alleles of a given locus that can be obtained, and will increase our knowledge of the way in which the information in the DNA is inherited, regulated, and ultimately expressed.

Yeast Spheroplasts Take Up Externally Added DNA

DNA can be introduced into yeast quite easily. The yeast cellulose cell wall is removed enzymatically to produce "spheroplasts." The spheroplasts

are then exposed to DNA, CaCl$_2$, and a polyalcohol (such as polyethyleneglycol) that makes the membrane permeable and allows entry of the DNA. The spheroplasts are suspended in agar and allowed to regenerate a new cell wall.

Expression of Yeast Genes in *E. coli*

A major breakthrough in the cloning of yeast genes for use in transformation was the discovery that some of them could complement mutations in *E. coli.* Thus, the leu2 gene of yeast, which codes for an enzyme (β-isopropylmalate dehydrogenase) in the leucine biosynthetic pathway, can complement the leuB mutation in *E. coli.* This is probably not because the *E. coli* RNA polymerase recognizes the yeast gene promoter, but rather because the *E. coli* polymerase occasionally transcribes stretches of yeast DNA at random. If one of these transcripts includes a yeast structural gene, the RNA can be translated into a functional enzyme (assuming that the yeast gene does not contain introns). About 30 percent of yeast genes are found to be functional in *E. coli,* and although this can be thought of as a cloning artifact, it is a very fortuitous one. The total DNA of yeast cells can be cloned into plasmids, which are then used to transform *E. coli* cells carrying appropriate mutations. Bacterial cells receiving plasmids carrying the corresponding yeast gene can then be selected. Several yeast genes coding for biosynthetic enzymes in the tryptophan, histidine, arginine, and uracil pathways (for example, trp1, his3, arg8, and ura3) have been directly cloned in this manner. The plasmid DNA can be extracted from the bacteria and then used to transform mutant yeast strains using the spheroplast method just described.

Shuttle Vectors

With the advent of efficient methods for transforming yeast spheroplasts, the cloning of yeast genes by complementation of mutations in *E. coli* as described above has been superseded by direct complementation of mutants in yeast itself. The method also depends upon so-called shuttle vectors, plasmids that contain both bacterial sequences that signal DNA replication in *E. coli* and sequences that signal DNA replication in yeast. After digestion with appropriate restriction en-

zymes, total yeast DNA is inserted into shuttle vectors and propagated in *E. coli.* The mixed population of plasmids is introduced into yeast spheroplasts. Any recombinant plasmid that complements a mutation in the recipient spheroplasts under selective conditions can be identified and reintroduced into *E. coli* and grown in large amounts (Figure 12-1). The important point is that direct complementation in yeast is not limited to genes for metabolic or biosynthetic enzymes. Any gene for which a mutation can be identified can, in principle, be cloned in this way.

Yeast Also Contains a Plasmid

Most strains of yeast contain an autonomously replicating ring of DNA called the 2μ circle

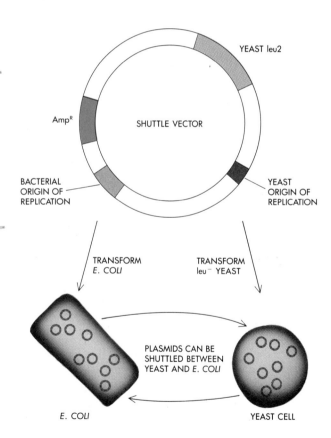

Figure 12-1
Shuttle vectors contain DNA sequences that allow replication in *E. coli* as well as sequences that allow replication in yeast. Such plasmids can thus be shuttled back and forth between the two organisms.

(Figure 12-2). This yeast plasmid, about 6300 bp in length, is present in the nucleoplasm of yeast at about 50 copies per cell. The 2μ DNA is packaged into nucleosomes that have a normal complement of histones. Like most bacterial plasmids, the 2μ circle contains a single origin of replication. In addition, the yeast plasmid itself encodes two so-called "REP" functions (presumably proteins, although they have not been identified) that promote amplification of the 2μ circles when the copy number is low. In this way the stability of the 2μ circle in yeast can be maintained even in the absence of attachment to the mitotic apparatus of the cell. Under normal conditions, the 2μ circle replicates at the same rate as the rest of the genome. When the copy number drops, however, the REP proteins can apparently override the normal coupling of plasmid replication to the cell cycle and initiate multiple rounds of independent replication of the 2μ circle, until the copy number is brought back to 30 to 50 per cell.

Increasing Transformation Efficiency by Addition of Replication Origins

Transforming DNA can be established in yeast either by integration into the chromosome or by autonomous replication as an episome. The presence or absence of ARS (autonomously replicating sequence) elements determines the fate of the introduced DNA. High-efficiency episomal transformation is achieved by including on the circular plasmid a DNA segment that contains an origin of DNA replication (an ARS), which allows replication in the yeast cell independently of the yeast chromosome (Figure 12-2). The presence of an ARS element on a plasmid frequently allows as much as 1 percent of a yeast spheroplast population to be transformed.

Such ARS segments have been isolated from the indigenous yeast plasmid (the 2μ circle) and from randomly cloned segments of the yeast chromosome. They contain at least 60 bp, are AT-rich

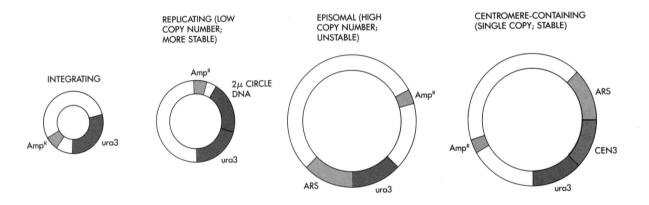

Figure 12-2
Four yeast plasmids. An integrating plasmid contains a yeast selectable marker but no yeast origin of replication. Such plasmids will be stable in yeast only by integration into the yeast chromosome. A replicating plasmid contains a yeast selectable marker and a segment of DNA from the yeast 2μ circle. The 2μ-circle DNA contains the origin of replication and also the "rep" genes, which stably maintain the plasmid extrachromosomally at a relatively low copy number. An episomal plasmid contains a segment of DNA (the ARS sequence) that allows the plasmid to replicate autonomously in yeast cells; however, there is no mechanism to *maintain* such extrachromosomal plasmids at high copy number during mitosis, and these ARS-containing plasmids are unstable. Stability is achieved by adding a CEN sequence, a segment of DNA from one of the yeast centromeres that bind to the spindle apparatus during mitosis. The CEN sequence ensures stable segregation of the extrachromosomal plasmid during mitosis.

(about 80 percent) and have within them the consensus sequence AAA$\frac{C}{T}$ATAAA. Interestingly, segments of DNA that have been cloned from maize, *Dictyostelium,* and *Drosophila* also function as ARS elements in yeast plasmids. While ARS elements from these various sources function as signals for extrachromosomal replication of DNA introduced into yeast, there is no firm evidence as yet that these sequences function as sites of initiation of *chromosomal* DNA replication in yeast or in any of the other species. So the presence of these sequences in the chromosomes of many diverse species may simply be the result of chance.

Stabilizing Yeast Plasmids with Yeast Centromere DNA

When recombinant bacterial plasmids such as pBR322 containing inserted ARS elements and foreign genes are introduced into yeast cells, the efficiency of transformation is high. But usually the plasmid is lost from the cells as they multiply if the selective pressure is removed. After ten generations only about 5 percent of the cells still have the plasmid. During cell division the plasmids apparently do not segregate regularly and uniformly between the two daughter cells. To overcome this difficulty, DNA segments containing sequences from the centromere (CEN) region of yeast chromosomes can be introduced into the plasmid (Figure 12-2). These sequences ensure the attachment of the chromosomes to the spindle fibers of the mitotic apparatus, and therefore effect the equal segregation of the chromosomes when the cell divides. Plasmids containing these CEN sequences are thus stably maintained by the same mechanism that ensures equal segregation of chromosomes.

CEN sequences were cloned, identified, and isolated by the chromosome-walking procedure. Two genetic markers that had been mapped by classical genetics and so were known to be close to, but on either side of, the centromere of chromosome 3 were selected. Clones from a yeast DNA library with segments of these genes were identified. Further clones overlapping the first pair were selected. The DNAs were then sequenced until the sequence of the complete segment from one gene to the other and spanning the centromere was obtained. The cloned DNA fragments were introduced individually into shuttle vectors, grown in *E. coli,* and then introduced into yeast. Plasmids carrying one particular segment of DNA were stably maintained in the host yeast cells and were presumed to contain the centromere of chromosome 3. DNA segments with stabilizing activity have also been isolated from chromosomes 4 and 11 by walking from centromere-linked markers on those chromosomes. Sequence comparison of the segments from chromosomes 3 and 4 have revealed several short blocks of sequence homology flanking an AT-rich region 80 to 90 base pairs in length.

In addition to ensuring equal segregation during mitosis, a *bona fide* centromere would be expected to obey the rules of centromeric segregation during meiosis leading to gametes. Plasmids containing long stretches of DNA from the centromeric regions (6 to 10 kb) behave as expected in meiosis. With few exceptions, a single CEN plasmid goes to one or the other pole in the first meiotic division and then segregates to both daughters in the second meiotic division. Smaller subcloned regions of this 6-to-10-kb segment are capable of stabilizing plasmids in mitosis, but they do not direct proper segregation at meiosis. This indicates that the fully functional centromere sequence may be longer than that deduced from experiments with only mitotically dividing cells.

With the knowledge that CEN sequences stabilize plasmids, it is now possible to isolate them by randomly cloning total yeast DNA into ARS-containing plasmids and selecting the recombinant plasmids that are stably maintained following transformation. Usually CEN-containing plasmids are maintained at a low copy number in transformants, at about an average of one plasmid per cell. For studies of gene regulation, it is, of course, desirable to have a system in which the gene dosage can be controlled; this is another advantage of CEN-containing plasmids.

In summary, then, the most efficient plasmid vector for yeast transformation would contain: sequences of *E. coli* plasmid pBR322, an ARS sequence, a CEN sequence, a selectable yeast marker gene (such as leu2), and one or several unique restriction enzyme sites to allow insertion of foreign DNA.

Hairpin Loops at the Ends (Telomeres) of Yeast Chromosomes

Yeast transformation has also been useful for defining the structure of telomeres, the ends of eukaryotic chromosomes. The mechanism by which the ends of linear DNA duplicate is not a trivial problem, because DNA polymerase only synthesizes DNA starting from an RNA primer. The replicated linear DNA chromosomes could thus end up with terminal gaps arising from excision of the RNA primers used to commence synthesis at their 5' ends. Some linear DNA viruses, like bacteriophage λ, avoid the gap problem by forming circular DNA intermediates. Other viruses having linear DNA molecules, like the phages T7 and T4, have the same sequence at the ends of their DNA molecules, so that their early replicating DNA intermediates can aggregate into long end-to-end concatamers which are later cut into complete genome-length molecules.

Another way out of the gap dilemma is to have the two strands of the linear DNA linked to one another at the terminus; that is, in a hairpin loop. When a replication fork proceeds to the end of such a structure, the end of the hairpin is now the center of symmetry of an inverted repeat, which, as we saw previously, can form a cruciform structure. Such a DNA molecule can then be resolved by nucleolytic cleavages on opposite strands to form two daughter DNA molecules, with a gap or a nick in each (Figure 12-3).

The amplified ribosomal DNA (rDNA) genes of the ciliate *Tetrahymena* were the first telomeric structures to be analyzed in detail. It was found that the ends of these extrachromosomal rDNA molecules are indeed hairpin structures. The rDNA telomeres consist of the repeating DNA unit CCCCAA. The number of these C_4A_2 units was found to vary from 20 to 70 in different rDNA molecules, giving length heterogeneity in the sizes of the restriction enzyme fragments that are generated from the ends of rDNA molecules. These *Tetrahymena* rDNA hairpins were ligated to the ends of a linear plasmid containing the yeast leu2 gene and an ARS sequence. When such a plasmid was introduced into yeast spheroplasts, it replicated and remained a linear DNA molecule. This indicated that *Tetrahymena* telomeres can function in yeast. Furthermore, the essential struc-

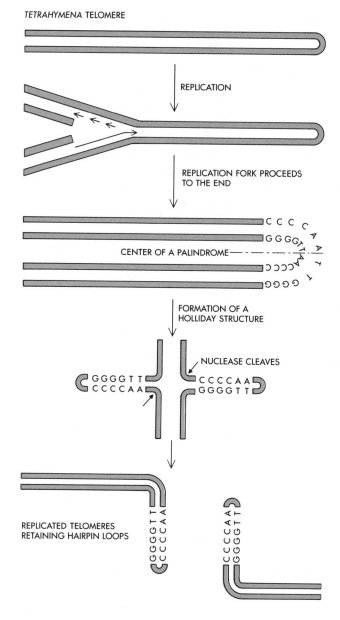

Figure 12-3
A model for telomere replication. The ends of *Tetrahymena* rDNA molecules are hairpin loops consisting of a variable number of CCCCAA repeating units on one strand and GGGGTT units on the other. The replication fork copies out to the end of the DNA molecule. What was the terminus of the molecule is now the center of an inverted repeat structure. The palindromic sequence can form a cruciform or Holliday structure that can be resolved by nucleolytic clips on opposite strands. The result is two replicated daughter molecules with hairpin loops.

ture of the rDNA telomere, that is, its hairpin loop, was maintained (Figure 12-4).

These *Tetrahymena* telomeres were then used to clone chromosomal telomeres from yeast. This was done by constructing a linear plasmid similar to the one just described, but with a *Tetrahymena* telomere at only *one* end. The other ends of the linear molecules were ligated to total yeast DNA that had been cleaved with the restriction enzyme

Figure 12-4
Cloning yeast telomeres. (a) The ends of *Tetrahymena* rDNA (telomeres) containing hairpin loops are ligated to each end of a linear plasmid containing the yeast leu2 gene and an ARS sequence. Such molecules replicate and remain linear when they are introduced into yeast cells. (b) A linear plasmid is constructed as in part (a), but with a *Tetrahymena* telomere at only one end. These molecules are ligated to total yeast DNA that has been cleaved with *Pvu* II. When they are introduced into yeast cells, plasmids that picked up a yeast telomere are able to replicate as linear plasmids.

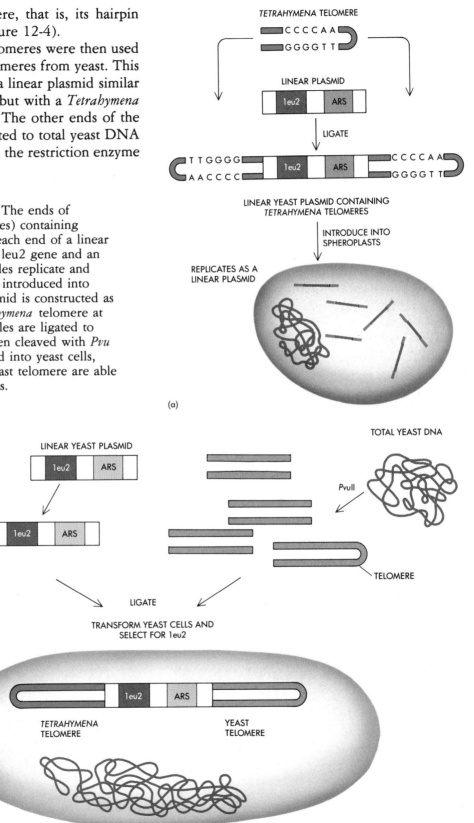

Pvu II, which generates about 2000 DNA fragments, of which 34 should be telomeres (assuming 17 chromosomes). This ligated DNA was then used to transform yeast cells. Several leu⁺ colonies were found to contain linear plasmids with yeast DNA at one end. Thirty to forty copies of this fragment are present in the total yeast genome, consistent with a model in which every yeast telomere is homologous to the copy that was cloned. Further analysis indicated that the different chromosomal telomeres in yeast are very similar to one another for about 3 or 4 kb from the end; the chromosomal sequences then diverge.

By using smaller and smaller pieces of yeast telomeric DNA, and by introducing specific mutations in this DNA, it should be possible to pinpoint those sequences or structures that are essential for telomere function.

Directed Integration of Cloned DNA into the Yeast Chromosome

DNA transformed into yeast spheroplasts can integrate into chromosomes. Almost all integration occurs by a crossover event between homologous sequences on the incoming DNA molecule and the yeast chromosome. If the DNA is introduced into the yeast cell as a circular molecule, integration turns out to be a very rare event, occurring on the order of 1 in 10⁶ cells, even if the region of homology with the chromosomal sequence is more than 10,000 base pairs. However, if the plasmid is first cut with a restriction enzyme and is then introduced into yeast spheroplasts, it integrates into the chromosome at a site homologous to the cut site about 100 times more frequently than when it is introduced as a circular molecule (Figure 12-5). So it is possible to direct an incoming plasmid to a specific site in the yeast chromosome by cutting it at an appropriate place with a restriction enzyme.

This ability to direct transforming DNA to a specific site in a yeast chromosome can be used to replace a wild-type gene with a mutant one. The procedure, called "allele replacement," allows the study of the effects of a specific gene mutation made *in vitro* on the expression of that gene. Be-

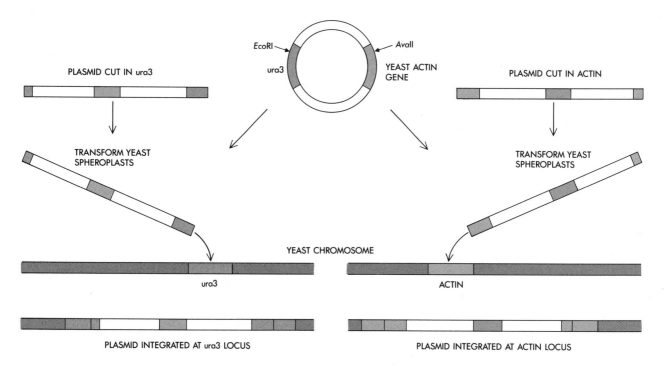

Figure 12-5
Directed integration into the yeast chromosome. An integrating yeast plasmid (that is, one without a yeast origin of replication) can be directed to integrate at a specific site by cleaving the plasmid with a restriction enzyme. For reasons that are unclear, a linear DNA molecule with ends homologous to a sequence in the yeast chromosome is hundreds of times more recombinogenic than a circular molecule. Cleaving a plasmid anywhere in the ura3 gene will cause integration into the ura3 locus in the yeast chromosome, whereas cleavage in the actin gene in the plasmid will ensure integration at the actin locus.

Figure 12-6
A yeast retriever vector. A wild-type yeast gene is cloned, along with its flanking sequences, into an integrating yeast plasmid (one without a yeast origin). The wild-type gene must be flanked by sites for two different restriction enzymes (here, *Hin*dIII and *Bam*HI). The plasmid also contains a yeast selectable marker (here, ura3). Such a clone can be used to retrieve a mutant allele of the cloned gene from the yeast chromosome: When a yeast spheroplast is transformed to ura$^+$ by one of these plasmids, integration will occur by a homologous crossover into the yeast chromosome at the mutant allele. If integration occurs *within* the gene, the procedure will not be successful; however, if the crossover occurs in the flanking sequences, the result will be a tandem arrangement of the wild-type and mutant alleles in the chromosome. As depicted here, subsequent cleavage of such DNA with *Hin*dIII and circularization will liberate the wild-type allele on a plasmid with the Amp^R gene, whereas cleavage with *Bam*HI and circularization will result in the mutant allele being retrieved. (If the integration had occurred in the 3' flanking sequences, then cleavage with *Hin*dIII and *Bam*HI would have given the opposite results.)

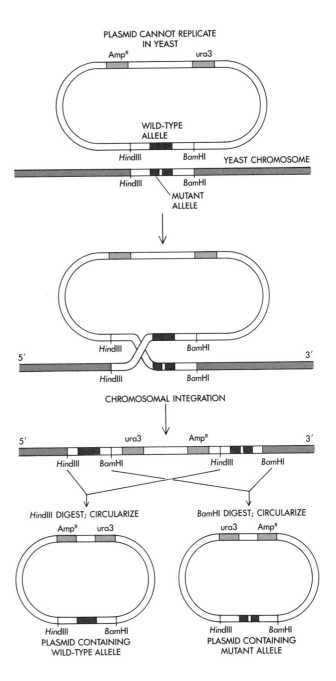

cause the altered gene is sitting in the yeast chromosome at its "correct" site, one can be more confident that any effects observed are actually due to the mutation, and not to changes in chromosomal location or changes in copy number. This technique was used to replace one of the yeast actin alleles with a mutant actin gene created by site-specific mutagenic procedures. The result was a recessive lethal mutation, indicating that functional actin is essential for the survival of yeast.

Retriever Vectors

The fact that integration of a plasmid into the yeast chromosome occurs by homologous recombination can be used to isolate naturally occurring mutant alleles of a yeast gene once the wild-type gene has been cloned. This can be done in either of two ways.

In one technique, a plasmid containing the wild-type yeast gene and flanking sequences is constructed in such a way that there are two unique restriction sites, one on each side of the yeast gene

(Figure 12-6). The plasmid also contains the ampicillin-resistance gene of pBR322 and a yeast selectable marker (such as ura3), but no yeast replication origin, so that the only way that yeast cells can be transformed to ura$^+$ is by integration of the plasmid into a chromosome. (Although this is a low-frequency event with a circular plasmid, it will occur.) If the recombination of the plasmid with the yeast chromosome occurs in the flanking sequences of the gene, then the chromosome of the transformed yeast cell now contains two copies of the gene in tandem: the newly integrated copy and the endogenous allele (Figure 12-6). Yeast chromosomal DNA from the transformed cells is

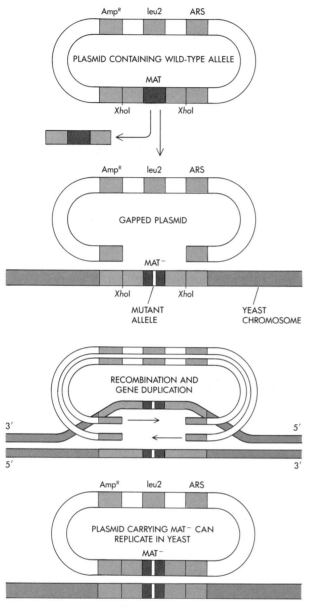

Xhol ... Xhol

MUTANT
ALLELE   YEAST
               CHROMOSOME

TRANSFORM *E. COLI* AND SCREEN FOR
PLASMIDS CARRYING MAT⁻ INSERT

Figure 12-7
Another type of retriever vector. A yeast gene is
cloned into a plasmid with a yeast origin of
replication. The yeast gene (here, MAT) is then
excised by using restriction enzymes, but 5′ and
3′ flanking sequences remain on the plasmid.
When this construction is used to transform yeast
cells, the only way it can replicate is to be
repaired back into a circle. This can occur if the
5′ and 3′ flanking sequences of the cloned gene
pair up with their homologous sequences in the
chromosome. DNA replication will result in the
mutant allele from the chromosome being
"copied onto" the plasmid, which can then be
used to transform *E. coli.*

then cleaved separately with each of the two restriction enzymes whose recognition sites flanked
the yeast gene in the plasmid used for transformation. Depending on which side of the resident
chromosomal gene the recombination occurred,
cleavage with one of the restriction enzymes followed by circularization will result in an ampicillin-resistance plasmid containing the allele that was
resident in the chromosome, while cleavage with
the other enzyme will result in an ampicillin-resistance plasmid with the original cloned gene (Figure 12-7).

A more efficient method of retrieving mutant
alleles from the yeast chromosome is to begin with
a plasmid containing a yeast selectable marker, an
origin of replication, and a yeast gene with its
flanking sequences. Restriction enzymes are used
to excise the yeast gene, leaving only the flanking
sequences. The result is a plasmid with a large gap
in it. When this gapped plasmid is transformed
into yeast cells, it must be repaired to a circle in
order to replicate; the easiest way for this to occur
is for the flanking sequences to recombine with
their homologous sequences in the chromosome.
These sequences in the chromosome, of course,
flank the gene that is to be retrieved. This double
recombination event results in the chromosomal
copy of the gene being replicated into the gapped
plasmid (Figure 12-8). The now circular plasmid
containing a copy of the chromosomal allele can
proceed to replicate in yeast cells and be isolated.

Gene Organization

By now, several yeast genes have been completely
sequenced, including those for the proteins alcohol dehydrogenase, glyceraldehyde 3-phosphate
dehydrogenase, actin, enolase, anthranilate isome-
rase, mating type (MAT *a*1 and MAT α2), α-
pheromone (sex hormone), and the two forms of
cytochrome *c*. Only one of these genes, that for
actin, contains an intron. If this sample of yeast
genes coding for proteins is representative of the
whole genome, introns must be much rarer in
yeast than in higher eukaryotes. In contrast, introns appear to be a regular feature of yeast tRNA
genes.

Outside the coding sequences are found sequences of TATAAAA and TATAAA, analogous
to the "TATA boxes" of higher eukaryotes. These
regions most likely serve as promoter elements.

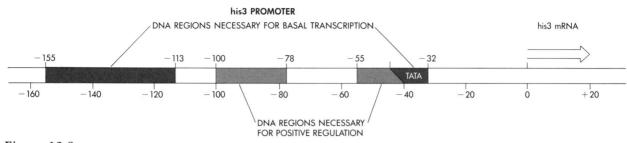

Figure 12-8
DNA sequences responsible for the regulation of the yeast his3 gene. Site-specific mutations were made throughout the promoter region of a cloned his3 gene, and the mutant genes were reintroduced into yeast at the normal his3 locus. It was found that two distinct regions were necessary for basal-level transcription, while two other distinct regions were necessary for the positive regulation of his3 transcription by amino acid starvation.

Much less clear is the nature of the sequences coding for the poly-A-addition sites. Some yeast genes may use AATAAA-like regions, whereas in other genes different sequences seem to play this role.

Regulation of Gene Expression in Yeast

Coordinately controlled genes in yeast are scattered throughout the genome on different chromosomes. This is unlike the situation in bacteria, where coordinately controlled genes are organized into operons (Chapter 4). For example, three of the yeast genes specifying the enzymes of the histidine biosynthetic pathway are on different chromosomes. Analysis of the DNA sequences on the 5′ side of these structural genes (the region in which regulatory signals might exist) revealed remarkably little homology; fewer than 14 base pairs of sequence are common to all three regions. The question now is how, given so little homology in supposed regulatory regions, a coordinated regulation of expression is achieved. Clearly, in eukaryotic organisms there seems to be no evolutionary pressure to keep coordinately controlled genes in contiguous blocks, or operons.

Studies of how the individual yeast genes are controlled have revealed that the sequences on the 5′ side of many regulated yeast genes, including cytochromes and the histidine biosynthetic enzymes, have *two* components necessary for normal regulated expression. Promoter sequences lie close to the transcription-initiation site and are necessary for basal levels of transcription. In addition, sites for the action of positive regulatory molecules lie more than 100 base pairs further upstream of the site of transcription initiation (Figure 12-8).

Another important result from studies of gene regulation in yeast by recombinant DNA methods is that at least some regulatory functions act at sites very distant from the beginning of the gene being regulated. For example, the HML and HMR silent mating-type genes are not expressed, even though they have completely functional genetic information, including a promoter. To be expressed, copies of the gene must be inserted in the MAT locus (Chapter 11). The two silent genes are kept silent by two sequences of DNA that are many hundreds of base pairs away from the gene. Comparison of the structure of the chromatin at the MAT locus, where expression occurs, with that of the two silent genes reveals significant differences. Apparently, the structure of the chromatin—in other words, the way the DNA is complexed with histones and other proteins—can alter the accessibility of the promoter to RNA polymerase. The same conclusion comes from studies of the yeast transposons, the Ty elements. Insertion of a Ty element hundreds of base pairs upstream from a transcriptional unit of a gene can cause the gene to be constitutively expressed; alternatively, such an insertion can completely shut off the gene's expression. These effects may also be attributed to alterations in the structure of the chromatin.

This correlation of chromatin structure, DNA sequence, and transcription with genetic analysis is greatly increasing our understanding of gene regulatory mechanisms in eukaryotes. The knowledge that such mechanisms exist in yeast, a simple eukaryote, will at the very least make the identification of similar mechanisms in higher eukaryotes much easier.

READING LIST

Books

Strathern, J. N., E. W. Jones, and J. R. Broach, eds. *The Molecular Biology of the Yeast Saccharomyces: Life Cycle and Inheritance.* Cold Spring Harbor Laboratory, Cold Spring Harbor, N.Y., 1981.

Strathern, J. N., E. W. Jones, and J. R. Broach, eds. *The Molecular Biology of the Yeast Saccharomyces: Metabolism and Gene Expression.* Cold Spring Harbor Laboratory, Cold Spring Harbor, N.Y., 1982.

Original Research Papers (Reviews)

YEAST TRANSFORMATION

Hicks, J. B., A. Ginnen, and G. R. Fink. "Properties of yeast transformation." *Cold Spring Harbor Symp. Quant. Biol.,* 43: 1305–1313 (1979).

Botstein, D., and R. W. Davis. "Principles and practice of recombinant DNA research with yeast." In J. N. Strathern, E. W. Jones, and J. R. Broach, eds., *The Molecular Biology of the Yeast Saccharomyces: Metabolism and Gene Expression.* Cold Spring Harbor Laboratory, Cold Spring Harbor, N.Y., 1982, pp. 607–636.

SHUTTLE VECTORS

Hicks, J. B., J. N. Strathern, A. J. S. Klar, and S. L. Dellaporta. "Cloning by complementation in yeast: The mating type genes." In J. K. Setlow and A. Hollaender, eds., *Genetic Engineering—Principles and Methods,* vol. 4. Plenum, New York, 1982, pp. 219–248.

PLASMIDS

Broach, J. R. "The yeast plasmid 2μ circle." In J. N. Strathern, E. W. Jones, and J. R. Broach, eds., *The Molecular Biology of the Yeast Saccharomyces: Life Cycle and Inheritance.* Cold Spring Harbor Laboratory, Cold Spring Harbor, N.Y., 1981, pp. 445–470.

Broach, J. R. "The yeast plasmid 2μ circle." *Cell,* 28: 203–204 (1982). (Review.)

REPLICATION ORIGINS

Broach, J. R., Y. Li, J. Feldman, M. Jayaram, J. Abraham, K. A. Nasmyth, and J. B. Hicks. "Localization and sequence analysis of yeast origins of DNA replication." *Cold Spring Harbor Symp. Quant. Biol.,* 47: 1165–1173 (1983).

Stinchcomb, D. T., M. Thomas, J. Kelly, E. Selker, and R. W. Davis. "Eukaryotic DNA segments capable of autonomous replication in yeast." *Proc. Natl. Acad. Sci. USA,* 77: 4559–4563 (1980).

CENTROMERE DNA

Bloom, K. S., and J. Carbon. "Yeast centromere DNA is in a unique and highly ordered structure in chromosomes and small circular minichromosomes." *Cell,* 29: 305–317 (1982).

Fitzgerald-Hayes, M., J.-M. Buhler, T. G. Cooper, and J. Carbon. "Isolation and subcloning analysis of functional centromere DNA (*CEN11*) from yeast chromosome XI." *Mol. Cell. Biol.,* 2: 82–87 (1982).

Fitzgerald-Hayes, M., L. Clarke, and J. Carbon. "Nucleotide sequence comparisons and functional analysis of yeast centromere DNAs." *Cell,* 29: 235–244 (1982).

TELOMERES

Szostak, J. W., and E. H. Blackburn. "Cloning yeast telomeres on linear plasmid vectors." *Cell,* 29: 245–255 (1982).

Szostak, J. W. "Replication and resolution of telomeres in yeast." *Cold Spring Harbor Symp. Quant. Biol.,* 47: 1187–1194 (1983).

Blackburn, E. H., M. L. Budarf, P. B. Challoner, J. M. Cherry, E. A. Howard, A. L. Katzen, W.-C. Pan, and T. Ryan. "DNA termini in ciliate macronuclei." *Cold Spring Harbor Symp. Quant. Biol.,* 47: 1195–1207 (1983).

DIRECTED INTEGRATION

Orr-Weaver, T. L., J. W. Szostak, and R. J. Rothstein. "Yeast transformation: A model system for the study of recombination." *Proc. Natl. Acad. Sci. USA,* 78: 6354–6358 (1981).

Shortle, D., J. E. Haber, and D. Botstein. "Lethal disruption of the yeast actin gene by integrative DNA transformation." *Science,* 217: 371–373 (1982).

GENE ORGANIZATION

Smith, M., D. W. Leung, S. Gillam, C. R. Astell, D. L. Montgomery, and B. D. Hall. "Sequence of the gene for iso-1-cytochrome c in Saccharomyces cerevisiae." *Cell,* 16: 753–761 (1979).

Holland, J. P., and M. J. Holland. "The primary structure of a glyceraldehyde-3-phosphate dehydrogenase gene from *Saccharomyces cerevisiae.*" *J. Biol. Chem.,* 254: 9839–9845 (1979).

Tschumper, G., and J. Carbon. "Sequence of a yeast DNA fragment containing a chromosomal replicator and the TRP1 gene." *Gene,* 10: 157–166 (1980).

Ng, R., and J. Abelson. "Isolation and sequence of the gene for actin in *Saccharomyces cerevisiae.*" *Proc. Natl. Acad. Sci. USA,* 77: 3912–3916 (1980).

Holland, M. J., J. P. Holland, G. P. Thrill, and K. A. Jackson. "The primary structures of two yeast enolase genes." *J. Biol. Chem.,* 256: 1385–1395 (1981).

Astell, C. R., L. Ahlstrom-Jonasson, M. Smith, K. Tatchell, K. A. Nasmyth, and B. D. Hall. "The sequence of the DNAs coding for the mating-type loci of Saccharomyces cerevisiae." *Cell,* 27: 15–23 (1981).

Zaret, K. S., and F. Sherman. "DNA sequence required for efficient transcription termination in yeast." *Cell,* 28: 563–573 (1982).

Bennetzen, J. L., and B. D. Hall. "The primary structure of the Saccharomyces cerevisiae gene for alcohol dehydrogenase I." *J. Biol. Chem.,* 257: 3018–3025 (1982).

Kurjan, J., and I. Herskowitz. "Structure of yeast pheromone gene (MF α): A putative α factor precursor contains four tandem copies of mature α factor." *Cell,* 30: 933–943 (1982).

REGULATION OF GENE EXPRESSION

Jones, E. W., and G. R. Fink. "Regulation of amino acid and nucleotide biosynthesis in yeast." In J. N. Strathern, E. W. Jones, and J. R. Broach, eds., *The Molecular Biology of the Yeast Saccharomyces: Metabolism and Gene Expression.* Cold Spring Harbor Laboratory, Cold Spring Harbor, N.Y., 1982, pp. 181–299.

Rose, M., M. Casadaban, and D. Botstein. "Yeast genes fused to β-galactosidase in *E. coli* can be expressed normally in yeast." *Proc. Natl. Acad. Sci. USA,* 78: 2460–2464 (1981).

Guarante, L., and M. Ptashne. "Fusion of *E. coli lacZ* to the cytochrome *c* gene of *S. cerevisiae.*" *Proc. Natl. Acad. Sci. USA,* 78: 2199–2203 (1981).

Faye, G., D. Leung, K. Tatchell, B. Hall, and M. Smith. "Deletion mapping of sequences essential for *in vivo* transcription of the iso-1-cytochrome *c* gene." *Proc. Natl. Acad. Sci. USA,* 78: 2258–2262 (1981).

Struhl, K. "Regulatory sites for *his3* gene expression in yeast." *Nature,* 300: 284–287 (1982).

Struhl, K. "The yeast *his3* promoter contains at least two distinct elements." *Proc. Natl. Acad. Sci. USA,* 79: 7385–7389 (1982).

Nasmyth, K. A. "The regulation of yeast mating type chromatin structure by SIR. An action at a distance affecting both transcription and transposition." *Cell,* 30: 567–578 (1982).

Beier, D. R., and E. T. Young. "Characterization of a regulatory region upstream of the *ADR2* locus of *S. cerevisiae.*" *Nature,* 300: 724–728 (1982).

Abraham, J., J. Feldman, K. A. Nasmyth, J. N. Strathern, A. J. S. Klar, J. R. Broach, and J. B. Hicks. "Sites required for position-effect regulation of mating type information in yeast." *Cold Spring Harbor Symp. Quant. Biol.,* 47: 989–998 (1983).

13

Genetic Engineering of Plants by Using Crown Gall Plasmids

Over the past 20 years, methods have been developed for redifferentiating whole mature plants from cells growing in culture. These developments have recently been coupled with the discovery that crown gall plasmids of certain soil bacteria naturally integrate into the chromosomes of the plant cells these bacteria infect. As a result, we have begun to exploit an entirely novel method of introducing individual genes into plants—a method that has great potential importance for plant breeding. Many other possible recombinant DNA techniques for plant genetic engineering are being investigated, but the crown gall plasmid system is currently the most advanced.

Conventional Plant Breeding Methodologies

Plant breeding, the production of more useful plants, has by now become a very sophisticated branch of applied Mendelian genetics. The yields of crops like wheat and corn have steadily increased over the last 50 years, and these rates of increase are being maintained. One area of plant breeding that might be amenable to recombinant technology is the transfer of simple traits like disease resistance from one variety or species to another. To introduce a particular desired gene or set of genes by conventional methods, the two lines are sexually crossed to give first-generation hybrids with a genetic constitution derived from both parents. The hybrids are then grown up and repeatedly back-crossed with one parent until a plant with the desired genetic makeup emerges. This plant will have most of the genes of the one parental variety with a few particular desired characteristics from the other; the process is called introgression. Such breeding is necessarily slow and usually spans several years, even when a range of genetic and breeding tricks are used to accelerate it. Furthermore, it is essentially restricted to sexually compatible species that can hybridize with each other; thus, it is limited by the natural species barriers to gene exchange.

Plant Cells in Culture

When a plant is wounded, a patch of soft cells called a callus grows over the wound and, with time, phenolic compounds accumulate in these cells and harden and effectively seal the wound. The hardened callus is the plant's equivalent of scar tissue. If a piece of young, still-soft callus is removed and placed in a culture medium containing salts, sugars, vitamins, amino acids, and the appropriate plant growth hormones, instead of hardening, the cells continue to divide, and give rise to a disorganized mass of relatively undifferentiated cells, a "callus culture." Pieces of tissue taken from inside a plant, or from young seedlings grown under sterile conditions, also give rise to similar cell cultures when the tissue pieces are placed in media containing plant hormones.

Redifferentiation of Whole Plants from Culture Plant Cells

When cultures of plant cells of certain species are exposed to media with appropriate growth hormones, some of the cells can be induced to redif-

ferentiate into shoots, roots, or whole plants, depending on the concentration and type of hormones present. At least some of the cells have developmental totipotency; that is, they have the potential to develop into a whole plant. A variety of factors in addition to the hormone balance influence this redifferentiation. The age of the culture is important: The longer the cells are passaged by replanting in fresh medium, the fewer the number that retain developmental totipotency. Physical factors such as humidity, light intensity, pH, and so on are also important to control. Only certain species of plants can be induced to redifferentiate: cereals, woody plants, and legumes, for example, have very rarely been induced to redifferentiate, and the few successes have been achieved with great difficulty. On the other hand, tobacco, petunia, carrot, tomato, and some other species readily redifferentiate.

Redifferentiation of callus cultures is useful to plant breeders because many genetically identical plants can be produced quickly under sterile conditions. This is particularly true in the floral industry. These days most commercially available orchids are produced in this way or by inducing rapid propagation in sterile conditions. Some florists sell small plants, "test-tube babies," that are each regenerated from a piece of callus culture.

Plant Protoplasts Can Regenerate Whole Plants

When the cellulose cell wall that surrounds each plant cell is digested with fungal cellulase enzymes, a "plant protoplast" is liberated. The response of some protoplasts to their nakedness is to synthesize and deposit a new cell wall and then to divide and grow into a callus culture. For a few species, including potato, tomato, *Datura* (Jimson weed), petunia, and tobacco, whole new plants have been regenerated from individual protoplasts via a callus. Because they lack a cell wall, fresh protoplasts of the same or different species can be induced to fuse with one another by addition of polyethyleneglycol and calcium ions (Ca^{++}) to the medium. The product is a somatic-cell hybrid protoplast that redevelops a cell wall and that may, depending on the species, begin to divide and regenerate. Protoplasts also may take up macromolecules, including DNA, that are normally unable to penetrate the cell wall barrier.

Making Hybrid Plants by Protoplast Fusion

Once somatic hybrid cells have been obtained via protoplast fusion, whole plants can sometimes be regenerated. Such asexually produced hybrids have been made by fusing the protoplasts of certain pairs of species within petunia (Figure 13-1, page 166), tobacco, carrot, and *Datura* genera, and by fusing potato and tomato protoplasts to create a new intergeneric hybrid, the "pomato," which unfortunately has no commercial value. To distinguish the hybrid fusion products from fused pairs of protoplasts of the same species, selection or screening procedures are often employed. For example, mutant cells that are albino because they cannot make chlorophyll may be fused. If the mutations in the cells of the two species complement each other, colonies from interspecific hybrid cells will turn green (Figure 13-2, page 167).

The cells of a callus culture begin as diploid because they are the descendants of somatic cells. However, the factors that maintain the chromosome composition strictly diploid *in vivo* are not present in culture, and typically with the passage of time more of the cells in a culture develop abnormal chromosome compositions and changes in ploidy. This is sometimes a serious drawback. In the case of plant cells, the capacity to regenerate a whole plant may well be lost with time in culture, due to changes in ploidy. The simplest way to minimize this problem is to use fresh or short-term cultures, in which most cells are diploid. Alternatively, it is possible that karyological variability may prove useful to the breeder if cells with unusual chromosome constitutions can be regenerated into plants. This has occurred with some interspecific barley hybrids that have been put into culture and then regenerated into plants.

When the protoplasts of two diploid cells are fused, the product is tetraploid. It has the diploid chromosomes of each parent. In contrast, pollen grains, the male sex cells of plants, are haploid. In a small number of species and under appropriate culture conditions, pollen grains in anthers can be induced to develop into haploid plants. By fusing such haploid protoplasts it is possible to regenerate fertile diploid plants.

No commercially valuable hybrid has yet been produced asexually by protoplast fusion. Fur-

Figure 13-1

Hybridization of different petunia species by fusion of somatic cells (protoplast fusion). (a) Protoplasts of cells from the leaves of *Petunia.* The protoplasts were obtained by digesting away the cellulose cell wall with enzymes. (b) A hybrid protoplast obtained by fusing a green-leaf protoplast of *Petunia parodii* with a colorless protoplast of an albino mutant of *Petunia hybrida.* In an appropriate culture medium, hybrid protoplasts resynthesize a cell wall and then begin to divide. (c) A somatic-cell hybrid plant regenerating from a callus (a mass of undifferentiated hybrid cells) produced by the multiplication of fused protoplasts. (d) Flowering plants of, from left to right: *P. parodii;* a hybrid produced by normal sexual hybridization of *P. hybrida* and *P. parodii;* a somatic-cell hybrid of *P. hybrida* and *P. parodii,* produced by the method illustrated in Figure 13-2; and *P. hybrida.* (Courtesy of J. B. Power and E. C. Cocking, Agricultural Research Council Group, University of Nottingham, England.)

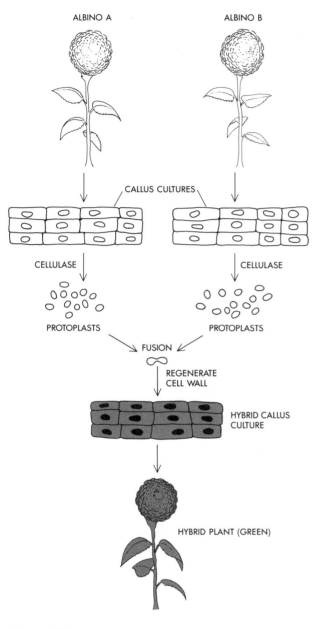

ALBINO A   ALBINO B

CALLUS CULTURES

CELLULASE   CELLULASE

PROTOPLASTS   PROTOPLASTS

FUSION

REGENERATE CELL WALL

HYBRID CALLUS CULTURE

HYBRID PLANT (GREEN)

Figure 13-2
A procedure for hybridizing two plant species by protoplast fusion. Protoplasts of mutant albino strains from two different species are liberated by digestion of the cell walls with fungal cellulase enzymes. The protoplasts are induced to fuse by addition of polyethyleneglycol and Ca^{++}. The fused protoplasts regenerate a cell wall and then divide and grow into a callus culture. Because the mutations in the two albino strains were complementary, the result is a green hybrid plant.

thermore, most of the fertile interspecies hybrid plants produced by protoplast fusion can also be produced by sexual crossing, and the products of both routes are identical. This gives confidence in the fusion method, but the aim of the asexual method is to *increase* the range of hybridization, not to duplicate sexual crossing. However, very occasionally, fertile hybrid plants *are* produced by protoplast fusion between species that cannot be crossed sexually. In other words, sexual incompatibility can sometimes be overcome by protoplast fusion. This achievement must still be followed by generations of introgressive breeding to obtain a useful new plant variety.

Genetically Engineering Plants

Both conventional plant breeding and protoplast fusion are in a sense genetic engineering; they can both be used to manipulate the genetic makeup of plants toward a desired combination of hereditary traits. They lack, however, the precision of the gene transfer that can be achieved in bacteria, especially now that recombinant DNA methods have been developed. What plant geneticists would like to be able to do is introduce single specific genes into already useful varieties of plants. That would involve two basic steps: first, obtaining particular genes in pure form and in useful amounts; and second, devising ways of inserting these genes into plant chromosomes so that they can function. The first step is no longer an insurmountable problem; using recombinant DNA methods, it is now possible to grow any segment of any DNA in bacteria. Unfortunately, it is extremely difficult to identify the particular segment of interest among a collection of clones, particularly for genes affecting such traits as crop yield, which are polygenic and biochemically undefined. The second step—reintroduction of cloned genes into plants—also presents unsolved problems. Recent work has, however, uncovered a natural gene vector that has evolved in soil bacteria; this vector allows the bacteria to genetically engineer plant cells for their own parasitic ends. We are just beginning to learn how we might exploit this so far unique example of natural genetic engineering.

Crown Gall Tumors

Among a group of soil bacteria known as the *Agrobacteria,* there are several species that can infect plants and cause a crown gall, a lump or callus of tumor tissue that grows in an undifferentiated way at the site of infection. That the bacteria are involved in the induction of galls was shown as long ago as 1911. Virtually all dicotyledons—that is, plants with two seed leaves—are susceptible, although the grasses and other monocotyledons are not. The cells of crown galls are in many ways analogous to animal cancer cells. They have acquired the properties of independent, unregulated growth. When crown gall cells are put into culture, they grow to form a callus culture even in media devoid of the plant hormones that must be added to induce normal plant cells to grow in culture. Moreover, once crown gall cells are rendered tumorous by *Agrobacteria* they remain transformed, even if all the bacteria are eliminated with antibiotics. Investigation of a particularly potent inducer of crown galls, *Agrobacterium tumefaciens,* has revealed that the tumor-inducing agent in the bacteria is a plasmid that functions through integrating some of its DNA into the chromosomes of its host plant cells (Figure 13-3).

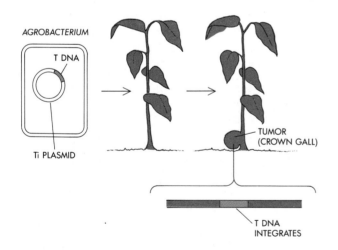

Figure 13-3
Certain species of *Agrobacteria* infect plants and cause crown galls, lumps or calli of tumor tissue. The tumor-inducing agent is a plasmid (the Ti plasmid) that integrates some of its DNA (the T, or transforming, DNA) into the host cell chromosome.

Tumor-Inducing (Ti) Plasmids

Cells of crown gall tumors induced by *A. tumefaciens* begin to synthesize unusual amino acids called opines, which are derived from arginine. Opines are never found in normal plants. Strains of *A. tumefaciens* that induce opines can use them as sources of carbon and nitrogen, and the ability both to induce and to metabolize opines is encoded by plasmids in the bacteria. Host plant cells cannot use these new amino acids, so the bacterial infection not only causes cells to become tumorous but also subverts the plant's metabolism to making amino acids that only the bacteria can use as a food. The two most commonly studied opines are octopine and nopaline.

Tumor induction, the induction of opine synthesis, and the capacity to metabolize opines all depend on the presence of the tumor-inducing or Ti plasmids in the respective bacteria. The Ti plasmids are circular DNA molecules with molecular weights of about 1.2×10^8 (3 to 5 percent of the *Agrobacter* chromosome); they exist in the bacterial cells as independently replicating genetic units. The Ti plasmids are classified according to the type of opine they induce. Most are either octopine or nopaline plasmids.

A. tumefaciens cells harbor only one sort of Ti plasmid, either a nopaline or an octopine plasmid. The base sequences of the two plasmid DNAs are not closely related, except for four regions of extensive homology, one of which includes the genes responsible for crown gall transformation. This suggests that the nopaline and octopine plasmids may have diverse evolutionary histories.

Ti Plasmid Mutants

Proof that Ti plasmid DNA sequences rather than bacterial chromosomal genes are responsible for the maintenance of the tumorous growth of crown gall cells came from DNA hybridization analysis, and from creation of strains of *Agrobacterium* that were completely cured of the Ti plasmid or that contained mutant plasmids.

Agrobacteria lacking Ti plasmids do not induce crown gall disease or the synthesis of opines in infected plants. Transposon-induced mutants of the Ti plasmid fell into three major classes. One class failed to synthesize opines but could still induce crown galls; a second class could no longer

induce tumors. A third class caused normal cells in the vicinity of the tumor cells to undergo abnormal differentiation—for example, excessive proliferation of roots or shoots. Very rarely, tumorous cells that could themselves differentiate into whole plants were found. Normally, crown gall cells cannot do this.

These genetic studies established that Ti plasmid DNA is responsible for tumorgenesis, opine synthesis, and the suppression of differentiation. Genes for these three functions map close together in one segment of the plasmid DNA called the transforming (T) DNA, which is the portion of the plasmid DNA that is transferred to host cell chromosomes during infection.

Integration of T DNA into the Plant Chromosome

When animal tumor viruses like SV40 and adenoviruses transform animal cells to malignancy, part or all of the viral DNA is integrated into the cell's chromosomes (Chapter 10). The DNA of crown gall cells was therefore probed for plasmid DNA by using Southern blot hybridization techniques; as hoped, copies of the T DNA segment (about 20,000 base pairs long) were found covalently integrated into the DNA of tumor cells. The T DNA segments of the octopine and nopaline Ti plasmids integrate at various, seemingly random places in the host chromosomes (but not in either their mitochondrial or chloroplast DNA). The integrated T DNA induces and maintains the tumorous state as well as the synthesis of opines. Several tumor cell lines have been maintained in hormone-free media for over 20 years with opines continually being synthesized.

The T DNA in octopine tumor cells has seven genes that specify distinct RNA transcripts in the plant. Most, if not all, of these genes are controlled by separate promoters. Four and possibly five of the genes suppress shoot and root formation by tumor cells, and another gene specifies the enzyme that synthesizes the particular opine made by the tumor cells. The products of two of the genes that suppress root formation may function in some ways like the plant hormone auxin, which promotes root growth; a gene that suppresses shoot formation may specify a product that mimics another plant hormone, cytokinin, which promotes

shoots. It is the balance between auxins and cytokinins that determines the pathway of differentiation in normal plants. Crown gall cells, which have high levels of the analogs to these hormones, can therefore grow in hormone-free media and remain undifferentiated calli.

Is T DNA a Transposon?

The only DNA that moves from the Ti plasmid to host plant cell chromosomes is the T DNA. The question thus arises whether T DNA is a transposable element that has specifically evolved to enable *Agrobacterium* to thrive within the environment of a suitable host plant. DNA sequence analysis, however, does not reveal T DNA to have the structure of any previously analyzed movable element. T DNA possesses neither the terminal inverted repeats of bacterial-type transposons nor the LTRs of retroviral DNA. The nopaline T DNA, however, does have 25-base-pair direct repeats at its borders with host cell DNA. A very similar sequence is found at a border of octopine T DNA, which suggests that these specific 25 base pairs do have some functional significance.

Complicating the matter is the finding that removal of the right-hand border decreases, but does not entirely destroy, the ability of the Ti plasmid to induce tumors. Moreover, exhaustive mutagenesis of T DNA through the creation of deletions or large insertions has so far failed to uncover any evidence for a transposase or any other recombination-promoting molecule coded by the T DNA. The enzyme(s) responsible for inserting T DNA into plant cell chromosomes must be coded either by genes on other segments of the Ti plasmid or by host chromosomal genes.

Mendelian Inheritance of T DNA

As we mentioned earlier, a mutant octopine Ti plasmid that is less tumorgenic than normal has been isolated; in the populations of tobacco cells it transforms, rare individual cells can be induced to redifferentiate into whole plants. The redifferentiated tobacco plants (virtually all the work with Ti plasmids has been done with tobacco plants and cells in culture) continued to make octopine in all their tissues because of the T DNA present. They were sexually crossed with normal tobacco plants and the resulting seed was grown.

Octopine synthesis, and therefore the T DNA, was inherited through both egg and pollen in a simple Mendelian fashion. Once it is inserted in a chromosome, the T DNA behaves like a normal Mendelian plant gene.

Ti Plasmid DNA Can Be Used as a Vector

Two properties of the T DNAs of Ti plasmids make them virtually ideal vectors for introducing foreign genes into plants. First, the host range of *Agrobacterium* is very broad; they are capable of transforming cells of virtually all dicotyledonous plants. Whether the range can eventually be broadened to include monocotyledons (the grasses) remains to be seen. Secondly, the integrated T DNA is inherited in a Mendelian way, and its genes have their own promoters to which foreign genes can be coupled and expressed.

The simplest way to introduce T DNA into plant cells is to infect them with *A. tumefaciens* containing the appropriate Ti plasmid, and let nature do the rest. Therefore, we need to be able to insert desired genes into the T regions of Ti plasmids. The entire Ti plasmid, however, is a larger molecule than is normally used in recombinant DNA work. To solve this size problem, the following strategy has been developed (Figure 13-4).

First, the T region is cut out of a Ti plasmid with restriction enzymes and introduced into one of the standard cloning vector plasmids that are used with *E. coli*. Large amounts of the vector carrying the T DNA can be grown in *E. coli* and then isolated. The next step is to use restriction enzymes and recombinant DNA techniques to insert a particular gene into the T DNA. This hybrid, containing the T DNA and the gene inserted into it, can be grown in large amounts in *E. coli* and then introduced into *A. tumefaciens* cells containing the corresponding entire Ti plasmid. Homologous genetic recombination between the T DNA segment of the native Ti plasmid and the cloned T DNA segment carrying the foreign gene results in transfer of the engineered T DNA to the Ti plasmid and the displacement of its normal T DNA. The outcome is *A. tumefaciens* with a Ti plasmid whose T region carries the desired foreign gene. The last step is to infect plants with these engineered *A. tumefaciens* bacteria. The crown gall cells that result will be transformed by the T DNA carrying the foreign gene, and the goal of introducing a desired gene into plant cells will have been reached (Figure 13-4).

Transformation of Plant Cells and Protoplasts

The usual way to transform plant cells with T DNA is to paint *Agrobacteria* that harbor Ti plasmids on a wound made in a plant shoot. However, with improvements in techniques for plant cell and protoplast culture, an often more convenient method that allows infection and transformation *in vitro* has been devised. Leaf cells are removed from plants, converted to protoplasts, and put into culture. At this stage, when the protoplasts have just regenerated a cell wall and begun to divide, the culture is infected with *Agrobacteria* and left for several hours. Antibiotics are then added to kill the bacteria, and the cells are grown in a medium containing plant hormones for a few weeks until they have formed small calli. At this stage, the medium is changed to one lacking plant hormones. Only the transformed cells will continue to survive and multiply. The transformed cells can then be tested for the presence of T DNA or its hallmark, the synthesis of opines. Cells from such cultures sometimes spontaneously regenerate into shoots or plants carrying T DNA and making opines.

With a much lower efficiency, it is also possible to transform protoplasts directly with Ti plasmid DNA. Freshly prepared protoplasts are exposed to plasmid DNA in a medium containing polyethyleneglycol and Ca^{++} ions—essentially the same medium that is used to induce protoplast fusion. T DNA is taken up by the protoplasts, which are then cultured in a medium with plant hormones to allow regeneration of cell walls and cell division. After a few weeks, when calli have developed, the medium is replaced with one lacking plant hormones. Only the transformed cells will survive and continue to multiply.

The fact that protoplasts can be transformed by pure Ti DNA alone proves that *Agrobacteria* are not essential for transformation. Their role is solely that of a vector to bring Ti DNA into plant cells.

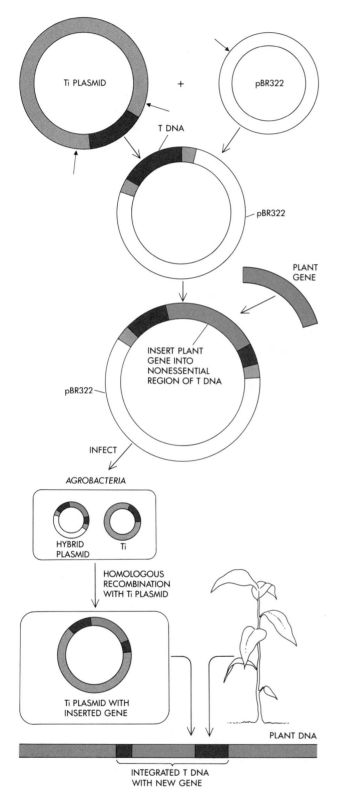

Figure 13-4
A procedure for using the Ti plasmid as a vector. First the T DNA of the Ti plasmid is cut out with restriction enzymes and cloned into pBR322. Next a foreign plant gene is inserted into the cloned T DNA region in the pBR322. The resulting hybrid plasmid is mixed with *Agrobacteria* colonies containing normal Ti plasmids; their T DNA recombines with that of the hybrid plasmids to form Ti plasmids carrying the foreign gene. The *Agrobacteria* carrying the foreign gene are used to infect plants, which incorporate the modified T DNA into their chromosomes.

Mobilization of T DNA by the *vir* Segment of the Ti Plasmid

One of the limitations of using the Ti plasmid for gene transfer in plants is its large size (approximately 180 kb). For this reason, most early attempts to use the Ti plasmid followed the sequence outlined above: cloning the T DNA region of the Ti plasmid into pBR322, inserting the desired gene into the T DNA, reintroducing this hybrid plasmid into *Agrobacteria* harboring an intact Ti plasmid, and then, through homologous recombination, inducing the chimeric plasmid to hop onto the intact Ti plasmid. Recent work, however, has shown that this procedure can be simplified: all that is needed for *Agrobacteria* to infect and transform plant cells is an intact T DNA region, as we mentioned above, and another region of the Ti plasmid called *"vir."* More important from a practical standpoint is that these two regions do not have to be on the same plasmid. If an *Agrobacterium* harbors a Ti plasmid containing the *vir* region and another plasmid containing the T DNA, the bacteria can transform plant cells, and the T DNA (and whatever other genes have been inserted into it) will be incorporated into the plant genome. No homologous recombination is necessary in the bacteria.

Attenuated T DNA Vectors Allow Regeneration of Whole Plants from Single Cells

Another limitation of using Ti plasmid T DNA as a vector is the fact that plant cells transformed with

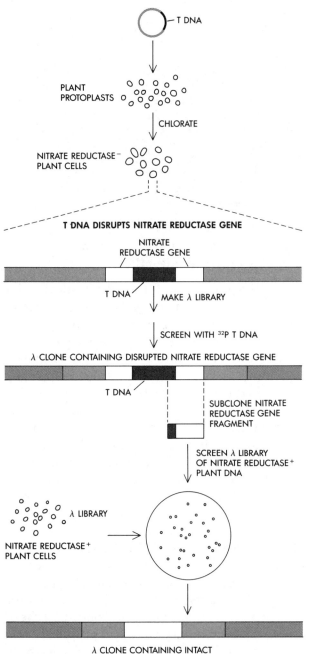

PLANT
PROTOPLASTS

CHLORATE

NITRATE REDUCTASE⁻
PLANT CELLS

T DNA DISRUPTS NITRATE REDUCTASE GENE

NITRATE
REDUCTASE GENE

T DNA   MAKE λ LIBRARY

SCREEN WITH ³²P T DNA

λ CLONE CONTAINING DISRUPTED NITRATE REDUCTASE GENE

T DNA

SUBCLONE NITRATE
REDUCTASE GENE
FRAGMENT

SCREEN λ LIBRARY
OF NITRATE REDUCTASE⁺
PLANT DNA

λ LIBRARY

NITRATE REDUCTASE⁺
PLANT CELLS

λ CLONE CONTAINING INTACT
NITRATE REDUCTASE GENE

Figure 13-5
T DNA insertion mutagenesis can be used to clone plant genes. Cloned T DNA or other plant transposons will integrate and disrupt plant genes. Plant protoplasts that have been transfected with T DNA can be treated with chlorate to select for nitrate reductase⁻ mutants. These will have had T DNA inserted into the nitrate reductase gene. A λ or cosmid library can be made from these mutants and screened with T DNA to obtain a recombinant containing the disrupted nitrate reductase gene. A fragment from this clone flanking the T DNA (and therefore containing part of the nitrate reductase gene) can be used as a probe to screen a λ library made from nitrate reductase⁺ plant DNA, to obtain a λ clone containing the intact nitrate reductase gene.

T DNA usually cannot be regenerated into whole plants. However, naturally occurring or transposon-induced T DNA mutants have been found which will transform plant cells but which do not block the redifferentiation of these cells into whole plants. These mutants were called "rooty" mutants and were mapped to a specific region of the T DNA. This discovery immediately suggested an experiment in which the T DNA would be specifically mutated at the rooty locus, where it would not interfere with plant regeneration; the "mutation" would be the insertion into the T DNA of the gene we wished to introduce into the plant cells. The gene coding for yeast alcohol dehydrogenase (ADH) was inserted into the rooty locus of cloned T DNA, and this "attenuated" T DNA was used to transform tobacco cells that were then used to regenerate entire tobacco plants. The tobacco plants were found to contain multiple copies of the ADH–T DNA in all their cells. The plants were fertile, moreover, and the seedlings also contained multiple copies of the chimeric T DNA. Preliminary results indicate that the yeast ADH gene is not expressed in the regenerated tobacco plants.

Finally, the latest advance in the use of T DNA as a vector for introducing genes into plants is the use of specific plant promotors to express the transferred genes. The Ti plasmid gene that codes for nopaline synthetase was isolated and sequenced, and its (putative) promotor region was identified. The structural genes for octopine synthetase or prokaryotic chloramphenicol acetyl-

transferase were cloned downstream from this promotor, and these hybrid genes were introduced into plant cells. These genes were found to be expressed in the plant cells, under the control of the nopaline synthetase promotor.

It is now well within our power, then, to introduce genes into plant cells and cause these cells to regenerate an entire plant in which the foreign genes are expressed. This technique has enormous potential as a tool for the study of gene regulation in higher plants and for the genetic engineering of plant species.

T DNA Insertion Can Be Used to Isolate Plant Genes

During transformation, T DNA is inserted into many different sites in the plant genome; indeed, as far as we know, the site of insertion is random for each transforming event. When T DNA is inserted in a structural gene for a plant enzyme, that gene is inactivated: The transformed cells are, in effect, insertion mutants. This can be exploited to isolate plant genes once the particular gene that has been inactivated is recognized. The T DNA, which can be recognized by hybridization is cut out of the plant DNA with appropriate restriction enzymes in such a way that flanking plant cell sequences remain attached. The T DNA with its flanks can then be used as a probe to screen a library of clones of the total DNA of the plant (Figure 13-5). The flanks of the T DNA segment will hybridize to, and therefore reveal, clones carrying normal copies of the gene that suffered the insertion. These can then be propagated and the normal gene obtained in large amounts so that it can be used for further manipulations.

Practical Applications of Plant Engineering Using Ti Plasmids

The long-term goals of all this research on the Ti plasmids of *Agrobacterium tumefaciens* and other strains of *Agrobacteria* are to provide plant breeders with a novel way of introducing single genes into plants, and to give plant molecular biologists molecular probes for studying plant development. Plant cell culture, gene cloning, and the discovery of the T DNA of Ti plasmids make it realistic to anticipate that eventually we will be able to exploit for practical plant breeding what are at present laboratory experimental systems. As we have already stressed, a serious limitation is that thus far it has proven impossible to induce the transformation of cells of monocotyledons (for example corn and wheat) with Ti plasmids. In addition, genes useful for increasing crop yields must be identified, cloned, and then properly inserted into the T DNA. This is a formidable task.

At present much effort is being devoted to the isolation of transposable elements from plants similar to the P element of *Drosophila* for use as generalized vectors for the introduction of desired genes into suitable recipient plants. One obvious system under development is the classical Ac (Ds) system of maize. Maize also contains other apparently transposonlike elements, one of which, the Robertson mutator (Mu), may be as short as 1.5 kb with 200-base-pair inverted sequences at its two ends and will perhaps prove to be a promising tool.

READING LIST

Original Research Papers (Reviews)

PLANT CELL CULTURE AND PROTOPLAST
ISOLATION

Murashige, T. "Plant propagation through tissue cultures." *Ann. Rev. Plant Physiol.,* 25: 135–166 (1974).

Helgeson, J. P. "Tissue and cell suspension culture." In *Nicotiana: Procedures for Experimental Use.* U.S. Dept. of Agriculture Technical Bulletin 1586, pp. 52–59 (1980). (Review.)

HAPLOID PLANT CULTURES

Melchers, G. "Haploid higher plants for plant breeding." *Zeitschrift Für Pflanzenzwechtung,* 67: 19–32 (1972). (Review.)

DeVreaux, M., and D. deNettancourt. "Screening mutations in haploid plants." In K. J. Kasha, ed., *Haploids in Higher Plants: Advances and Potential.* University of Guelph Press, Ontario, 1974, pp. 309–322. (Review.)

Nitsch, C. "Pollen culture." In K. J. Kasha, ed., *Haploids in Higher Plants: Advances and Potential.* University of Guelph Press, Ontario, 1974, pp. 123–135. (Review.)

SOMATIC-CELL HYBRIDIZATION

Constable, F. "Somatic hybridization in higher plants." *In Vitro,* 12: 743–748 (1972).

Carlson, P. S., H. H. Smith, and R. Dearing. "Parasexual interspecific plant hybridization." *Proc. Natl. Acad. Sci. USA,* 74: 5109–5112 (1972).

Gamborg, O. L. "Advances in somatic cell hybridization in higher plants." *Stadler Genet. Symp.,* 7: 37–46 (1975).

Mastrangelo, I. A. "Protoplast fusion and organelle transfer." In *Nicotiana: Procedures for Experimental Use.* U.S. Dept. of Agriculture Technical Bulletin 1586, pp. 65–73 (1980). (Review.)

TUMOR INDUCING (Ti) PLASMIDS OF
AGROBACTERIUM

Chilton, M.-D., M. H. Drummond, D. J. Merlo, D. Sciaky, A. C. Montoya, M. P. Gordon, and E. W. Nester. "Stable incorporation of plasmid DNA into higher plant cells: The molecular basis of crown gall tumorigenesis." *Cell,* 11: 263–271 (1977).

Schell, J., M. Van Montagu, M. De Beuckeleer, M. De Block, A. Depicker, M. De Wilde, G. Engler, C. Gentello, J. P. Hernalsteens, M. Holsters, J. Seurinck, B. Silva, F. Van Vilet, and R. Villarroel. "Interactions and DNA transfer between *Agrobacterium tumefaciens,* and the Ti plasmid and the plant host." *Proc. R. Soc. London B,* 204: 251–266 (1979).

Thomashow, M. F., R. Nutter, A. L. Montoya, M. P. Gordon, and E. W. Nester. "Integration and organization of Ti plasmid sequences in crown gall tumors." *Cell,* 19: 729–739 (1980).

Zambruski, P., M. Holsters, K. Kruger, A. Depicker, J. Schell, M. Van Montagu, and H. M. Goodman. "Tumor DNA structure in plant cells transformed by *Agrobacterium tumefaciens.*" *Science,* 209: 1385–1391 (1980).

Thomashow, M. F., R. Nutter, K. Postle, M.-D. Chilton, F. R. Blattner, A. Powell, M. P. Gordon, and E. W. Nester. "Recombination between higher plant DNA and the Ti plasmid of *Agrobacterium tumefaciens.*" *Proc. Natl. Acad. Sci. USA,* 77: 6448–6452 (1980).

Chilton, M.-D., R. K. Saiki, N. Yadav, M. P. Gordon, and F. Quetier. "T-DNA from *Agrobacterium* Ti plasmid is in the nuclear DNA of crown gall tumor cells." *Proc. Natl. Acad. Sci. USA,* 77: 4060–4064 (1980).

Willmitzer, L., M. De Beuckeleer, M. Lemmers, M. Van Montagu, and J. Schell. "DNA from Ti plasmid is present in the nucleus and absent from the plasmids of crown gall plant cells." *Nature,* 287: 359–361 (1980).

Lemmers, M., M. De Beuckeleer, M. Housters, P. Zambryski, A. Depicker, J. P. Hernalsteens, M. Van Montagu, and J. Schell. "Internal organization, boundaries and integration of Ti plasmid DNA in nopaline crown gall tumors." *J. Mol. Biol.,* 144: 353–376 (1980).

Garfinkel, D. J., R. B. Simpson, L. W. Ream, F. F. White, M. P. Gordon, and E. W. Nester. "Genetic analysis of crown gall: Fine structure map of T-DNA by site-directed mutagenesis." *Cell,* 27: 143–153 (1981).

Van Montagu, M., and J. Schell. "The Ti plasmids of *Agrobacterium.*" *Curr. Top. Microbiol. Immunol.,* 96: 237–254 (1982). (Review.)

Leemans, J., C. Shaw, R. Deblaere, H. De Greve, J. P. Hernalsteens, M. Maes, M. Van Montagu, and J. Schell. "Site-specific mutagenesis of *Agrobacterium* Ti plasmids and transfer of genes to plant cells." *J. Molec. Applied Genet.,* 1: 149–163 (1982).

Chilton, M.-D., D. A. Tepfer, A. Petit, C. David, F. Casse-Delbart, and J. Tempe. "*Agrobacterium rhizogenes* inserts T-DNA into the genomes of host plant root cells." *Nature,* 295: 432–434 (1982).

Bevan, M. W., and M.-D. Chilton. "T-DNA of the *Agrobacterium* TI and RI plasmids." *Ann. Rev. Genet.,* 16: 357–384 (1982).

White, F. F., D. J. Garfinkel, G. A. Huffman, M. P. Gordon, and E. W. Nester. "Sequences homologous

to *Agrobacterium rhizogenes* T-DNA in the genomes of uninfected plants." *Nature,* 301: 348–350 (1983).

Willmitzer, L., P. Dhaese, P. H. Schreier, W. Schmalenbach, M. Van Montagu, and J. Schell. "Size, location and polarity of T-DNA-encoded transcripts in nopaline crown gall tumors; common transcripts in octopine and nopaline tumors." *Cell,* 32: 1045–1056 (1983).

Joos, H., D. Inzé, A. Caplan, M. Sormann, M. Van Montagu, and J. Schell. "Genetic analysis of T-DNA transcripts in nopaline crown galls." *Cell,* 32: 1057–1067 (1983).

GENETIC ENGINEERING OF Ti PLASMIDS

Matzke, A. J. M., and M.-D. Chilton. "Site-specific insertion of genes into T-DNA of the *Agrobacterium* tumor-inducing plasmid: An approach to genetic engineering of higher plant cells." *J. Molec. Applied Genet.,* 1: 39–50 (1981).

De Greve, H., J. Leemans, J.-P. Hernalsteens, L. Thia-Toong, M. De Beuckeleer, L. Willmitzer, L. Otten, M. Van Montagu, and J. Schell. "Regeneration of normal and fertile plants that express octopine synthase, from tobacco crown galls after deletion of tumor-controlling functions." *Nature,* 300: 752–755 (1982).

Barton, K. A., A. N. Binns, A. J. M. Matzke, and M.-D. Chilton. "Regeneration of intact tobacco plants containing full length copies of genetically engineered T-DNA to R1 progeny." *Cell,* 32: 1033–1043 (1983).

Hoekema, A., P. R. Hirsch, P. J. J. Hooykaas, and R. A. Schilperoort. "A binary plant vector strategy based on separation of *vir-* and T-region of the *Agrobacterium tumefaciens* Ti-plasmid." *Nature,* 303: 179–189 (1983).

Herrera-Estrella, L., A. Depicker, M. Van Montagu, and J. Schell. "Expression of chimaeric genes transferred into plant cells using a Ti-plasmid-derived vector." *Nature,* 303: 209–213 (1983).

IN VITRO TRANSFORMATION OF PROTOPLASTS

Marton, L., G. J. Wullems, L. Molendijk, and P. A. Schilperoort. "In vitro transformation of culture cells of *Nicotiana tabacum* by *Agrobacterium tumefaciens.*" *Nature,* 277: 129–131 (1979).

Cocking, E. C., M. R. Davey, D. Pental, and J. B. Power. "Aspects of plant genetic manipulation." *Nature,* 293: 265–269 (1981). (Review.)

Wullems, G. J., L. Molendijk, G. Ooms, and R. A. Schilperoort. "Retention of tumor markers in F1 progeny plants in *in vitro* induced octopine and nopaline tumor tissues." *Cell,* 24: 719–727 (1981).

Wullems, G. J., L. Molendijk, G. Ooms, and R. A. Schilperoort. "The expression of tumor markers in intraspecific somatic hybrids of normal and crown gall cells of *Nicotiana tabacum.*" *Theor. Applied Genet.,* 56: 203–208 (1981).

Wullems, G. J., L. Molendijk, and R. A. Schilperoort. "Differential expression of crown gall tumor markers in transformants obtained after *in vitro* transformation of *Nicotiana tabacum* protoplasts by *Agrobacterium tumefaciens.*" *Proc. Natl. Acad. Sci. USA,* 78: 4344–4348 (1981).

Muller, A., and R. Grafe. "Isolation and characterization of cell lines of *Nicotiana tabacum* lacking nitrate reductase." *Mol. Gen. Genet.,* 161: 67–76 (1978).

TRANSPOSONS AS TOOLS FOR GENETIC ENGINEERING

Robertson, D. S. "Characterization of a mutator system in maize." *Mutation Res.,* 51: 21–28 (1978).

Strommer, J. N., S. Hake, J. Bennetzen, W. C. Taylor, and M. Freeling. "Regulatory mutants of the maize *Adh1* gene caused by DNA insertions." *Nature,* 300: 542–544 (1982).

Marx, J. L. "A transposable element of maize emerges." *Science,* 219: 829–830 (1983). (News report.)

14

Transferring Genes
into Mammalian Cells

Our understanding of the control of gene expression in prokaryotes has been based largely on the introduction of genes or other defined segments of DNA into bacteria and the assessment of the genes' ability to function normally. These procedures have allowed the unambiguous identification of cis-acting and trans-acting regulatory elements (Chapter 4, p. 47) around a given gene, and have demonstrated that the ability of such elements to control gene expression is not dependent on the location of the gene or operon in the chromosome but is inherent in the DNA itself.

A similar approach to the study of gene expression in higher eukaryotes was impossible until recently. Now, however, we are able routinely to introduce cloned genes into the genomes of a variety of mammalian cells.

Ca++ Stimulates the Uptake of DNA by Vertebrate Cells

Tumor virus DNA that is free of viral coat proteins is able to induce viral multiplication and give rise to progeny virus particles when it is added to susceptible host cells, and is thus able to transform normal cells into their cancerous equivalents. However, for many years the efficiencies of these events were so low that they could not be effectively studied. A great technical advance was made with the discovery that purified DNA from adenovirus, when precipitated with Ca++ and added to a monolayer of normal rat cells growing in culture, led to the appearance of many more transformed foci than when Ca++ was absent. The mechanism of the preferential uptake of Ca++-precipitated DNA by cells remains unclear, but apparently the cells actually phagocytize the DNA granules, and a small fraction of the DNA molecules later stably integrate into the cell's chromosomal DNA.

Given that all double-helical DNA was known to have roughly the same morphology, there seemed good reason to believe that the functional uptake (or transfection) of any eukaryotic DNA would also be stimulated by Ca++. Since, however, only a very small fraction of the added DNA finally becomes functionally integrated into cellular DNA, "marker" techniques had to be developed to allow cells with integrated DNA to be selectively multiplied and easily identified against a background of much larger numbers of unmodified cells.

Thymidine Kinase (Tk) Has Served as the Archetypal Selective Marker for Transfection Experiments

Several well-defined selectable genes (markers) have already been used in genetic studies in eukaryotic cells. The best characterized so far is the gene for thymidine kinase (tk), an enzyme used in the salvage pathway of pyrimidine biosynthesis. The enzyme takes thymidine that has been formed by the degradation of DNA and phosphorylates it to dTMP, which by the addition of two or more phosphorylated groups becomes dTTP and gets reincorporated into DNA. The "normal" pathway for dTTP biosynthesis, however, is through dCDP. The presence of tk is thus not essential for cells to survive, and cells completely lacking tk activity can live perfectly well.

Tk− mutants can easily be isolated by feeding cells bromodeoxyuridine (BrUdr). This nucleoside analog, when incorporated into DNA, is lethal to the cell. BrUdr, however, is incorporated into DNA only if it is first phosphorylated by the

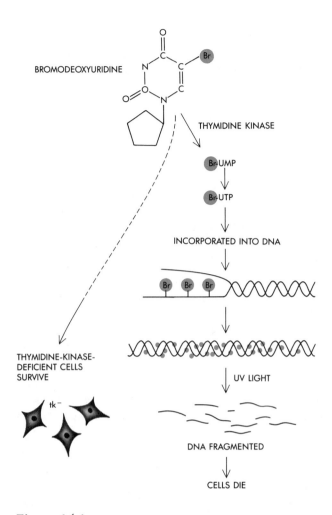

Figure 14-1
A procedure used to obtain tk⁻ mutants. Cells with functional thymidine kinase enzyme phosphorylate bromodeoxyuridine (BrUdr) to BrUMP, which is subsequently incorporated into the cells' DNA. Brominated DNA is extremely sensitive to UV light and the cells die. The rare tk⁻ mutant survives such treatment.

action of tk. Cells with functional tk activity are thus killed by BrUdr, whereas the rare tk⁻ mutant will survive (Figure 14-1).

Tk⁻ cells, however, will die in "HAT," a medium containing hypoxanthine, aminopterin (which blocks the normal dCDP → dTDP step), and thymidine. In this case, the only source of dTTP for DNA biosynthesis is through thymidine kinase. In 1977 it was first found that exposure of tk⁻ mouse cells to Ca^{++}-precipitated DNA containing the thymidine kinase gene from herpes simplex virus resulted in the survival of resistant cell clones in HAT medium. Southern blotting confirmed that these survivors had indeed taken up and stably integrated the herpes tk gene (Figure 14-2). This meant that the tk gene could be used as a selectable marker for the integration of other genes linked to it.

Dominant-Acting Markers for Transformation of Normal Cells

The use of a cloned tk gene as a vector for gene transfer into eukaryotic cells obviously requires that the recipient cells be tk⁻. Though in principle it should be possible to render any eukaryotic cell tk⁻, in practice this process is often very time-consuming, since both of the tk alleles of diploid cells must be inactivated; furthermore, the tk⁻ cells used for gene transfer must have an extremely low frequency of reversion to tk⁺. Recently, therefore, much effort has been given to developing gene markers that work on normal cells as opposed to tk⁻ mutant cells.

Two such versatile vectors now exist. Both use prokaryotic genes that have been linked to eukaryotic control signals (Figure 14-3, page 178). One contains a bacterial gene that lets bacteria utilize xanthine as a source of purine nucleosides by coding for the enzyme xanthine-guanine phosphoribosyl transferase (XGPRT). The corresponding mammalian enzyme HGPRT uses hypoxanthine; it can use xanthine only very inefficiently. This vector was made by inserting the cloned bacterial XGPRT gene between the promoter and poly-A-addition sites of the SV40 large T antigen gene. The SV40 promoter is extremely

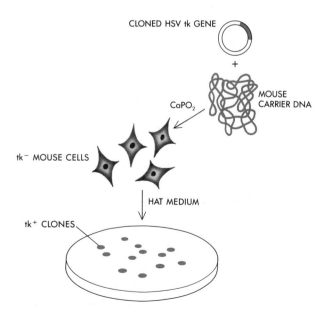

Figure 14-2
Transfection of the herpes simplex virus (HSV) thymidine kinase gene into tk⁻ mouse cells.

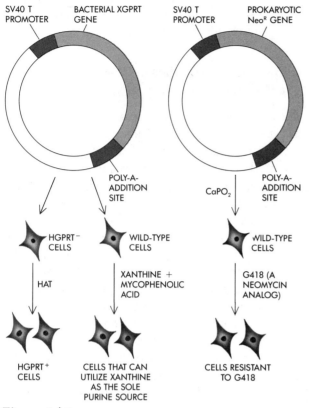

CHAPTER 14

Figure 14-3
Dominant-acting vectors for gene transfer in
eukaryotes. Both vectors use prokaryotic
structural genes under the control of eukaryotic
promoters.

efficient, and consequently results in the expression of a large amount of bacterial XGPRT enzyme. This "SVgpt" vector not only transforms HGPRT⁻ mammalian cells to the HGPRT+ phenotype, but, more importantly, it can be used as a dominant-acting vector (that is, one that can transform wild-type cells) if the selection medium is a mixture of mycophenolic acid and xanthine. Mycophenolic acid blocks the HGPRT enzyme, making xanthine the sole source of purine nucleosides. So only cells that have taken up the SVgpt vector with the bacterial gene can survive, by using the xanthine in the medium and the acquired bacterial gene to metabolize it.

The second dominant-acting vector consists of the prokaryotic neomycin-resistance gene (NeoR gene) ligated into the SV40 early region. This gene codes for an enzyme that phosphorylates and thus inactivates neomycin, which is toxic to ribosomes. Eukaryotic cells are sensitive to a neomycin

analog called G418, which is also inactivated by the product of the NeoR gene. The SVneo vector can thus be used to transform mammalian cells to resistance to G418.

Cotransformation Following Intracellular Ligation

The existence of selectable markers led to the idea that it should be possible to introduce any gene into mammalian cells by ligating that gene to the cloned selectable marker. Prior ligation outside of cells, however, turned out not to be necessary: Analysis of mouse cells that had taken up the tk gene revealed that they had also incorporated other DNA sequences that were included with the tk gene in forming the Ca⁺⁺ precipitate. (For technical reasons, to form a Ca⁺⁺ phosphate coprecipitate with DNA, a minimum mass of DNA must be present. Thus, in gene transfer experiments, a large excess of "carrier" DNA is usually added to the cloned tk gene to form the precipitate.) The finding that cells transformed to the tk+ phenotype had taken up some of the carrier DNA was at first attributed to the fact that the tk selection was simply identifying subpopulations of cells that were competent to take up and integrate DNA, and it was assumed that the other non-tk sequences would be integrated randomly throughout the genome. However, it was later found that within the mouse cells, the exogenously added DNA is actually ligated into a large concatamer (a series of units linked together as in a chain) that can contain up to 800 to 1000 kilobases. Apparently this large structure is integrated randomly as a unit into a chromosome. Thus the DNA added with the tk gene ("cotransformed" DNA) is physically linked to the tk gene in the mouse cell (Figure 14-4).

Using this technique of cotransformation, virtually any cloned segment of DNA can be easily introduced into eukaryotic tissue culture cells, if they are competent for transformation and if an appropriate selection is available, simply by including the cloned DNA along with the selectable marker when forming the Ca⁺⁺ precipitate.

Microinjection of DNA into Mammalian Cells

DNA can also be stably introduced into tissue culture cells by its direct microinjection into the nu-

clei of the cells, using a glass micropipette that has been drawn out to an extremely thin diameter (from 0.1 to 0.5 microns). Such a procedure requires some fairly sophisticated equipment (a micropipette puller for making the needles, and a micromanipulator to position the needles correctly for injections), but given this equipment and enough practice, one can inject 500 to 1000 cells per hour with DNA, and have up to 50 percent of the injected cells stably integrate and express the injected genes. The advantage of this procedure is that in principle *any* piece of DNA can be introduced into *any* cell; no selective pressure need be applied to maintain the transferred gene. The disadvantage of microinjection is, as was mentioned above, the expensive equipment that is required, the extensive practice needed to master this tedious technique, and the relatively small number of cells that can be treated in any one experiment.

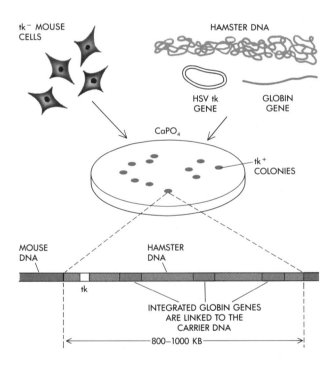

Figure 14-4
Cotransformation during gene transfer. When tk is transfected into eukaryotic cells by Ca^{++} precipitation, part of the carrier DNA and any other DNA present in the precipitate are ligated together into a large (800-to-1000-kb) structure and inserted into the chromosome.

Semistable Inheritance of Transfected Methylated DNA

Genomic DNA contains methylated bases that arise by postreplicative modification (Chapter 2). In vertebrates, the only methylated base is 5-methylcytosine, which is found almost exclusively at the sequence $\frac{(5')CG(3')}{(3')GC(5')}$. In birds and mammals, 50 to 70 percent of all such dinucleotides are modified by methylation. Methylation of newly replicated DNA chains is believed to be carried out by an enzyme (methylase) that acts exclusively on hemimethylated $\frac{(5')Me-CG}{GC}$ sites (Figure 14-5, page 180). By such a scheme, a pattern of methylation, once established, would tend to be inherited. Now there is increasing speculation that functionally active genes are less methylated than their inactive counterparts. For this to be true, there must exist some mechanisms that selectively (1) demethylate key control sites, perhaps by preventing their modification following DNA replication, (2) maintain given methylation patterns accurately, and (3) add methyl groups to previously unmethylated sites (Figure 14-6, page 181).

Support for the concept that specific methylation can be inherited comes from recent experiments in which previously unmethylated bacteriophage DNA (ϕX174) was methylated *in vitro* and then transfected into mouse cells. Twenty-five cell generations later, the methylation sites of the transferred phage DNA sequences were examined, and most were found to retain the methyl groups that had been enzymatically added. Methyl-C inheritance, however, was not perfect, with apparently a small fraction of such sites failing to be methylated at each round of chromosome replication. Equally interesting was the consequence of methylation of eukaryotic (chicken) tk DNA that had been cloned in *E. coli*. Methylation of these previously unmethylated DNA sequences led to reduced transfection frequencies, in comparison to the unmethylated tk control samples. These results and others to be discussed (Chapter 16, page 205) make it most tempting to believe that the lower rates of transfection with the enzymatically methylated DNA reflect the methylation of key control sequences, which blocks their function.

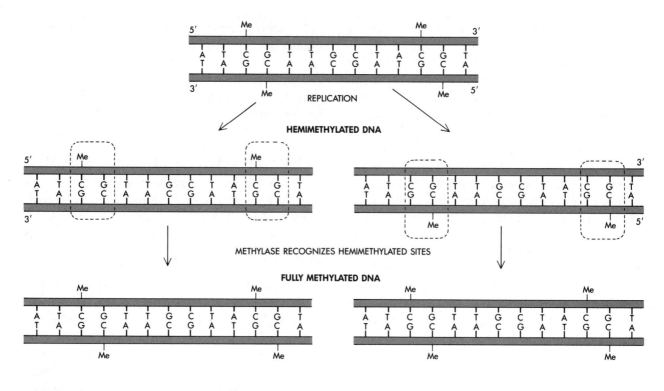

Figure 14-5
The eukaryotic "maintenance" methylase recognizes hemimethylated sites.

Isolation of Transferred Genes

The ability to transfer genes into mouse cells also allows isolation of those genes. To accomplish this, two different approaches, which may be termed "screening" and "rescue," have been devised. In the screening strategy (Figure 14-7), the donor DNA is first cleaved with a number of restriction enzymes until one or a few enzymes that do not cut the gene (that is, do not destroy the gene's ability to transfer the phenotype) are found. Total DNA is cleaved with this enzyme and ligated to a specific type of heterologous DNA (often plasmid pBR322 is used). In this way, all of the eukaryotic DNA fragments, including the one containing the selectable marker, are "tagged" with a prokaryotic marker. Gene transfer is carried out as usual. The selected gene is now sitting in the mouse cell next to the tag (that is, the pBR322) sequence. In practice, DNA from this "primary recipient" is isolated and used to transform a second population of cells. This is necessary because all the eukaryotic DNA fragments in the original DNA were tagged, and a primary recipient cell may have ran-

domly incorporated *several* pieces of tag DNA in addition to the one next to the selectable marker. The secondary transformant, however, will usually contain the tag sequence only next to the selected gene. A library of this DNA is then constructed in phage or in cosmids. The library is screened with the tag sequence, which, it is hoped, will be included on the same phage or cosmid containing the selectable gene. This approach was first used successfully to isolate the hamster APRT (adenine phosphoribosyl transferase) gene (Figure 14-7, page 182).

If human DNA is used as the donor, a DNA tag need not be applied, because of the presence in the human genome of the Alu sequence (Chapter 7, page 100), which is repeated several hundred thousand times and which does not cross-hybridize appreciably with any sequences in mouse cells. This Alu sequence is so ubiquitous in the human genome that the probability of finding one or several Alu sequences near a given gene is essentially 100 percent. In other words, the human genes have a natural tag. Human DNA is used to transform mouse cells, and then a second round of transformation is carried out, for the reasons de-

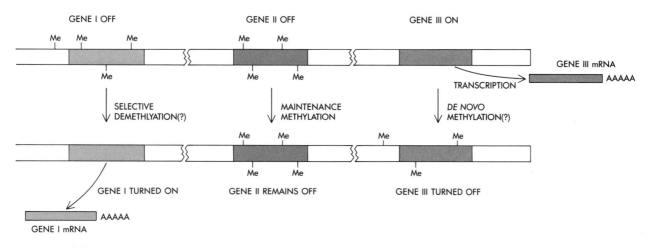

GENE I OFF   GENE II OFF   GENE III ON

Figure 14-6
A model for the regulation of gene expression by methylation. In addition to the maintenance methylase, which recognizes hemimethylated sites on a gene (gene II), there may be other enzymes that can remove methyl groups from a specific gene (gene I), thus activating it, or that can methylate a previously unmethylated gene (gene III), turning it off.

scribed above. DNA from the secondary transformant is made into a library, and the library is screened using a cloned Alu sequence. Again, the Alu sequence should be on the recombinant phage that contains the selectable marker.

The "rescue" approach for isolating transferred genes is based on the addition of a *functional* DNA tag to the DNA initially used for transfer. This can be a bacterial drug-resistance marker (Figure 14-8, page 183), or a prokaryotic suppressor tRNA. As outlined above, the first step is to find restriction enzymes that do not cleave the gene to be transferred. Once these are identified, the DNA is cut with the enzyme, and the population is ligated either to pBR322 DNA that contains the intact drug-resistance marker and that has been cut with the same enzyme, or to a cloned *E. coli* suppressor tRNA called "supF." The DNA is introduced into mouse cells, and the initial transformants are isolated and their DNA is used for a second round of transformation. By now the transferred eukaryotic gene should be next to the prokaryotic selectable marker. If pBR322 was the initial vector, DNA from the secondary transformant is cut and then circularized with DNA ligase. The transferred gene should be contained on a molecule with the drug-resistance gene from the plasmid. The total DNA population is used to transform *E. coli,* and the appropriate drug resistance is monitored. The only transformants obtained should contain plasmids with the drug-resistance gene and the selectable eukaryotic gene.

If the suppressor tRNA was the vector in the original ligation, DNA from the secondary transformant is made into a library with a λ phage vector that contains an amber mutation (a stop codon) in the lysis gene. These phage require a functional supF gene to complete the lytic cycle. Thus, although all the DNA from the secondary transformant can be packaged, phage growth and lysis following transduction into bacteria will occur only with phage that contain the supF gene, which should be next to the transferred gene.

Regulation Following Gene Transfer

The ability to introduce virtually any piece of DNA into mammalian cells has allowed the identification of DNA sequences responsible for controlling the expression of certain eukaryotic genes. In general, when a cloned gene is introduced into eukaryotic cells it continues to respond to the signals that normally control the expression of the gene *in vivo*. As in prokaryotes, the control of transcription of specific eukaryotic genes utilizes cis-acting elements (such as promoters) that usually lie quite close to their respective genes. Modulation of the genes' activity is often carried out by trans-acting molecules that, by binding to the control elements, either turn on or turn off the func-

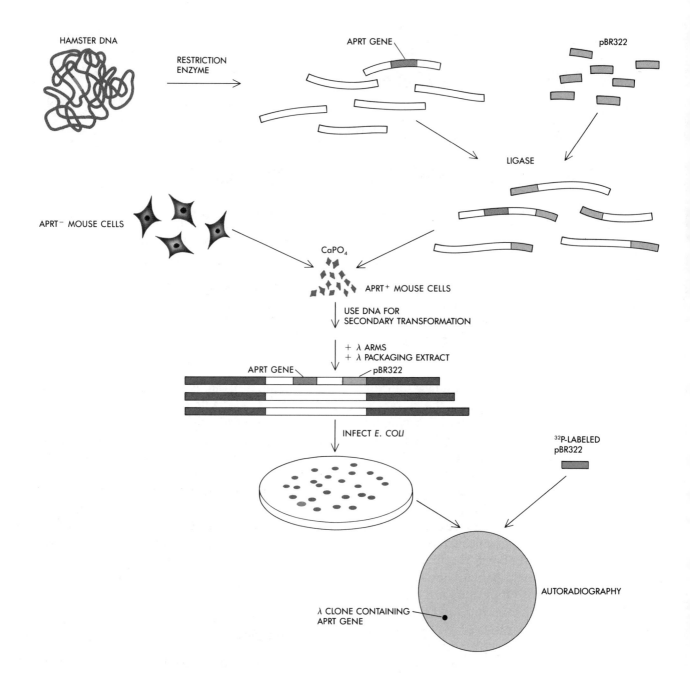

Figure 14-7
A screening method for isolation of a transferred gene. Hamster DNA is cut with a restriction enzyme that does not cleave the APRT gene. The total digest is then ligated to pBR322 that has been cut with the same enzyme. Enough pBR322 DNA is used in the reaction to ensure that every piece of hamster DNA is "tagged" with pBR322. This DNA is used to transfect APRT⁻ mouse cells to APRT⁺. DNA from these APRT⁺ cells is used for a second round of transformation; the DNA from the resulting cells is made into a library. The hamster APRT gene should be adjacent to the pBR322 tag sequence, hopefully on the same clone. This would mean that the library could be screened with labeled pBR322 to obtain the clone containing the APRT gene.

tions of those genes. Many eukaryotic genes, for example, are turned on or off by various steroid hormones. The transcription of the mouse mammary tumor virus (MMTV) genome, for instance, is positively controlled by glucocorticoid hormones. A cloned MMTV provirus, when introduced into mouse cells that contain a receptor for glucocorticoid hormones, continues to respond to the hormone, and transcription of the transferred MMTV gene is increased by a factor of five to ten by addition of the hormone (Figure 14-9, page 184). A cloned gene for rat α_{2u}-globulin, under the control of several hormones *in vivo,* also responds to insulin and glucocorticoids when it is introduced into mouse cells. In both of these cases, very little flanking DNA is necessary for the gene to respond normally. When small (300-base-pair) segments of 5′ flanking DNA from MMTV or α_{2u} are ligated to the structural gene for herpes simplex tk, tk expression becomes inducible by hormones. Other genes that have been found to be regulated normally in a heterologous environment include the human α-interferon gene, which can be induced by viral infection; the mouse metallothionine gene, which responds to Cd^{++} ions following gene transfer; and, most surprisingly, the *Drosophila* heat-shock gene p70, which, when introduced into mouse cells, is induced if the mouse cells are exposed to high temperatures. In all of these cases, the regulatory elements seem to lie quite close to (and in some cases possibly in) the gene.

Establishing the Existence of Specific Human Cancer Genes by DNA Transfer Experiments

As we outlined above, the Ca^{++}-precipitate technique for gene transfer was first used with tumor virus DNA to transform normal mammalian cells to the malignant phenotype. Recently, gene transfer techniques have been employed to identify and clone genes that may be involved in the formation of human cancers. These experiments started with the finding that high-molecular-weight DNA isolated from chemically transformed mouse cells or from primary cultures of human solid tumors (such as bladder cancers) can transform normal mouse fibroblasts (NIH 3T3 cells) to the malignant phenotype. To show this, tumor DNA was

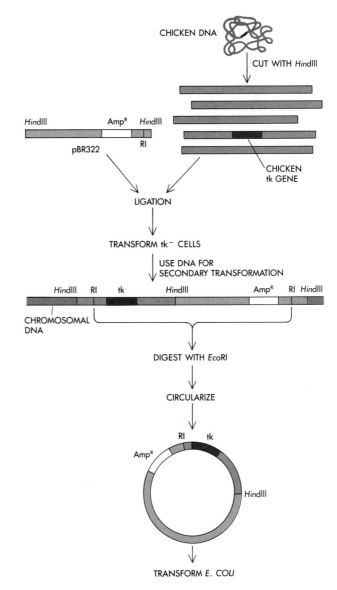

Figure 14-8
Plasmid rescue of the chicken tk gene. Chicken DNA is cleaved with an enzyme (here, *Hin*dIII) that does not cut in the tk gene. The *Hin*dIII-cleaved chicken DNA is ligated to *Hin*dIII-cleaved pBR322, and the ligation product is used to transform tk⁻ cells to tk⁺. DNA from these tk⁺ cells is used in a second round of transformation. The DNA from the secondary transformants is cut with another enzyme that does not cut the tk gene (here, *Eco*RI). The fragments are then circularized with DNA ligase and used to transform *E. coli* to ampicillin resistance. The tk gene should be contained on a plasmid with a functional Amp^R gene.

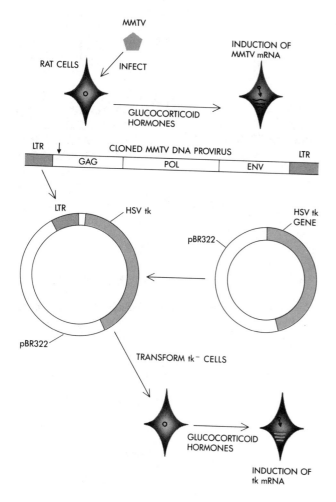

a single gene. The latter statement is based on the fact that the efficiency of gene transfer into mammalian cells by using total genomic DNA as a donor is so low that, if the simultaneous transfer and expression of even *two* genes were required to transform the cells, transformation would never be seen.

Cutting tumor cell DNA with various restriction enzymes generally inactivates its cancer-causing ability. The slopes of these inactivation curves vary from one tumor cell line to another, suggesting that a variety of different genetic changes can generate cancer cells (Table 14-1). The different patterns of human Alu repetitive sequences found in secondary transformants (induced by DNA from mouse cells made cancerous by exposure to human tumor cell DNA) support this conclusion.

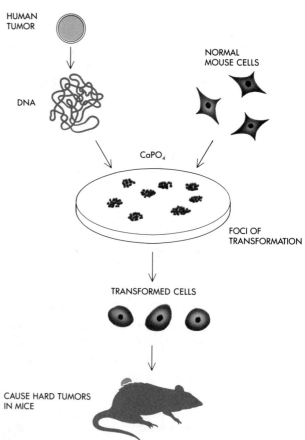

Figure 14-9
The transcription of mouse mammary tumor virus (MMTV) proviral DNA is induced by glucocorticoid hormones. The site for this regulation was shown to be in the LTR by fusing the LTR to a cloned HSV tk gene. When this construct is transfected into mouse cells, tk mRNA becomes inducible by glucocorticoids.

added to a dish of 3T3 mouse cells in the form of a Ca^{++} precipitate. The treated cells were placed in a low-serum medium for several weeks, after which small "foci" of cancerlike cells appeared; these were not found in cultures not exposed to such DNA. When the cells from these foci were injected into mice, they caused the formation of hard tumors (Figure 14-10). This indicated that (1) for these tumors at least, the transformed phenotype is dominant (that is, it is apparently due to the presence and not the absence of a specific gene product), and (2) the transformed phenotype in 3T3 cells can be induced by the expression of

Figure 14-10
Transfer of human oncogenes into mouse cells. When DNA from cell lines obtained from human tumors is transfected into mouse cells, small "foci" of transformed mouse cells appear in a few weeks. When such cells are injected into mice, hard tumors result.

The only human Alu sequences that such secondary transformants should contain are those located adjacent to or within the introns of the transferred human cancer gene. Now it seems clear that the Alu sequences transferred along with the bladder cancer, colon cancer, and neuroblastoma genes are all unique, and this implies that these three tumors are caused by three different cancer genes. Equally importantly, the cancer gene present within the DNA of lung cancer cells cannot be distinguished from that found within the DNA of colon cancer cells, so these two tumors may have a common genetic basis. Definitive proof for these conclusions, however, must wait for the cloning and subsequent sequencing of the respective genes.

Cloning Human Oncogenes

By now three putative human oncogenes have been cloned by using the Alu screening (Figure 14-11) and tRNA rescue (Figure 14-12, page 186) techniques we have described. Already four different labs have independently cloned cancer genes present in bladder cancer cell lines. All these genes appear to be the same (approximately 5.4 kb in size), but this similarity probably reflects the fact, only just being realized, that these supposedly different cell lines had a common origin. The neuroblastoma gene has also now been completely cloned by use of the tRNA rescue method, revealing a gene of approximately 13.5 kb. Cloning of the colon (lung) cancer gene proved more difficult to achieve since it is too large to be cloned in one piece in phage λ. Thirty-five different λ phages,

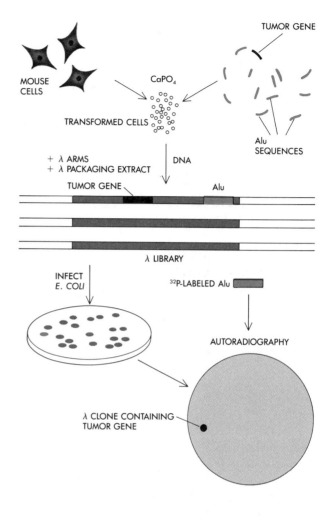

Figure 14-11
Cloning an oncogene by using Alu screening. This method is very similar to that depicted in Figure 14-7, except that no tag sequence is needed to identify the human gene being transferred into mouse cells. The Alu sequences present in all human DNA provide natural tags.

Table 14-1

Ability of Various Restriction Enzymes to Cleave Different Tumor Genes

Restriction Enzyme	Bladder Carcinoma DNA	Neuroblastoma DNA	Colon Cancer DNA
None	+	+	+
BamHI	+	−	−
HindIII	+	−	−
EcoRI	+	−	−
SalI	+	+	−

+ = DNA gives foci on 3T3 cells
− = DNA does not give foci

each containing overlapping partial sequences, were first isolated and then analyzed using Alu screening techniques and homologies to the Kirsten sarcoma virus ras gene sequences (see below). Gene walking procedures were then employed to establish its 45-kb gene structure.

Despite their wide variation in size, the bladder cancer, neuroblastoma, and colon (lung) cancer genes have the same basic exon–intron arrangement, with four exons used to code for similar but distinct proteins of molecular weights of about 21,000 (p21) each. The p21 proteins are found attached in small numbers to the outer

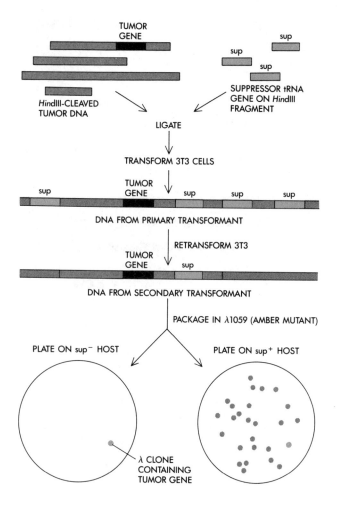

Figure 14-12
Suppressor rescue of a human oncogene. In this technique, tumor DNA is cut with an enzyme that does not cut the oncogene. The cleaved DNA is ligated to a cloned *E. coli* gene coding for a suppressor tRNA called supF. The DNA is used to transform mouse cells; the DNA from a second round of transformation is used to make a λ library. The arms that are used in the packaging, however, contain one or two amber mutations in vital genes. Therefore, the only λ phage that can grow in *E. coli* are those that have incorporated the supF gene, which will hopefully be contained on the same piece of DNA as the tumor gene.

plasma membrane of the cancer cells, and they are closely homologous to the products of cancer genes previously found in oncogenic retroviruses. The human bladder cancer gene is very similar to the ras oncogene of the Harvey sarcoma virus, whereas the lung cancer gene is very similar to the ras oncogene of the Kirsten sarcoma virus. Like the retroviral oncogenes, these human cancer genes have their normal cell equivalents; indeed, the cancer genes are thought to be derived from their normal cell equivalents by mutations that somehow convey oncogene potential to the protein products. The next obvious step was to sequence both the human oncogenes and their normal equivalents. The first such results have shown that the cancer gene in human bladder carcinoma cells differs from its counterpart in normal cells by a single point mutation. This mutation converts a glycine residue at position 12 in the normal protein product (the p21 protein) to a valine in the

protein of the carcinoma cells. However, there is at present no proof that this simple change is the sole cause of bladder carcinoma, nor do we know the role of the p21 protein in either its normal or mutated state.

Clearly the recombinant DNA analysis of the genetic basis of cancer is still in its infancy. The next few years should witness the arrival of powerful insights into the genetic origins of cancer, as well as set us on the right path to finding out how the respective protein products give rise at the molecular level to cancerous phenotypes. These discoveries may help first in cancer diagnosis, but ultimately they should affect the treatment of cancer. For example, it may already be possible to develop DNA probes that will allow detection of mutated "cancer" genes such as those found in the human bladder carcinoma cells. In Chapter 17 we shall discuss further the use of DNA probes for detecting nonmalignant genetic diseases.

READING LIST

Original Research Papers (Reviews)

CA++-STIMULATED DNA UPTAKE

Graham, F. L., P. J. Abrahams, C. Mulder, H. L. Meijneker, S. Warnaar, F. A. J. de Vries, W. Fiers, and A. J. van der Eb. "Studies on in vitro transformation by DNA and DNA fragments of human adenoviruses and simian virus 40." *Cold Spring Harbor Symp. Quant. Biol.,* 39: 637–650 (1974).

TK AS A SELECTIVE MARKER

Wigler, M., S. Silverstein, L.-S. Lee, A. Pellicer, Y.-C. Cheng, and R. Axel. "Transfer of purified herpes virus thymidine kinase gene to cultured mouse cells." *Cell,* 11: 223–232 (1977).

Maitland, N. J., and J. K. McDougall. "Biochemical transformation of mouse cells by fragments of herpes simplex virus DNA." *Cell,* 11: 233–241 (1977).

Wigler, M., A. Pellicer, S. Silverstein, and R. Axel. "Biochemical transfer of single-copy eucaryotic genes using total cellular DNA as donor." *Cell,* 13: 725–731 (1978).

Wigler, M., A. Pellicer, S. Silverstein, R. Axel, G. Urlaub, and L. Chasin. "DNA-methylated transfer of the adenine phosphoribosyltransferase locus into mammalian cells." *Proc. Natl. Acad. Sci. USA,* 76: 1373–1376 (1979).

DOMINANT-ACTING VECTORS

Southern, P. J., and P. Berg. "Transformation of mammalian cells to antibiotic resistance with a bacterial gene under control of the SV40 early region promoter." *J. Molec. Applied Genet.,* 1: 327–342 (1982).

Mulligan, R., and P. Berg. "Selection for animal cells that express the *Escherichia coli* gene coding for xanthine–guanine phosphoribosyltransferase." *Proc. Natl. Acad. Sci. USA,* 78: 2072–2076 (1981).

COTRANSFORMATION

Perucho, M., D. Hanahan, and M. Wigler. "Genetic and physical linkage of exogenous sequences in transformed cells." *Cell,* 22: 309–317 (1980).

Robins, D. M., S. Ripley, A. S. Henderson, and R. Axel. "Transforming DNA integrates into the host chromosome." *Cell,* 23: 29–39 (1981).

Goodenow, R. S., M. McMillan, M. Nicolson, B. T. Sher, K. Eakle, N. Davidson, and L. Hood. "Identification of the class I genes of the mouse major histocompatibility complex by DNA-mediated gene transfer." *Nature,* 300: 231–237 (1982).

METHYLATED DNA

Wigler, M., D. Levy, and M. Perucho. "The somatic replication of DNA methylation." *Cell,* 24: 33–40 (1981).

ISOLATION OF TRANSFERRED GENES

Perucho, M., D. Hanahan, L. Lipsich, and M. Wigler. "Isolation of the chicken thymidine kinase gene by plasmid rescue." *Nature,* 285: 207–210 (1980).

Lowy, I., A. Pellicer, J. F. Jackson, G. K. Sim, S. Silverstein, and R. Axel. "Isolation of transforming DNA: Cloning of the hampster aprt gene." *Cell,* 22: 817–823 (1980).

REGULATION FOLLOWING GENE TRANSFER

Kurtz, D. T. "Hormonal inducibility of rat α_{2u} globulin genes in transfected mouse cells." *Nature,* 291: 629–631 (1981).

Buetti, E., and H. Diggelmann. "Cloned mouse mammary tumor virus DNA is biologically active in transfected mouse cells and its expression is stimulated by glucocorticoid hormones." *Cell,* 23: 335–345 (1981).

Corces, V., A. Pellicer, R. Axel, and M. Meselson. "Integration, transcription and control of a *Drosophila* heat shock gene in mouse cells." *Proc. Natl. Acad. Sci. USA,* 79: 7038–7042 (1981).

Lee, F., R. Mulligan, P. Berg, and G. Ringhold. "Glucocorticoids regulate expression of dihydrofolate reductase cDNA in mouse mammary tumour virus chimaeric plasmids." *Nature,* 294: 228–232 (1981).

Hamer, D. H., and M. J. Walling. "Regulation *in vivo* of a cloned mammalian gene: Cadmium induces the transcription of a mouse metallothionein gene in SV40 vectors." *J. Molec. Applied Genet.,* 1: 273–288 (1982).

HUMAN CANCER GENES

Krontiris, T. G., and G. M. Cooper. "Transforming activity of human tumor DNAs." *Proc. Natl. Acad. Sci. USA,* 78: 1181–1184 (1981).

Perucho, M., M. Goldfarb, K. Shimizu, C. Lama, J. Fogh, and M. Wigler. "Human-tumor-derived cell lines contain common and different transforming genes." *Cell,* 27: 467–476 (1981).

Murray, M. J., B.-Z. Shilo, C. Shih, D. Cowing, H. W. Hsu, and R. A. Weinberg. "Three different human tumor cell lines contain different oncogenes." *Cell,* 25: 355–361 (1981).

Goldfarb, M., K. Shimizu, M. Perucho, and M. Wigler. "Isolation and preliminary characterization of a

human transforming gene from T24 bladder carcinoma cell." *Nature,* 296: 404–409 (1982).

Parada, L. F., C. J. Tabin, C. Shih, and R. A. Weinberg. "Human EJ bladder carcinoma oncogene is homologue of Harvey sarcoma virus *ras* gene." *Nature,* 297: 474–478 (1982).

Weinberg, R. A. "Fewer and fewer oncogenes." *Cell,* 30: 3–4 (1982). (Review.)

Cooper, G. M. "Cellular transforming genes." *Science,* 217: 801–806 (1982). (Review.)

Der, C. J., T. G. Krontiris, and G. M. Cooper. "Transforming genes of human bladder and lung carcinoma cell lines are homologous to the *ras* gene of Harvey and Kirstein sarcoma viruses." *Proc. Natl. Acad. Sci. USA,* 79: 3637–3640 (1982).

Ellis, R. W., D. DeFeo, T. Y. Shih, M. A. Gonda, H. A. Young, N. Tsuchida, D. R. Lowy, and E. M. Scolnick. "The p21 *src* genes of Harvey and Kirsten sarcoma viruses originate from divergent members of a family of normal vertebrate genes." *Nature,* 292: 506–511 (1981).

Tabin, C. J., S. M. Bradley, C. I. Bargmann, R. A. Weinberg, A. G. Papageorge, E. M. Scolnick, D. Dhar, D. R. Lowy, and E. H. Chang. "Mechanism of activation of a human oncogene." *Nature,* 300: 143–149 (1982).

Reddy, E. P., R. K. Reynolds, E. Santos, and M. Barbacid. "A point mutation is responsible for the acquisition of transforming properties by the T24 human bladder carcinoma oncogene." *Nature,* 300: 149–152 (1982).

Taparowsky, E., Y. Suard, O. Fasano, K. Shimizu, M. Goldfarb, and M. Wigler. "Activation of the T24 bladder carcinoma transforming gene is linked to a single amino acid change." *Nature,* 300: 762–765 (1982).

Shimizu, K., M. Goldfarb, M. Perucho, and M. Wigler. "Isolation and preliminary characterization of the transforming gene of a human neuroblastoma cell line." *Proc. Natl. Acad. Sci. USA,* 80: 383–387 (1983).

Shimuzu, K., D. Birnbaum, M. A. Ruley, O. Fasano, Y. Suard, L. Edlund, E. Taparowsky, M. Goldfarb, and M. Wigler. "The structure of the K-*ras* gene of the human lung carcinoma cell line calu-1." *Nature* (1983) (in press).

15

Viral Vectors

The molecular biology of DNA and RNA tumor viruses (Chapter 10) had been extensively studied even before the advent of recombinant DNA technology. Because so much was known about the life cycles of these viruses and their entries into and exits from animal cells, and because, depending on the cells they infected, they could either replicate to give progeny or integrate their DNA into host chromosomes, they were obvious candidates for use as vectors for the introduction of foreign DNA into animal cells.

Moreover, virus infection can be a much more efficient way of getting DNA into some cells than the transfection procedures described in Chapter 14. With virus infection, it is possible to ensure that each recipient cell has many copies of the foreign gene. And since the viral genome includes strong promoters, it is possible to ensure efficient expression of foreign genes inserted into the viral DNA. As a result, the amounts of gene products, both RNAs and proteins, are much higher and therefore more accessible to study.

SV40 Vectors

SV40 has been the most commonly used vector for introducing foreign genes into animal cells; its DNA is a covalently closed circle that has about 5200 base pairs and that can be divided into "early" and "late" regions. The early region is expressed throughout the lytic cycle, whereas expression of the late genes occurs only after viral DNA replication has begun. Between the early and late regions there is a DNA sequence containing the origin of viral DNA replication. The early-region gene products, the T antigens, are responsible for malignant transformation of

"nonpermissive" cells (cells in which the virus cannot complete its replication), as well as for initiation of viral DNA replication in permissive cells (Chapter 10). The late region encodes the viral structural proteins that form the protein capsid or coat of the virus particles.

Two approaches have been exploited to utilize SV40 as a vector system. The first is to make recombinants between SV40 DNA and a piece of foreign DNA, and then to pass the hybrid molecules through permissive cells eventually to produce encapsidated virions containing the recombinant DNA. Alternatively, recombinants between SV40 and foreign DNA can be constructed and not packaged into virions. The recombinant DNA either replicates transiently as a plasmidlike entity independent of the cellular DNA, or it is integrated into a chromosome.

SV40 Virions as Vectors

To produce SV40 virus particles containing recombinant DNA, either the early or the late region of the SV40 DNA is replaced with a segment of foreign DNA. The recombinant molecules cannot synthesize essential viral functions (either the T antigens if the early region has been replaced, or the capsid proteins if the late region has been replaced). Therefore, in order to obtain virus particles containing the foreign DNA, the missing viral functions are frequently supplied by complementation using coinfecting "helper" virus particles.

SV40 Late-Region Replacement

When the late region of SV40 is replaced with foreign DNA, a helper SV40 that has a deletion

mutation in its early genes is used. Permissive cells are simultaneously transfected with the SV40 molecules containing the foreign gene and DNA extracted from the mutant helper virus particles. The SV40 recombinant produces the early gene products, while the coinfecting helper mutant provides the viral late proteins for new capsids. As a result, a mixed population of progeny virus particles is produced, some containing the foreign DNA, others the mutant viral genome (Figure 15-1). Often, there is no way to separate the two classes of virions in the mixture, but the total population can be used to infect either permissive or nonpermissive cells. These procedures were first used to insert the genes for the mouse α- and β-globins and the gene for a prokaryotic amber-suppressor tRNA into SV40. Cells infected with these recombinant viruses were found to produce large amounts of functional RNA transcribed from the inserted sequence.

SV40 Early-Region Replacement

When the early region of SV40 RNA is replaced by foreign DNA, production of recombinant virus particles depends on the supply of early gene proteins, the T antigens. One way to obtain T antigens

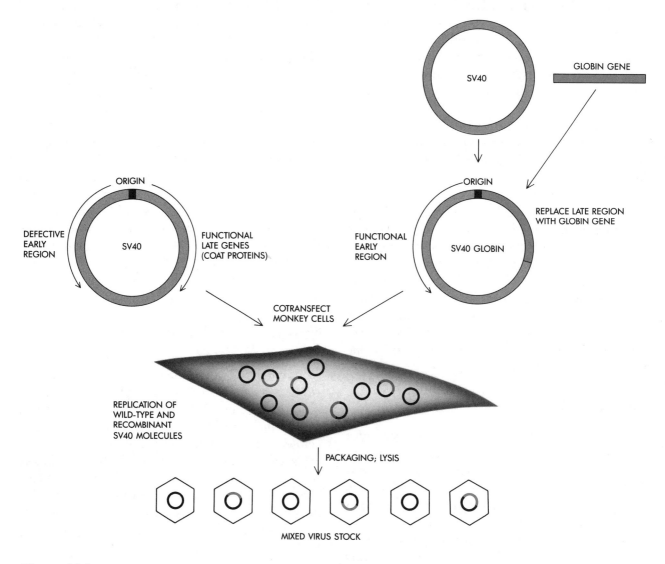

Figure 15-1
The late region of SV40 is replaced with a cloned globin gene. The hybrid construct is then used to transform monkey cells along with SV40 DNA that has a mutation in the early region. The globin–SV40 hybrid molecule provides early functions, while the coinfecting helper virus produces viral coat proteins. The result is a mixed virus stock consisting of the defective-early-region SV40 and globin–SV40 hybrid DNA, now packaged into virions.

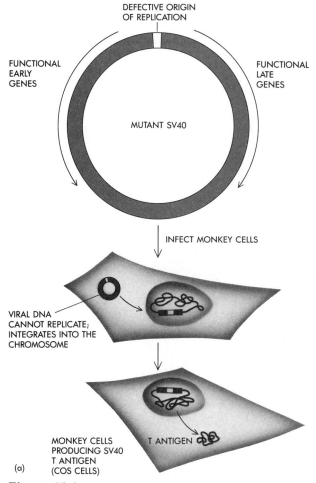

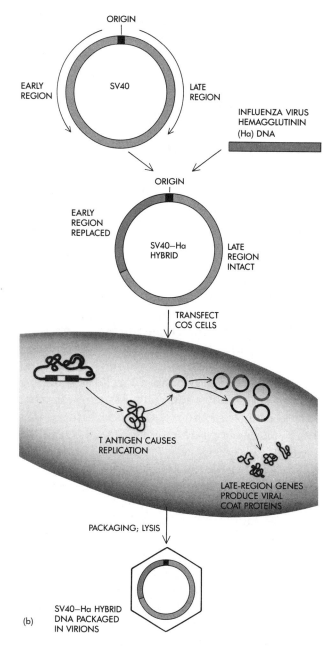

Figure 15-2
SV40 early-region replacement. (a) Monkey cells
are infected with a mutant SV40 that possesses
functional early and late genes but that lacks a
functional origin of replication. Such molecules
cannot replicate, and therefore they integrate
into the chromosome of the monkey cell, where
they continue to produce functional SV40 T
antigen. These cells are called "COS" cells.
(b) The early region of SV40 is replaced with a
cloned influenza virus hemagluttinin (Ha) gene.
When this hybrid construct is transfected into
COS cells, the T antigen being produced in the
cells causes the SV40–Ha DNA to replicate, and
the functional late region makes viral coat
proteins. The SV40–Ha molecules are packaged
into virions.

while avoiding coinfection with a mutant helper
virus is to introduce the recombinant DNA into a
monkey cell line called COS. COS cells have been
transformed by SV40 DNA that contains a functional early gene region but that has a defective
origin of viral DNA replication. As a result, the
SV40 DNA integrated into the host chromosome
specifies functional viral T antigens but the viral
DNA, lacking an origin of replication, cannot rep-

licate independently of the cell's chromosomes
(Figure 15-2a). When any SV40 molecule that
lacks functional early genes but that has an SV40
origin of DNA replication is introduced into these
cells, it will replicate, because the T antigens provided by the cells will recognize the origin of replication in the incoming viral DNA. Virions containing recombinant DNA in which the influenza
virus hemagglutinin (Ha) gene replaces the SV40

early region have been produced in this way (Figure 15-2b). The advantage of replacing early genes by foreign DNA and propagating virions in COS cells is that *all* progeny virus particles should have the recombinant genome; there are no contaminating helper viruses.

Analysis of Cloned Surface Antigen Genes

A gene inserted into an SV40 vector and introduced into animal cells should be efficiently and correctly expressed. The protein that is made should be fully functional; it should undergo any posttranslational modification that it normally undergoes, and, if it is a protein that is ordinarily transported through the intracellular membrane system to the cell surface, this transport process should be completed. Cloning in SV40 of a gene coding for such a protein provides an opportunity not only to study transcription, mRNA splicing, and translation, but also the intracellular transport mechanisms that sort and deliver the protein to its correct cellular location, and the mechanisms that insert and anchor it in membranes. Through the use of *in vitro* mutagenesis (Chapter 8), amino acid codons in the gene can be selectively mutated and the effects of such mutations can be assessed. In this way it should be possible to define the functional domains of the gene and its corresponding protein.

For example, the flu virus hemagglutinin (Ha) gene cloned into SV40 expression vectors is efficiently expressed in infected monkey cells (Figure 15-3). The hemagglutinin is fully glycosylated and transported in a normal way to the cell surface membrane. There it remains in a biologically and antigenically active form, anchored through a carboxyl-terminal domain containing hydrophobic amino acids that lie in the lipid bilayer of the membrane.

When cells were infected with the SV40 vector carrying an Ha gene from which the sequences that specify the carboxyl-terminal hydrophobic segment had been deleted, the hemagglutinin was synthesized and glycosylated, but was then secreted into the culture medium rather than being anchored to the membrane. This neatly and unambiguously proved that the C-terminal hydrophobic domain of the hemagglutinin is responsible for

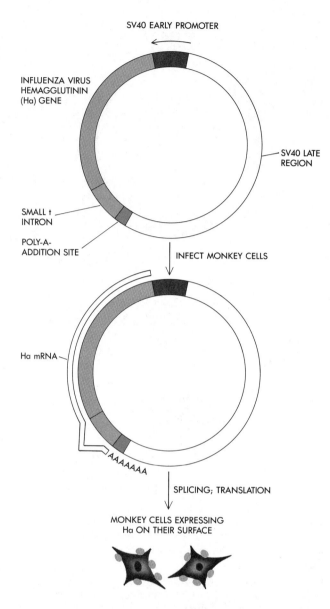

Figure 15-3
The SV40 early promoter can be used to express large amounts of flu virus hemagglutinin.

maintaining it as an integral membrane protein. As we have mentioned (Chapter 7), the amino-terminal segment of both secreted and integrated outer-surface membrane proteins in eukaryotic cells is also hydrophobic, and acts as a leader to translocate the nascent polypeptide across the membranes of the intracellular transport system. During this translocation, the amino-terminal leader is cleaved from the growing polypeptide. Removal of the 5′ end of the cloned Ha gene, which codes for the

hemagglutinin leader, resulted in hemagglutinin that was not glycosylated and was not translocated across intracellular membranes.

Experiments such as these would not be possible without eukaryotic cloning vectors. Cloning in bacteria would allow study of the early steps in gene expression, but not of posttranslational modifications such as glycosylation and transport through intracellular membrane systems. How proteins are sorted within cells and delivered to their correct destinations is now a central preoccupation of cell biologists. Gene cloning provides a way of tackling the problem.

Plasmidlike Replication of DNA in COS Cells

The SV40 vectors we have just discussed depend on the completion of the replication cycle of the virus. In COS cells, *any* piece of DNA that includes an SV40 origin of replication will replicate because of the presence of SV40 T antigen in the cells. The foreign DNA will, at least transiently, replicate independently of the cellular DNA.

Recombinant DNAs containing the SV40 origin and a foreign gene will, when introduced into COS cells, replicate to an enormously high copy number. The transcription of the foreign gene from the plasmidlike DNA molecules can then be studied. In one such experiment, transcription of the β-globin gene was found to initiate correctly and to produce a precisely spliced mRNA (Figure 15-4). Because so many gene copies are present in each cell and because the total amount of RNA made is correspondingly high, this system provides a quick way of screening mutations for their effects on transcription and RNA processing.

Rescue of Integrated SV40 DNA by COS Cell Fusion

Nonpermissive mouse cells transfected with a plasmid containing the SV40 origin of replication linked to some selectable marker will incorporate the DNA into the chromosomes. If these cells are then fused with COS cells (fusion can be effected with polyethyleneglycol), the T antigen being produced by the COS cells will diffuse to the nuclei of the mouse cells and initiate DNA replication at the SV40 origins in the mouse cell chromosomes.

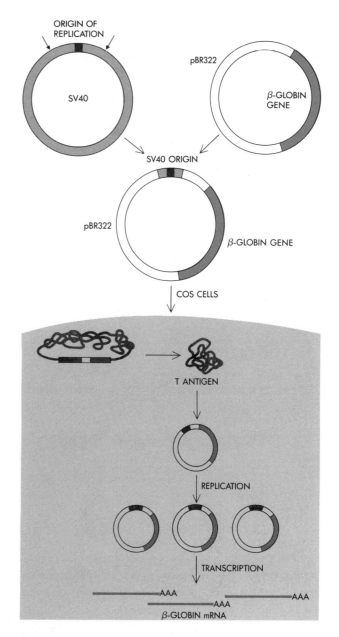

Figure 15-4
A foreign gene cloned on a plasmid with the SV40 origin replicates to a very high copy number when transfected into COS cells. The high copy number allows efficient transcription of the foreign gene.

The replication proceeds bidirectionally and eventually produces circular molecules that pop out of the chromosomes. These molecules can easily be purified away from the bulk of the chromosomal DNA. If plasmid pBR322 sequences are located near the SV40 origin, the circular molecules produced in this manner can be propagated directly in

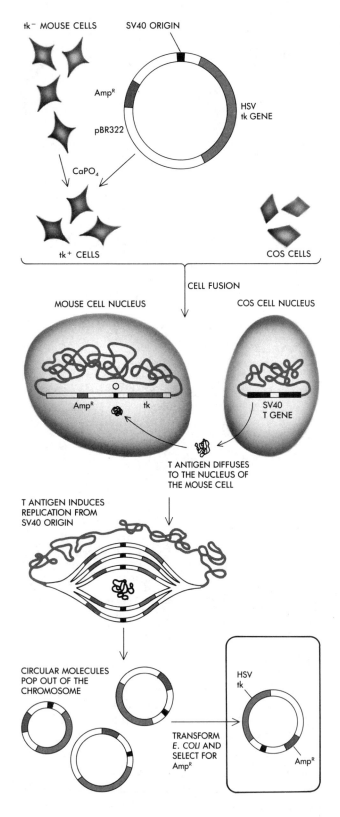

Figure 15-5
Rescue of SV40-origin-containing DNA from mouse cells. The HSV tk gene is cloned into a plasmid containing the SV40 origin and the *E. coli* ampicillin-resistance gene. The hybrid is used to transfect tk⁻ mouse cells to tk⁺. When these tk⁺ mouse cells are fused with COS cells, the T antigen being produced by the COS cells causes the SV40 origin in the mouse chromosome to begin to replicate bidirectionally. The replicating DNA pops out of the mouse chromosome and circularizes. This DNA can be used to transform *E. coli* to ampicillin resistance.

E. coli. Thus the vector can replicate in both bacteria and animal cells (Figure 15-5). This system is, therefore, similar to the shuttle vectors we described for moving plasmids between yeast and bacteria (Chapter 12).

Discovery of Enhancer Sequences Using SV40 Vectors

In the course of constructing recombinant molecules containing the SV40 early-region promoter and origin of replication, the discovery was made that certain sequences upstream from the promoter of the SV40 T antigen greatly increase transcription from virtually any promoter contained nearby on the same plasmid molecule. These SV40 "enhancer sequences" were found to be 72-base-pair segments of DNA that are repeated in a tandem fashion. When these 72-base-pair repeats are cloned on a plasmid with the rabbit β-globin gene and the recombinant plasmid is introduced into human cells, large amounts of β-globin mRNA are produced. Plasmids without the enhancer sequences produce essentially no β-globin messenger (Figure 15-6). Since their discovery in SV40, enhancer sequences have also been found in murine leukemia virus, polyoma, and bovine papilloma viruses.

It is still unclear how the enhancer sequences exert their effects. Surprisingly, it was discovered that these sequences seem to have no polarity; that is, they can enhance transcription from a promoter whether they are cloned 5' or 3' relative to the promoter. Also, these enhancer sequences can increase transcription from "illegitimate" promo-

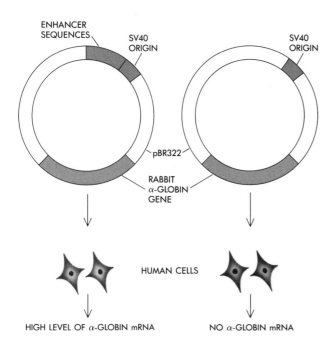

Figure 15-6
Enhancer sequences. Specific DNA sequences upstream of the SV40 promoter greatly increase the transcription from any promoter contained nearby on the same plasmid.

ters. Often, when cloned DNA is introduced into cells, RNA will be transcribed from DNA sequences that are not normally used for transcription but that, by chance, have sequences the RNA polymerase can recognize (CAT- or TATA-box-like sequences). Enhancers increase transcription even from such sequences.

One model that would explain these findings is that enhancer sequences keep the DNA in an open, protein-free conformation, and thus provide entry sites for RNA polymerase. In this scheme, the RNA polymerase molecules bind to the open DNA at the enhancer and then slide along the DNA, scanning for a sequence that they recognize as a promoter sequence. Support for such a model comes from the observation that if two promoters are cloned in tandem downstream from an enhancer sequence, an increase in transcription is found only from the first promoter, that is, the one closer to the enhancer. According to the model, this would mean that the polymerase enters at the enhancer, moves along the DNA until it reaches

the first promoter, and then begins transcribing; it never reaches the second promoter to initiate there. If no legitimate promoter exists on the cloned DNA molecule, then the RNA polymerase would start transcribing when it found a DNA sequence that approximates a legitimate promoter.

Further support for this model comes from an old observation that SV40 minichromosomes, circular SV40 DNA molecules that are arranged in nucleosomes, have a region near the origin that is free of nucleosomes. This observation was originally interpreted as having to do with SV40 replication, but now it seems that this open area of SV40 DNA (which can also be detected by its hypersensitivity to nucleases) is, in fact, the enhancer region. If the enhancer sequence is cut out from its normal position in the SV40 molecule and recloned at new positions, the nucleosome-free regions in minichromosomes derived from such molecules are now found at these new positions in the SV40 DNA.

Still unknown is whether enhancers are specific to certain viruses that may have evolved a mechanism to allow extremely efficient transcription from their promoters, or whether enhancers will be found in all eukaryotic DNA.

Papilloma Virus DNA Replicates like a Plasmid in Mouse Cells

Another DNA virus that is being used as a vector in eukaryotic cells is bovine papilloma virus (BPV). This virus, which causes warts in cattle, has a genome of 8.0 kilobases. The circular BPV DNA has the ability to transform certain mouse cell lines to the malignant phenotype. In these transformed cells, the BPV DNA remains circular and extrachromosomal, at about 30 to 100 copies per cell. This is the only well-documented example of a "plasmid" being stably maintained in higher eukaryotes. Foreign DNA can be cloned into the BPV genome and introduced into mouse cells in this way. The recombinant molecules usually remain extrachromosomal. For example, the rat insulin gene has been inserted into BPV DNA. In transformed mouse cells, 30 to 50 copies of the recombinant are maintained in the plasmidlike state, and high levels of insulin mRNA and protein

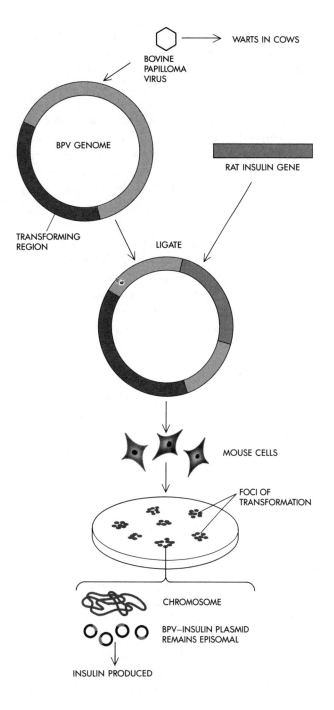

are produced (Figure 15-7). If pBR322 sequences are also introduced into BPV DNA, the resulting recombinant can be shuttled back and forth between *E. coli* and mouse cells.

RNA Tumor Viruses Can Be Used as Vectors

Retroviruses can also be used as vectors in eukaryotes. As we discussed earlier (Chapter 10), the genome of a retrovirus consists of an RNA molecule that resembles an mRNA in that it contains a methylated cap at the 5′ end and a poly-A tract at the 3′ end. The RNA genome is copied into a double-stranded DNA molecule that integrates into the genome of the cell. Once integrated, the provirus is transcribed very much like a cellular

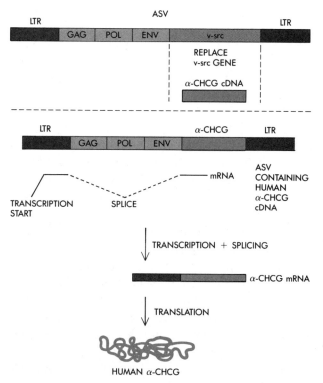

Figure 15-7
Bovine papilloma virus DNA replicates autonomously in mouse cells and remains stable at about 50 to 200 copies per cell. The rat insulin gene contained on such a replicating plasmid produces large amounts of insulin mRNA and protein.

Figure 15-8
Retroviral vectors. The LTRs of most retroviruses are extremely active promoters; foreign DNA (in this example, the cDNA for the α subunit of human chorionic gonadotropin, α-CHCG) cloned downstream from a viral LTR will usually be transcribed very efficiently.

gene. The resultant RNA may or may not be packaged into viral particles. The production of virions does not necessarily kill the host cell; the virions bud from the plasma membrane, and this is usually not cytopathic.

The integrated provirus contains the same DNA sequence at each end, the LTR sequence. The only known promoter for viral gene expression is contained in the LTR, and the production of the various RNAs coding for the different proteins is regulated at the level of RNA splicing. Cloned proviruses can therefore be used as vehicles for introducing genes into animal cells; this is done by replacing one of the viral genes with a foreign sequence (Figure 15-8). Cells infected with such a recombinant will produce large amounts of the new mRNA, with transcription initiating in the LTR. If a helper virus is included, the RNA copy of the recombinant virus will be packaged into viral particles. If the original provirus recombinant contains a heterologous gene (one with introns), the RNA copy of this recombinant should have the introns spliced out. However, the first experiments to test this idea have shown that the removal of introns from foreign genes inserted into cloned retroviral proviruses is a slow process that requires many generations of viral replication to achieve. The first-generation progeny virus particles do have introns in their RNA genomes. This implies that during the replication of retroviruses, the proviral RNA transcripts that are destined to enter progeny virus particles are somehow protected from the splicing mechanisms that trim the RNA that is destined to act as messenger.

READING LIST

Books

Gluzman, Y., ed. *Eukaryotic Viral Vectors.* Cold Spring Harbor Laboratory, Cold Spring Harbor, N.Y., 1982.

Gluzman, Y., and T. Shenk, eds. *Enhancers and Eukaryotic Gene Expression.* Cold Spring Harbor Laboratory, Cold Spring Harbor, N.Y., 1983 (in press).

Original Research Papers (Reviews)

SV40 VECTORS

Hamer, D. H., and P. Leder. "Expression of the chromosomal mouse β^{maj}-globin gene cloned in SV40." *Nature,* 281: 35–40 (1979).

Hamer, D. H., K. D. Smith, S. H. Boyer, and P. Leder. "SV40 recombinants carrying rabbit β-globin gene coding sequences." *Cell,* 17: 725–735 (1979).

Mulligan, R. C., B. H. Howard, and P. Berg. "Synthesis of rabbit β-globin in cultured monkey kidney cells following infection with an SV40 β-globin recombinant genome." *Nature,* 277: 108–114 (1979).

Gruss, P., and G. Khoury. "Expression of simian virus 40-rat preproinsulin recombinants in monkey kidney cells: Use of preproinsulin processing signals." *Proc. Natl. Acad. Sci. USA,* 78: 133–137 (1981).

Gluzman, Y., R. J. Frisque, and J. Sambrook. "Origin-defective mutants of SV40." *Cold Spring Harbor Symp. Quant. Biol.,* 44: 293–300 (1980).

Gluzman, Y. "SV40-transformed simian cells support the replication of early SV40 mutants." *Cell,* 23: 175–182 (1981).

SURFACE ANTIGEN GENES

Gething, M.-J., and J. Sambrook. "Cell-surface expression of influenza haemagglutinin from a cloned DNA copy of the RNA gene." *Nature,* 293: 620–625 (1981).

Liu, C. C., D. Yansura, and A. D. Levinson. "Direct expression of hepatitis B surface antigen in monkey cells from an SV40 vector." *DNA,* 1: 213–221 (1982).

Sveda, M. M., L. J. Markoff, and C.-J. Lai. "Cell surface expression of the influenza virus haemagglutinin requires the hydrophobic carboxy-terminal sequences." *Cell,* 30: 649–656 (1982).

Gething, M.-J., and J. Sambrook. "Construction of influenza haemagglutinin genes that code for intracellular and secreted forms of the protein." *Nature,* 300: 598–603 (1982).

ENHANCER SEQUENCES

Banerji, J., S. Rusconi, and W. Shaffner. "Expression of a β-globin gene is enhanced by remote SV40 DNA sequences." *Cell,* 27: 299–308 (1981).

Benoist, C., and P. Chambon. "*In vivo* sequence requirements of the SV40 promoter region." *Nature,* 290: 304–310 (1981).

De Villiers, J., and W. Shaffner. "A small segment of polyoma virus DNA enhances the expression of a cloned β-globin gene over a distance of 1400 basepairs." *Nuc. Acids Res.,* 9: 6251–6264 (1981).

Levinson, B., G. Khoury, G. Vande Woude, and P. Gruss. "Activation of the SV40 genome by the 72 base-pair tandem repeats of Moloney Sarcoma Virus." *Nature,* 295: 568–572 (1982).

Conrad, S. E., and M. R. Botchan. "Isolation and characterization of human DNA fragments with nucleotide sequence homologies with the simian virus 40 regulatory region." *Mol. Cell. Biol.,* 2: 949–965 (1982).

PAPILLOMA VIRUS

Law, M.-F., D. R. Lowy, I. Dvoretzky, and P. M. Howley. "Mouse cells transformed by bovine papillomavirus contain only extrachromosomal viral DNA sequences." *Proc. Natl. Acad. Sci. USA,* 78: 2727–2731 (1981).

Sarver, N., P. Gruss, M.-F. Law, G. Khoury, and P. M. Howley. "Bovine papilloma virus deoxyribonucleic acid: A novel eucaryotic cloning vector." *Mol. Cell. Biol.,* 1: 486–496 (1981).

Binétruy, B., G. Meneguzzi, R. Breathnach, and F. Cuzin. "Recombinant DNA molecules comprising bovine papilloma virus type 1 DNA linked to plasmid DNA are maintained in a plasmidial state both in rodent fibroblasts and in bacterial cells." *EMBO J.,* 1: 621–628 (1982).

DiMaio, D., R. Treisman, and T. Maniatis. "Bovine papilloma virus vector that propagates as a plasmid in both mouse and bacteria cells." *Proc. Natl. Acad. Sci. USA,* 79: 4030–4034 (1982).

RETROVIRUS VECTORS

Shimotohno, K., and H. M. Temin. "Formation of infectious progeny virus after insertion of herpes simplex thymidine kinase gene into DNA of an avian retrovirus." *Cell,* 26: 67–77 (1981).

Wei, C.-M., M. Gibson, P. G. Spear, and E. M. Scolnick. "Construction and isolation of a transmissible retrovirus containing the *src* gene of Harvey Sarcoma

Virus and the thymidine kinase gene of herpes simplex virus type 1." *J. Virology,* 39: 935–944 (1981).

Tabin, C. J., J. W. Hoffman, S. P. Goff, and R. A. Weinberg. "Adaption of a retrovirus as a eucaryotic vector transmitting the herpes simplex virus thymidine kinase gene." *Mol. Cell. Biol.,* 2: 426–436 (1982).

Sorge, J., and S. Hughes. "The splicing of intervening sequences introduced into an infectious retrovirus vector." *J. Molec. Applied Genet.,* 1: 547–559 (1982).

16

The Introduction of Foreign Genes into Fertilized Mouse Eggs

As we have already discussed, the transfection of cells in tissue culture with cloned purified genes is proving a most valuable way of studying how genes in higher organisms are regulated. This approach is, for the most part, limited to cells that are already partially or fully differentiated. Therefore, it cannot directly reveal the molecular mechanisms responsible for the temporal and tissue-specific regulation of gene expression during early embryonic development. One way to approach this problem is to introduce foreign genes—which because they are foreign can readily be identified—into zygotes or early embryos, and then analyze their pattern of expression in the tissues of the embryos or the animals they give rise to.

Before the advent of recombinant DNA techniques, the only sources of large amounts of pure genes were viruses. In the early 1970s, DNA of SV40 was microinjected into very early mouse embryos at the blastocyst stage. The injected blastocysts were then reimplanted into the uteri of foster mothers and allowed to develop. About 40 percent of the mice that resulted had SV40 DNA in some of their cells. The mice were mosaic: In each individual tissue, some cells had SV40 DNA in their chromosomes and other cells did not. This proved that foreign DNA injected in very early embryos is incorporated stably into the chromosomes of some embryonic cells and maintained in them as they differentiate into adult tissues.

Chromosomal Integration of Foreign Genes

Once genes were purified by cloning and characterized by sequence analysis, there was every incentive to microinject them into mouse embryos

and follow their fate. Reflecting many improvements over the past decade in the techniques of handling early embryos and eggs, the usual procedure nowadays is to microinject the cloned genes into one of the two pronuclei of freshly fertilized mouse eggs. The male pronucleus, contributed by the fertilizing sperm, is usually chosen since it is larger. The injected fertilized egg is then either immediately implanted into the oviduct of a foster mother, or it is allowed to develop in culture to the blastocyst stage before being implanted into the uterus (Figure 16-1).

This type of experiment has been done with several cloned genes, including human interferon and insulin genes, rabbit β-globin genes, the thymidine kinase gene of herpes simplex virus, and mouse leukemia virus cDNA. In different experiments, the number of foreign DNA molecules inserted into each egg has ranged from about 100 to 30,000, and the size of the individual DNA molecules has ranged from 5,000 to about 50,000 base pairs. The percentage of eggs that survive the manipulation and develop to term varies, but it is usually between 10 and 30 percent. The proportion of these mice that have the foreign DNA integrated into their chromosomes is highly variable, from a few percent to about 40 percent. Thus, the present state of the art allows foreign DNA to be introduced into mice in this way with an overall efficiency of about 10 percent—a statistic that is slowly improving.

Integration of Foreign Genes Is Not Chromosome-Specific

In experiments done so far with different foreign

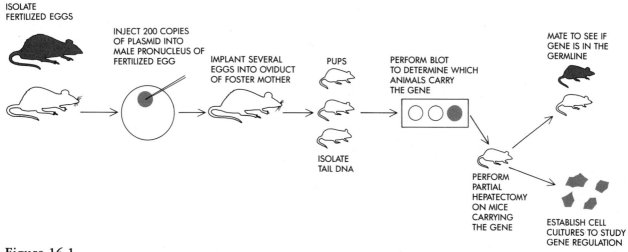

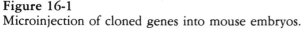

Figure 16-1
Microinjection of cloned genes into mouse embryos.

genes injected into oocytes, multiple copies of the foreign DNA are usually found integrated into one chromosome of the diploid set in the adult mouse. In any one experiment, the chromosome carrying the foreign DNA varies from mouse to mouse, so the integration event does not appear to be specific in any way. The number of copies of the foreign gene present in different animals resulting from eggs treated in identical ways can range from a few to over a hundred. The foreign genes are usually in a head-to-tail tandem array, with most copies intact. The amount of DNA that can be integrated and tolerated is quite substantial; in one experiment, for example, mice with as many as 30 to 50 tandem copies of a DNA insert 50,000 base pairs long were produced. In such situations it is easy to localize the integrated DNA on the recipient chromosome by *in situ* hybridization: One simply uses cells that have been blocked in cell division in such a way that each chromosome is clearly resolvable in the light microscope. These "metaphase squashes" are hybridized to a radioactively labeled copy of the foreign DNA. Autoradiography will produce grains over the location of the integrated genes.

Foreign DNA Is Stably Integrated into the Germ-Line Cells

Following microinjection into the male pronucleus of a fertilized egg, the foreign DNA is usually found in both somatic and germ-line cells (egg and sperm). It is thus transmitted through egg and sperm as a Mendelian trait (Figure 16-2, page 202). This implies that the integration event occurs very early in development, perhaps even during the first cell division of the zygote, before the germ cell population destined to give rise to eggs or sperm is segregated from the primordial somatic cells. In several cases the foreign DNA has been transmitted through three generations of mice. It is therefore stably integrated into the recipient mouse's genome, even though the foreign genes confer no selective advantage. As we shall see, however, in the passage from one generation to the next, the foreign DNA may suffer complex and as yet ill-defined modifications.

Expression of Foreign DNA in Mice

Once it was shown that foreign genes can be stably integrated into the somatic and germ cells of adult mice, the next step was to determine if these genes are expressed and whether their expression is regulated. Clearly, in the experiments we have been describing there is a very strong selection for the DNA integration events that are compatible with subsequent normal embryonic development. Any integration event that disrupts normal development will kill the embryo. It would not be surprising, therefore, if the injected eggs that develop to term were ones in which the foreign DNA is integrated in such a way that it is either genetically silent and unexpressed—a passive passenger—or normally regulated in its expression.

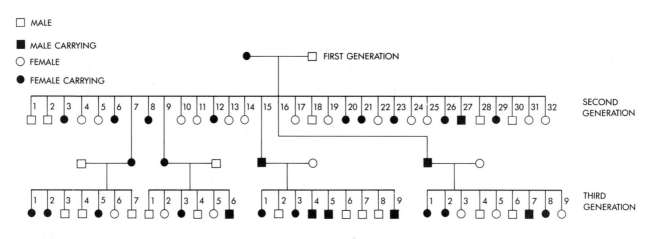

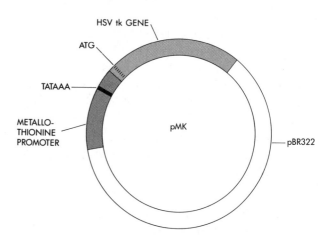

Figure 16-2
Microinjected foreign DNA can be transmitted through the germ line. Mice that had incorporated the pMK fusion gene were mated with mice that had not received the gene. Several offspring were found to have retained the pMK gene through the third generation.

The results to date furnish examples of both situations. In some of the experiments with cloned rabbit β-globin genes, no expression of the multiple tandem integrated copies of rabbit β-globin genes was found; no trace of β-globin mRNA or of β-globin polypeptide was detected either in erythroid red blood cell tissues or in other tissues. However, in other experiments with rabbit β-globin genes that had been cloned in a different vector, there was indirect evidence of a small amount of rabbit β-globin in the erythroid cells of some mice. Following injection of β-globin genes that had been cloned in still another vector, expression in muscle cells but not in erythroid cells was observed. Moreover, in a fetus that had integrated the herpes simplex thymidine kinase gene, the viral enzyme was detected at a low level, representing about a third of that of the endogenous mouse thymidine kinase. In these examples, there seems to be very little evidence for tissue-specific or regulated expression of the foreign gene.

Far more compelling evidence for the tissue-specific expression of integrated foreign genes was obtained in two other sets of experiments, one with a so-called fusion gene, and the other with mouse leukemia virus DNA.

Expression of the MK Fusion Gene Following Microinjection

The "MK" fusion gene was constructed by attaching the regulatory sequences of a mouse gene coding for the metal-binding protein metallothionine I in front of the coding sequence for the thymidine kinase of herpes simplex virus (Figure 16-3). It was hoped that the expression of this gene following microinjection would be regulated by the molecular signals that induce the synthesis of metallothionine in mouse cells, but that instead of producing metallothionine the cloned sequences would produce viral thymidine kinase. These expectations were borne out when the fusion gene was microinjected into the nuclei of mouse eggs in a culture that, several hours later, was exposed to cadmium ions. Viral thymidine kinase activity was

Figure 16-3
The pMK fusion plasmid was created by ligating the promoter region of the mouse metallothionine gene to the structural gene for herpes virus thymidine kinase.

detectable in the mouse eggs. (Cadmium is one of several heavy metals that, together with glucocorticoid hormones, are the natural inducers of metallothionine synthesis. The physiological role of metallothionine is to bind ions of heavy metals such as cadmium and mercury to protect the body from poisoning, and also to deliver zinc and copper ions to enzymes that need them.)

The Tissue-Specific Expression of the MK Gene

When the fusion gene was microinjected into the nuclei of fertilized eggs that were then implanted into foster mothers, some 10 to 15 percent of the progeny mice had the fusion gene integrated into a chromosome. In several of these mice, synthesis of viral thymidine kinase could be induced by injections of cadmium sulfate solution (Figure 16-4). The amount of induced enzyme was higher than the normal level of endogenous mouse thymidine kinase by a factor ranging from 7 to 100, but there was no correlation between the amount of enzyme and the number of tandem integrated copies of the fusion gene, which ranged in different animals from 1 or 2 to over 100. The important finding was that the induced expression of the fusion gene was tissue-specific, being greatest in the liver, considerably less in the kidney, and absent in the brain. This is precisely the pattern of tissue-specific expression of the endogenous metallothionine gene in mice.

Altered Expression of the MK Gene in Progeny

Breeding experiments proved that in each mouse, the MK genes were integrated into only one chromosome of the diploid set. Only half the offspring of matings between normal mice and mice with the MK gene inherited the latter. The responses of the first- and second-generation MK+ mice to cadmium sulfate were complex. In some offspring the amount of viral thymidine kinase activity that could be induced was greater than the induced level in the parent, whereas in other offspring no induction was observed. Apparently, during the passage of the genes through egg and sperm and embryonic development, they undergo changes that may result in either the extinction or the en-

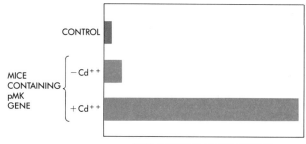

LIVER THYMIDINE KINASE ACTIVITY

Figure 16-4
Regulation of HSV tk activity by Cd^{++}. When mice that were found to have incorporated the pMK fusion gene into their liver DNA were injected with Cd^{++}, the normal inducer of the metallothionine promoter, the hepatic tk activity was greatly increased.

hancement of their expression. We do not at present know what these crucial changes are, but the fact that they occur may provide the opportunity to discover how, during passage through the embryo, the potential of a gene to be expressed is permanently enhanced or suppressed. That is, of course, the essence of tissue differentiation. The challenge now is to identify changes at the DNA level—for example, changes in the base sequence of regulatory regions, in the pattern and extent of methylation, or in the packaging of the MK genes in nucleosomes, or even changes in the position of the genes along the chromosome—and to correlate these changes with alterations in the expression of the gene.

Functional Expression of the MGH Fusion Gene

Even more exciting are the very recent experiments using a fusion gene containing the 5′ regulatory sequences of the mouse metallothionine (MT) gene attached to the coding sequences of the rat growth hormone (GH) gene. Insufficient amounts of growth hormone lead to dwarfism and excessive amounts lead to gigantism. Growth hormone is normally synthesized in the pituitary gland and travels to the liver, where it promotes the synthesis of a second set of hormones, the somatomedins, which in turn stimulate the growth of mesodermal tissues such as muscle, cartilage, and bone. Dwarfism can be partially reversed by the administration of growth hormone, and pre-

sumably, excessive exogenously added growth hormone would lead to abnormally large individuals.

A portion of a cloned rat growth hormone gene lacking its 5' regulatory region was fused to the MT regulatory region to generate the plasmid pMGH (Figure 16-5). The fusion was done in such a way that it led to the synthesis of an RNA molecule containing a stretch of MT untranslated bases, followed by the growth hormone AUG initiation codon and the remainder of the growth hormone gene, including its five intervening sequences and its poly-A-addition site. For injection into mouse eggs, a linear 5-kilobase fragment of the pMGH plasmid containing the MGH fusion gene was used, since there has been a growing consensus that linear DNA fragments containing free ends integrate into host DNA more effectively than circular plasmids (Chapter 12, page 158). Each male pronucleus was injected with about 600 copies of the fusion gene, and 170 fertilized eggs were inserted into the uteri of foster-mother mice. From these eggs, 21 animals developed.

The mice were checked for the presence of the MGH DNA by Southern blot hybridization, using DNA extracted from portions of their tails. Seven of the mice were positive, and most possessed multiple MGH copies. Six of these mice grew faster than their normal litter mates, even before heavy metals were added as inducers (Figure 16-6). (As yet, the effect of the heavy metals is not known.) As was expected from their fast

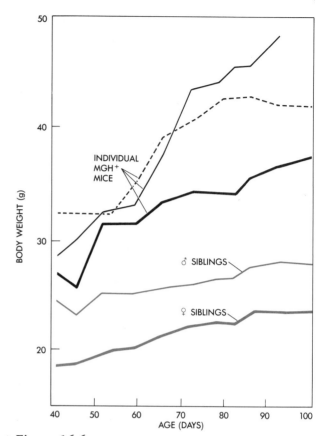

Figure 16-6
Mice that had incorporated the pMGH plasmid grew much faster than their normal litter mates.

growth, elevated levels of growth hormone mRNA were found in these animals, and analyses of growth hormone protein revealed levels 100 to 800 times those found in control mice. There thus seems no doubt that the increased growth of these transgenic mice resulted from the integration and expression of the metallothionein growth hormone fusion gene.

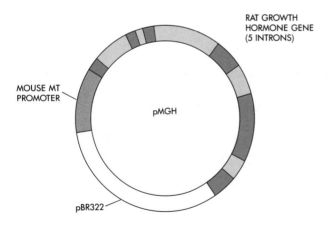

Figure 16-5
The pMGH fusion plasmid was created by ligating the promoter region of the mouse metallothionein gene to the structural gene for rat growth hormone.

Integration of MuLV DNA

The genome of the weakly oncogenic murine (mouse) leukemia virus (MuLV) DNA has also been introduced into the germ and somatic cells of mice and expressed in a tissue-specific manner. The infectious particles of MuLV, like the other retroviruses (Chapter 10), carry their genetic information in single-stranded RNA molecules. Infection involves the synthesis and chromosomal integration of a double-stranded DNA provirus. Three sorts of experiments have been done with MuLV and mouse eggs or embryos—infection

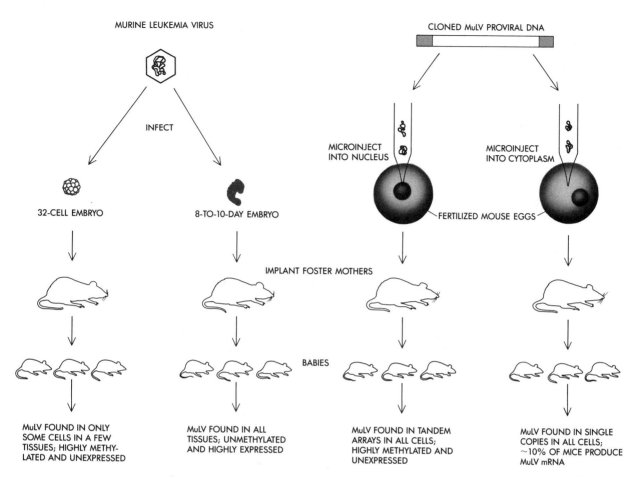

Figure 16-7
Different fates of murine leukemia viral DNA, depending upon the time and method of its introduction into embryos.

with the virus, cytoplasmic microinjection of proviral DNA, and nuclear microinjection of proviral DNA—with strikingly different results (Figure 16-7).

Early Embryos Infected with MuLV

When very early mouse embryos (at a 32-cell stage) are infected with MuLV particles and reimplanted in foster mothers, up to half of the progeny mice have MuLV proviral DNA in their somatic and germ cells. The animals are invariably mosaic; only some cells in each tissue have the viral DNA, and in the cells that do, there is only a single copy of it in one chromosome. The site of integration varies from animal to animal, but the DNA is always highly methylated. In the majority of animals the viral DNA is never expressed, but in some it is expressed during embryogenesis, shortly after birth, or very late in life. At other sites ex-

pression is completely extinguished. Apparently, therefore, the chromosomal location of a methylated gene is important for its activation (by demethylation?) during development.

In contrast to these results obtained by infection of very early embryos, when 8-to-10-day-old embryos are infected *in utero* with MuLV, highly expressed unmethylated DNA is found in all tissues, with every cell making progeny virus. By this stage in embryonic development, the virus acts as a straightforward infectious agent that the late-stage embryonic cells can no longer methylate or genetically suppress. This suggests that methylation is at least part of the mechanism of genetic suppression.

Microinjection of Proviral DNA

MuLV proviral DNA with flanking mouse chromosomal DNA sequences was cloned from the

cells of one of the mice from the experiments just described. (The mouse was one in which activation of its MuLV DNA occurred soon after birth.) The proviral DNA and flanking mouse sequences were microinjected into the nuclei of fertilized eggs; the mice obtained had 20 to 40 tandem integrated copies of the injected DNA, flanking sequences and all, at single sites on single chromosomes in all cells. The integrated DNA was highly methylated in all the tissues tested and no expression of the viral genome occurred.

By contrast, microinjection into the *cytoplasm* of fertilized eggs resulted in mice with just single copies of MuLV DNA integrated into all cells. The flanking mouse sequences of the microinjected DNA were not detectable. These mice were not mosaic; every cell had one copy of MuLV DNA, in striking contrast to the mosaic animals that resulted when 32-cell embryos were infected with MuLV particles. Apparently the microinjected DNA in the cytoplasm gives rise to viral RNA, which in turn gives rise to a new provirus that is integrated, perhaps at the 2-cell stage.

Only 3 of the 30 mice produced MuLV after cytoplasmic injection; they became viremic three weeks after birth. In each case, the MuLV DNA was at different chromosomal sites. One of the three early-viremic mice was checked at 5½ months; viral RNA was found in all tissues tested, including the brain and testis, but in very different amounts. In this animal, muscle tissue had at least five times more viral RNA than any other tissue. Apparently, therefore, the site of chromosomal integration influences the expression of the viral DNA in a tissue-specific manner.

First Implications of Implanting Genes in Fertilized Eggs

In summary, although the study of the fate of foreign genes introduced into mice by microinjecting DNA into fertilized eggs has hardly begun, we can conclude that:

1. Genes can be introduced in this way into somatic and germ-line cells.
2. The site at which the DNA integrates in a chromosome, as well as the extent of its methylation, influences its activation dur-

ing development. (Absence of reduction of methylation correlates with gene expression.)
3. The extent of the expression of foreign genes is also influenced by tissue differentiation and can be tissue-specific.
4. Appropriate foreign genes can be placed under the normal regulatory mechanisms of the host cells.

Finally, our ability to obtain any segment of DNA by gene cloning, coupled with the techniques that allow manipulation of eggs and injection of genes into them, may well prove to have a direct practical application. Such techniques are highly likely to be used in animal breeding to introduce desired genes into cattle, sheep, or other domestic animals. It would not be surprising if implantation of genes into embryos became an adjunct of conventional animal breeding during the next decade.

"Cloning" Animals

Throughout this book we have frequently used the words "cloned," "cloning," and "clone"; recently they have taken on new and, in some quarters, highly emotive meanings. The word "clone" was first used to describe a population of cells or organisms all derived from a single cell or organism by asexual multiplication in such a way that all the individuals in the clone have the same genetic constitution. When it became possible by recombinant DNA techniques to take a gene or any other segment of DNA and multiply it in bacteria, the term "gene cloning" seemed appropriate.

The clones that molecular biologists deal with are usually of bacteria or other microorganisms, cells in tissue culture, and, lately, DNA molecules. Gardeners and plant breeders, on the other hand, regularly handle and produce cloned higher organisms, the plants that they propagate asexually by cutting, grafting, splitting tubers and rhizomes, and so on. Higher plants naturally lend themselves to asexual reproduction and to cloning; for many species in the wild, asexual propagation is more important than sexual propagation. In contrast, higher animals do not naturally reproduce asexually. To clone an animal, it is necessary to remove surgically, or totally inactivate with radiation, the nucleus in a fertilized egg, and replace it with a

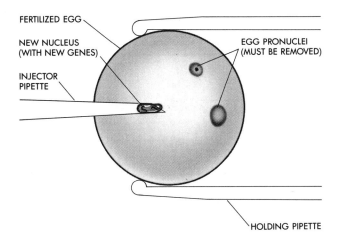

Figure 16-8
Transplanting nuclei to "clone" a frog. A fertilized egg is obtained and its two pronuclei are removed surgically or are inactivated with x-rays. A nucleus from another frog is then inserted into the egg.

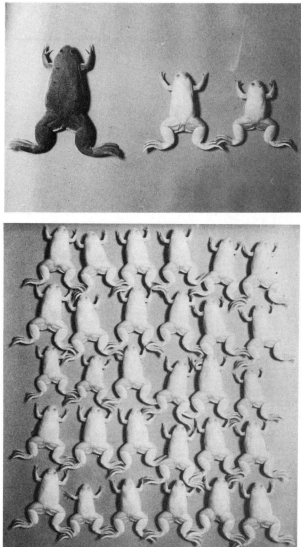

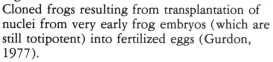

Figure 16-9
Cloned frogs resulting from transplantation of nuclei from very early frog embryos (which are still totipotent) into fertilized eggs (Gurdon, 1977).

nucleus taken from another individual. It requires transplantation of a whole intact and developmentally competent nucleus (Figure 16-8).

Experiments of this sort were first done with frogs' eggs, large cells that are easy to obtain and relatively easy to manipulate. The results showed that as the donor cells from which nuclei are taken become progressively more committed to particular developmental pathways, their nuclei lose the capacity to replace the fertilized egg nucleus. Thus, nuclei transplanted from cells of very early frog embryos that are still totipotent can give rise to adult frogs (Figure 16-9). On the other hand, nuclei transplanted from the cells of *adult* frogs have so far never promoted the development of an adult animal; the developmental process always fails at some embryonic or larval stage.

Nuclear transplantation with frogs' eggs was first achieved in 1952, but of course it would be much more interesting to be able to reproduce mammals asexually, rather than frogs. The technical problems of mammal reproduction by nuclear transplantation, however, are much greater simply because it is extremely difficult to manipulate mammalian eggs without damaging them. Figure 16-10 (page 208) outlines the sorts of steps that would be necessary. In 1981, one such set of ex-

periments with mice was reported, but it has not yet been repeated and independently confirmed. Until the methods can be made reproducible, they will not contribute significantly to our understanding of mammalian development.

In the immediate future there is little likelihood of nuclear transplantations being attempted with any other mammalian species. If the efficiency and reproducibility can be improved, the method

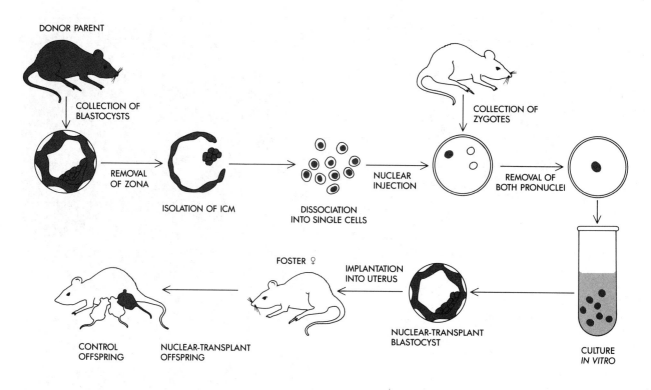

Figure 16-10
A potential scheme for cloning mice by nuclear transplantation. Totipotent mouse embryo nuclei must be collected from a pregnant female mouse. Blastocysts are obtained, and nuclei are removed from these cells by using a micropipette and injected into another fertilized mouse egg. The genome (the male and female pronuclei) of the recipient egg is then removed. The transplanted embryos are cultured until the blastocyst stage and then implanted into foster mothers.

may, however, find a place in animal breeding. In theory it could be attempted with human eggs and embryonic cells, but for what reason? There is no practical application. And it needs to be stressed that it has not proved possible, even with frogs, to produce a cloned adult by transplanting into an egg the nucleus of an *adult* cell. The gothic plot of the aged millionaire persuading doctors to clone several copies of himself by transplanting nuclei of his cells into fertilized eggs and then implanting them into women remains pure fantasy. This cannot even be done for old toads.

READING LIST

Original Research Papers (Reviews)

INJECTION OF SV40 DNA INTO EARLY EMBRYOS

Jaenish, R., and B. Mintz. "Simian Virus 40 DNA sequences in DNA of healthy adult mice derived from preimplantation blastocysts injected with viral DNA." *Proc. Natl. Acad. Sci. USA,* 71: 1250–1254 (1974).

INTEGRATING CLONED GENES

Gordon, J. W., G. A. Scangos, D. J. Plotkin, J. A. Barbos, and F. H. Ruddle. "Genetic transformation of mouse embryos by microinjection of purified DNA." *Proc. Natl. Acad. Sci. USA,* 77: 7380–7384 (1980).

Costantini, F., and E. Lacy. "Introduction of a rabbit β-globin gene into the mouse germ line." *Nature,* 294: 92–94 (1981).

Gordon, J. W., and F. H. Ruddle. "Integration and stable germ line transmission of genes injected into mouse pronuclei." *Science,* 214: 1244–1246 (1981).

Wagner, E. F., T. A. Stewart, and B. Mintz. "The human β-globin gene and a functional viral thymidine kinase gene in developing mice." *Proc. Natl. Acad. Sci. USA,* 78: 5016–5020 (1981).

Wagner, T. E., P. C. Hoppe, J. D. Jollick, D. R. Scholl, R. L. Hodinka, and J. B. Gault. "Microinjection of a rabbit β-globin gene into zygotes and its subsequent expression in adult mice and their offspring." *Proc. Natl. Acad. Sci. USA,* 78: 6376–6380 (1981).

Stewart, T. A., E. F. Wagner, and B. Mintz. "Human β-globin sequences injected into mouse egg, retained in adult, and transmitted to progeny." *Science,* 217: 1046–1048 (1982).

THE MK FUSION GENE

Brinster, R. L., H. Y. Chen, M. Trumbauer, A. W. Senear, R. Warren, and R. D. Palmiter. "Somatic expression of herpes thymidine kinase in mice following injection of a fusion gene into eggs." *Cell,* 27: 223–231 (1981).

Brinster, R. L., H. Y. Chen, R. Warren, A. Sarthy, and R. D. Palmiter. "Regulation of metallothionein-thymidine kinase fusion plasmids injected into mouse eggs." *Nature,* 296: 39–42 (1982).

Palmiter, R. D., H. Y. Chen, and R. L. Brinster. "Differential regulation of metallothionein-thymidine ki-nase fusion genes in transgenic mice and their offspring." *Cell,* 29: 701–710 (1982).

Palmiter, R. D., R. L. Brinster, R. E. Hammer, M. E. Trumbauer, M. G. Rosenfeld, N. C. Birnberg, and R. M. Evans. "Dramatic growth of mice that develop from eggs microinjected with metallothionein-growth hormone fusion genes." *Nature,* 300: 611–615 (1982).

MuLV DNA

Jaenisch, R. "Germ line integration and Mendelian transmission of the exogenous Moloney leukemia virus." *Proc. Natl. Acad. Sci. USA,* 73: 1260–1264 (1976).

Jaenisch, R. "Moloney leukemia virus gene expression and gene amplification in preleukemic and leukemic BALB/Mo mice." *Virology,* 93: 80–90 (1979).

Breindl, M., J. Doehmer, K. Willecke, J. Dausman, and R. Jaenisch. "Germ line integration of Moloney leukemia virus: Identification of the chromosomal integration site." *Proc. Natl. Acad. Sci. USA,* 76: 1928–1942 (1979).

Jaenisch, R., D. Jahner, P. Nobis, I. Simon, J. Lohler, K. Harbers, and D. Grotkopp. "Chromosomal position and activation of retroviral genomes inserted into the germ line of mice." *Cell,* 24: 519–529 (1981).

Harbers, K., D. Jahner, and R. Jaenisch. "Microinjection of cloned retroviral genomes into mouse zygotes: Integration and expression in the animal." *Nature,* 293: 540–542 (1981).

Jahner, D., H. Stuhlmann, G. L. Stewart, K. Harbers, J. Lohler, I. Simon, and R. Jaenisch. "*De novo* methylation and expression of retroviral genomes during mouse embryogenesis." *Nature,* 298: 623–628 (1982).

"CLONING" ANIMALS

Gurdon, J. B. "The developmental capacity of nuclei taken from intestinal epithelium cells of feeding tadpoles." *J. Embryol. Exptl. Morphol.,* 10: 622–640 (1962).

Gurdon, J. B., and B. Uehlinger. " 'Fertile' intestine nuclei." *Nature,* 210: 1240–1241 (1966).

Gurdon, J. B. "Changes in somatic cell nuclei inserted into growing and maturing amphibian oocytes." *J. Embryol. Exptl. Morphol.,* 20: 401–414 (1968).

Gurdon, J. B., F.R.S. "Egg cytoplasm and gene control." *Proc. Roy. Soc. Lond. B,* 198: 211–247 (1977).

DiBerardino, M. A. "Genetic stability and modulation of metazoan nuclei transplanted into eggs and

oocytes." *Differentiation,* 17: 17–30 (1980). (Review.)

Stewart, T. A., and B. Mintz. "Successive generations of mice produced from an established culture line of euploid teratocarcinoma cells." *Proc. Natl. Acad. Sci. USA,* 78: 6314–6318 (1981).

Illmensee, K., and P. C. Hoppe. "Nuclear transplantation in Mus musculus: Developmental potential of nuclei from preimplantation embryos." *Cell,* 23: 9–18 (1981).

Marx, J. L. "Three mice 'cloned' in Switzerland." *Science,* 211: 375–376 (1981). (News report.)

Edward, R. G. "Test-tube babies." *Nature,* 293: 253–256 (1981). (Review.)

17

Recombinant DNA and Genetic Diseases

One of the major medical applications of gene cloning that we can safely predict will be in the diagnosis of genetic diseases in fetuses, so that abortion can be offered to prevent the birth of incurably sick children. In the very distant future, it may also prove possible to treat some genetic diseases by transplanting normal genes into the cells of genetically sick people. Several genetic diseases are already being diagnosed at the DNA level, and the potential for further developments in this area is enormous.

Mendelian Inheritance

Sickle-cell anemia, muscular dystrophy, and cystic fibrosis are but three examples of the over 500 known genetic diseases that result from recessive mutations in single genes. Typically they afflict babies and young people, causing physically and often mentally crippling symptoms, and usually killing their victims before they reach sexual maturity. Although most genetic diseases are rare, together they represent an enormous burden of suffering. At a rough estimate, about half a million afflicted children are born each year throughout the world.

Single-gene genetic diseases run in families, and the Mendelian pattern of their inheritance over three or more generations proves that each is caused by a recessive mutation in an individual gene. For example, in Caucasian populations, about 1 in 20 people carry on one chromosome the recessive mutation that gives rise to cystic fibrosis. These heterozygous carriers do not suffer from cystic fibrosis because they also have the corresponding dominant normal gene on the other homologous chromosome. The problems arise when two carriers have children. In accordance with Mendel's rules, each pregnancy involving a pair of heterozygous carriers has a one-in-four chance of giving rise to a baby with cystic fibrosis. If we assume random partnering, about 1 in 400 marriages is at risk ($\frac{1}{20} \times \frac{1}{20}$) and 1 in 1600 babies will have the disease ($\frac{1}{400} \times \frac{1}{4}$). In fact, in populations of Caucasian stock, the incidence of the disease is about 1 in 2000 live births.

Inborn Errors of Metabolism

The recessive mutations that cause Mendelian genetic diseases inactivate enzymes or other essential proteins such as hemoglobin, growth hormone, or blood-clotting factors. In about 200 of the more than 500 single-gene diseases that have been identified by studies of family pedigrees, we know precisely which protein is defective. In sickle-cell anemia it is the β-globin chain of hemoglobin; in phenylketonuria, the enzyme phenylalanine hydroxylase; in Tay-Sachs disease, the enzyme hexosaminidase A; in classic hemophilia, blood-clotting factor VIII; and so on (Table 17-1, page 212). On the other hand, we are still ignorant of the primary biochemical lesion in over half of these diseases, including cystic fibrosis and muscular dystrophy.

The immediate consequence of inactivating an enzyme is, as Archibald Garrod realized as early as 1908 when he coined the term "inborn errors of metabolism," the failure of the metabolic path-

Table 17-1

Some Common Genetic Diseases*

Inborn Errors of Metabolism	Approximate Incidence Among Live Births	
1. Cystic fibrosis (mutated gene unknown)	$\frac{1}{1600}$	Caucasians
2. Duchenne muscular dystrophy (mutated gene unknown)	$\frac{1}{3000}$	boys (X-linked)
3. Gaucher's disease (defective glucocerebrosidase)	$\frac{1}{2500}$	Ashkenazi Jews, $\frac{1}{75,000}$ others
4. Tay-Sachs disease (defective hexosaminidase A)	$\frac{1}{3500}$	Ashkenazi Jews, $\frac{1}{35,000}$ others
5. Essential pentosuria (a benign condition)	$\frac{1}{2000}$	Ashkenazi Jews, $\frac{1}{50,000}$ others
6. Classic hemophilia (defective clotting factor VIII)	$\frac{1}{10,000}$	boys (X-linked)
7. Phenylketonuria (defective phenylalanine hydroxylase)	$\frac{1}{5000}$	among Celtic Irish, $\frac{1}{15,000}$ others
8. Cystinuria (mutated gene unknown)	$\frac{1}{15,000}$	
9. Metachromatic leukodystrophy (defective arylsulfatase A)	$\frac{1}{40,000}$	
10. Galactosemia (defective galactose-1-phosphate uridyl transferase)	$\frac{1}{40,000}$	

Hemoglobinopathies	Approximate Incidence Among Live Births	
1. Sickle-cell anemia (defective β-globin chain)	$\frac{1}{400}$	U.S. blacks. In some West African populations the frequency of heterozygotes is 40%.
2. β-Thalassemia (defective β-globin chain)	$\frac{1}{400}$	among some Mediterranean populations

*Although the vast majority of the over 500 recognized recessive genetic diseases are extremely rare, in combination they represent an enormous burden of human suffering. As is consistent with Mendelian mutations, the incidence of some of these diseases is much higher in certain racial groups than in others.

way in which that enzyme functions. The substrate of the inactivated enzyme accumulates and is very often shunted into alternative metabolic routes. The situation is analogous to a complex road network at rush hour; an accident blocking one road can disrupt the whole system, as commuters crowd onto alternative roads in an attempt to avoid the traffic jam. The normally integrated and balanced metabolism of the cell is disrupted. Some metabolites accumulate in abnormally high amounts, whereas others are severely or completely deficient. The body begins to poison itself in this way, and the manifold and characteristic symptoms of each disease develop.

Treatment of Inborn Errors of Metabolism

A handful—but only a handful—of these genetic diseases can be successfully treated or managed because we understand their biochemistry. Therapeutic strategies include supplying deficient metabolites (for example, the vitamin biotin in certain diseases that inactivate carboxylase enzymes), supplying proteins (such as clotting factor VIII in classic hemophilia), and strict dieting to regulate the intake of nutrients that can no longer be metabolized properly (for instance, phenylalanine in the case of phenylketonuria, and galactose

in galactosemia). However, the vast majority of these diseases—even those in which we know the primary biochemical lesion—remain untreatable. Either we cannot deliver the missing enzyme to the appropriate cells, or we still do not know exactly which of the many abnormal metabolites that accumulate causes the symptoms. In most cases the prognosis for afflicted children remains bleak.

Early Diagnosis and Abortion

In the absence of successful therapies, and because these diseases cause such suffering to the victims and their families, much effort has gone into the alternative approach, the early diagnosis of affected fetuses and the offer of abortion. When both parents are known to be heterozygous carriers—and all too often this discovery is made only after the birth of a genetically sick child—further pregnancies can be monitored closely if the parents wish to prevent the birth of another diseased child.

The basic approach to fetal diagnoses is to obtain fetal cells—either fibroblasts or blood cells taken by amniocentesis (Figure 17-1)—and to test them for biochemical defects. To date, over 40 different recessive genetic diseases have been

diagnosed prenatally. Currently the tests are ⎵⎵⎵ for the enzymes or other proteins suspected of being defective; that is, the tests are for the gene products rather than for the mutated genes themselves (Table 17-2, page 214). Of course, if the enzyme in question is not expressed in the fibroblasts or blood cells that are obtained by amniocentesis, it cannot be assayed. Moreover, tests for deficiency of an enzyme tell us nothing about the nature of the underlying mutation; for example, we don't know when a mutation is in a regulatory sequence or a transport protein rather than in the structural gene that codes for the amino acid chain of the enzyme.

DNA Analysis of Inherited Disorders

Gene cloning gives us the potential to diagnose these lethal recessive mutations directly. Once we have cloned a normal gene and determined its nucleotide sequence, we can map all the sites at which the gene will be cut by the many different restriction endonucleases. We can also use the cloned normal gene as a probe to fish out the corresponding gene from the DNA of the fetal cells being tested. Armed with an abundance of the normal gene, and given knowledge of its se-

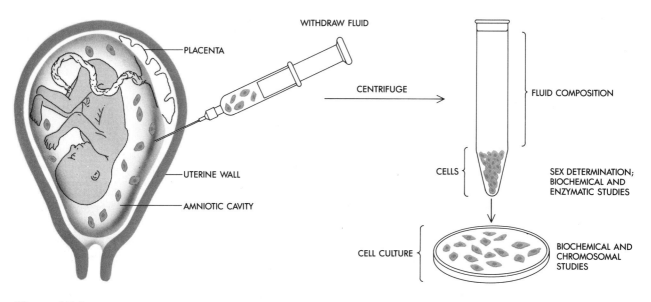

Figure 17-1
Amniocentesis. A sample of amniotic fluid (mostly fetal urine and other secretions) is taken by inserting a needle into the amniotic cavity during or around the 16th week of gestation. The fetal cells are separated from the fluid by centrifugation. The cells are then cultured so that a number of biochemical, enzymatic, and chromosomal analyses can be made.

Table 17–2

Genetic Diseases That Can Currently
Be Diagnosed Prenatally

Enzymopathies*	Hemoglobinopathies
Diseases of lipid metabolism	Sickle-cell anemia
There are eight, including:	β-Thalassemia
Tay-Sachs disease	
Gaucher's disease	
Krabbe's disease	
Fabry's disease	
Diseases of amino acid metabolism	
There are seven in all.	
Diseases of carbohydrate metabolism	
There are four, including:	
Galactosemia	
Diseases of mucosaccharide and mucolipid metabolism	
There are ten, including:	
Hurler's syndrome	
Hunter's syndrome	
San Filippo disease	
Other metabolic errors	
There are five, including:	
Lesch-Nyhan syndrome	
Xeroderma pigmentosum	

*Among the enzymopathies (another name for inborn errors of metabolism), by far the largest number of tests have been made for Tay-Sachs disease in Ashkenazi Jewish populations. The incidence of cases among the tested fetuses proves to be 25 percent, as is consistent with Mendel's laws.

quence and therefore a map of its restriction sites, we can devise screening procedures for mutations.

β-Thalassemias

The inherited disorders of hemoglobin synthesis, the thalassemias, have been the most extensively studied at the DNA level. As we have seen in Chapter 7 (page 99), the human hemoglobin polypeptides are encoded by two gene clusters: the α-like genes on the short arm of chromosome 10, and the β-like cluster on the short arm of chromosome 11. These regions have been cloned and large portions of them have been sequenced. Thalassemia is the name given to a condition in which there is a low level of, or sometimes a complete lack of, synthesis of one of the globin polypeptides. A defect in the rate of synthesis of a given protein could theoretically be due to an aberration in any one of the many steps in the pathway of gene expression: transcription, RNA processing, mRNA stability, or translation. Such aberrations may be due to point mutations leading to frameshift or nonsense mutations, or to deletion of all or part of the gene itself. Each of these types of defects has been found in naturally occurring β-thalassemias (Table 17-3).

Nonsense and Frame-Shift Mutations

β^0-Thalassemia, in which there is a complete absence of β-globin synthesis, has been found to be caused by a point mutation either in amino acid codon 17 or in amino acid codon 39. In both cases the mutation gives rise to a translation-termination codon. In addition to being untranslatable, these "nonsense" globin mRNAs tend to be extremely unstable *in vivo*. Other types of β^0-thalassemia have been found to be due to deletions or insertions in the β-globin coding sequence; these have resulted in a shift in reading frame (Table 17-3).

Transcription Mutations

Only two naturally occurring β-thalassemias that are due to mutations affecting transcription have been identified. One is a nucleotide substitution at −87 relative to the mRNA cap site. Patients with this mutation have an extremely low level of globin mRNA. This mutant gene, when transfected into human cells (Chapter 15, page 194), shows greatly lowered transcriptional activity compared to the normal gene. Another transcriptional mutant that has been found has a mutation in the "TATA box" (Chapter 7). This sequence, which is located at about −30 in virtually all eukaryotic genes, has a role in the initiation of transcription (Table 17-3).

RNA-Processing Mutations

Perhaps the most interesting class of β-thalassemias that has been discovered are those that are due to aberrant splicing of β-globin transcripts. The human β-globin gene has two introns that must be spliced out (Chapter 7). The dinucleotide sequence GT is always found at the 5′ (the donor) splice junction. Two different β^0-thalassemias in which this GT has been changed to AT, at the beginning of either the first or the second intron, have been identified (Table 17-3). In both cases, splicing at the correct site is prevented, and other donorlike sequences in the RNA are used. This, of course, leads to an incorrectly spliced (and untrans-

Table 17–3

Mutations in β-Thalassemia

Mutation Class	Type (β⁰ or β⁺)	Ethnic Origin
1. Nonsense mutation		
Codon 17	β^0	Chinese
Codon 39	β^0	Mediterranean
2. Frame-shift mutation		
-2 at codon 8	β^0	Turkish
-1 at codon 16	β^0	Indian
-1 at codon 44	β^0	Kurdish
-4 at codons 41/42	β^0	Indian
$+1$ at codons 8/9	β^0	Indian
3. Splice-junction substitution		
Intron 1, position 1: G→A	β^0	Mediterranean
Intron 2, position 1: G→A	β^0	Mediterranean
4. Consensus substitution		
Intron 1, position 5: G→C	β^+	Indian
Intron 1, position 6: T→C	β^+	Mediterranean
5. Internal intron substitution		
Intron 1, position 110	β^+	Mediterranean
Intron 2, position 705	β^+	Mediterranean?
Intron 2, position 745	β^+	Mediterranean
6. Transcriptional mutation		
-87 C→G	β^+ (probable)	Mediterranean
-28	?	Kurdish
7. Coding-region substitution affecting RNA processing		
Codon 26: G→A	β^ϵ	Southeast Asian
Codon 24: T→A	β^+	Black

latable) mRNA. Mutations can occur *within* an intron to create a new 3′ (acceptor) splice site that can compete with the correct 3′ site and cause aberrant splicing. Finally, mutations can occur in what would be the normal protein-coding region of the mRNA to create a new 5′ donor splice site that can compete with the normal one. Alterations in RNA splicing virtually always lead to mRNAs that cannot be translated into the correct protein.

Sickle-Cell Anemia

Sickle-cell anemia results from a mutation that changes a glutamic acid residue, coded by the triplet GAG, for a valine residue, coded by GTG, at position 6 in the β-globin chain of hemoglobin. As a result, sickle hemoglobin tends to crystallize in red blood cells, the cells become less flexible and are removed by the spleen, and anemia results. Fortunately, the mutation of A to T in the base

sequence of the β-globin gene eliminates a restriction site for the enzyme *Dde*I (as well as sites for other restriction enzymes). The sickle hemoglobin mutation can, therefore, be detected by digesting sickle-cell and normal DNA with *Dde*I and performing Southern blot hybridization. Normal DNA will generate two *Dde*I fragments of 201 and 175 base pairs, whereas sickle-cell DNA will generate only one fragment of 376 base pairs (Figure 17-2, page 216).

Although the restriction enzyme *Dde*I can be used to detect the sickle-cell mutation, it is not ideal for routine hospital screening procedures because there are too many *Dde*I restriction sites in the β-globin gene and its immediately neighboring DNA. As a result, digestion of the DNA produces a rather larger number of small DNA fragments that are relatively difficult to separate by gel electrophoresis. However, another convenient enzyme, *Mst*II, has been discovered. *Mst*II cuts at the

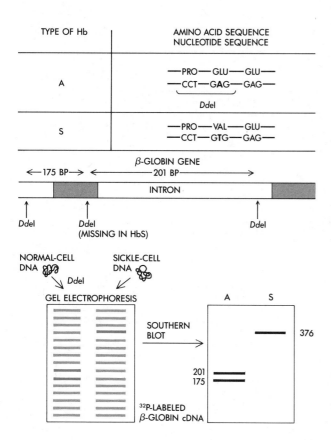

TYPE OF Hb	AMINO ACID SEQUENCE NUCLEOTIDE SEQUENCE
A	—PRO— GLU— GLU— —CCT— GAG— GAG— *Dde*I
S	—PRO— VAL— GLU— —CCT— GTG— GAG—

Figure 17-2
Detection of the sickle-cell globin gene by
Southern blotting. The base change (A → T)
that causes sickle-cell anemia destroys a *Dde*I site
that is present in the normal β-globin gene. This
difference can be detected by Southern blotting.

sequence CCTNAGG, which occurs less frequently than the *Dde*I site CTNAG. *Mst*II generates from normal DNA a 1.1-kilobase β-globin
gene fragment; in sickle-cell DNA this is replaced
by a 1.3-kilobase fragment. Fragments of this size
are more easily separated and recognized than the
considerably shorter ones *Dde*I produces. For that
very practical reason, *Mst*II has become the enzyme of choice for the direct detection of the sickle
globin mutation.

Diagnosis of α₁-Antitrypsin Deficiency with a Synthetic Oligonucleotide

Individuals who are deficient in the synthesis of a
protein called α$_1$-antitrypsin are greatly predis-
posed to developing pulmonary emphysema or in-
fantile liver cirrhosis. α$_1$-Antitrypsin, produced in

the liver, is somewhat misnamed, since its primary
physiological role is to inhibit a protease called
elastase, which is produced in polymorphonuclear
leukocytes. The balance between the levels of α$_1$-
antitrypsin and elastase is normally carefully controlled. In individuals with α$_1$-antitrypsin deficiency, this balance is upset and the elastic fibers of
the lung are slowly destroyed by elastase. As a
result, such individuals are 30 to 40 times more
prone to develop pulmonary emphysema. The α$_1$-
antitrypsin genes have been cloned from both normal individuals and an α$_1$-antitrypsin-deficient individual. Sequence analysis has revealed that the
mutant gene has a single base change (G → A) that
leads to an amino acid substitution (GLU →
LYS) at residue 324 (Figure 17-3). This apparently produces a nonfunctional protein. Unfortunately, the G → A substitution does not create or
destroy a restriction enzyme site, as is the case with
the sickle-cell anemia mutation, so prenatal screening using Southern blot analysis of restriction fragments to search for an altered restriction site is
impossible. However, a 19-base-long synthetic
oligonucleotide that is complementary to the
normal α$_1$-antitrypsin sequence around the mutation site can be prepared (Chapter 6, page 63).
The synthetic oligonucleotide will have one mismatch with the mutant gene. This oligonucleotide
can be labeled and used as a probe to distinguish
normal from mutant genes by raising the stringency of hybridization to a level at which the
oligonucleotide will hybridize stably to the normal
gene, to which it is perfectly complementary, but
not to the mutant gene, with which it has the single
mismatch (Figure 17-3).

As more mutant genes that are responsible for
genetic defects are cloned and sequenced, this
technique could prove to be very versatile for
prenatal diagnosis. It must be emphasized that this
type of analysis will be useful only if it can be
determined that a particular disease is always due
to the same underlying point mutation (as in sickle-cell anemia). In the case of the thalassemias,
which as we have seen are due to many (about 60)
different mutations, oligonucleotide probes specifically tailored to many segments of the wild-type
β-globin gene will have to be synthesized. Fortunately, only one form of thalassemia effectively
exists within a given geographic region, and here
specific oligonucleotides have already proved very

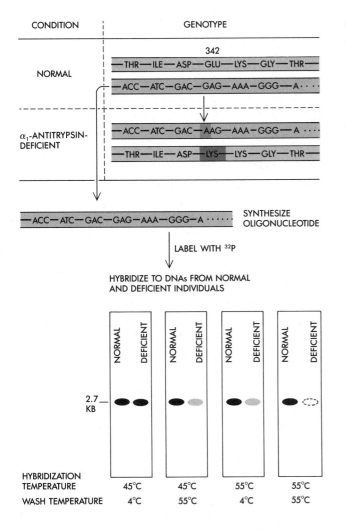

Figure 17-3
Diagnosis of α_1-antitrypsin deficiency by using a synthetic oligonucleotide. DNA from a patient with α_1-antitrypsin deficiency was analyzed and found to have a single base change (G → A) in the protein-coding region. A 19-base oligonucleotide was synthesized chemically to be complementary to the normal gene (and therefore to have one mismatch with the mutant gene). This oligonucleotide can distinguish between the normal and mutant genes when it is used as a probe in Southern blot analysis (Chapter 6).

useful. In Sardinia, the frequency of babies born with thalassemia has already drastically declined.

Citrullinemia Is Correlated with Aberrant Argininosuccinate Synthetase mRNA

The disease citrullinemia, characterized by ammonia poisoning, mental retardation, and early death, is due to the absence of the enzyme argininosuccinate synthetase, a urea cycle enzyme that catalyzes the conversion of citrulline and aspartate to argininosuccinate. Human cell lines resistant to the arginine analog canavanine were found to produce very high levels of argininosuccinate synthetase and correspondingly high levels of its mRNA. This made cloning of the cDNA relatively easy. The cloned cDNA was used to analyze DNA from citrullinemia patients using Southern blot hybridization. No systematic differences have been found between these individuals and normal patients. Cultured skin fibroblasts from citrullinemia patients show extremely low or undetectable argininosuccinate synthetase enzyme activity, but northern blot analysis of mRNA from these cells using the cloned cDNA as a probe revealed that they all contained an mRNA species that hybridized to the cDNA probe (Figure 17-4, page 218). In some of the mutant lines, however, the mRNA was shorter than normal; in others, the message was apparently aberrantly spliced. This situation is reminiscent of that with the β-thalassemias, and it may well be found that several different mutations in the gene for argininosuccinate synthetase may lead to citrullinemia. In all, five out of every six citrullinemia patients had a detectable alteration in the argininosuccinate synthetase mRNA. It is possible that such analysis of mRNA may prove feasible for prenatal diagnosis of this disease.

Diagnosing Mutations by Linkage

In the U.S. black population, the sickle mutation in the β-globin gene is associated 60 percent of the time with another mutation in the DNA near the β-globin gene. This second mutation eliminates a *Hpa*I restriction site that is present in people without the sickle trait. As a result, when the DNA of U.S. blacks with sickle-cell disease is digested with the enzyme *Hpa*I, the β-globin gene occurs in a

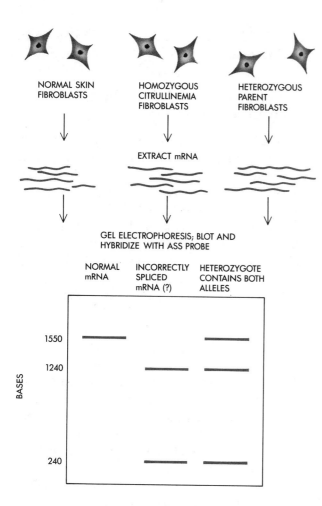

Figure 17-4
Altered argininosuccinate synthetase (ASS) mRNA in a citrullinemia patient. A comparison of mRNA from normal vs. mutant skin fibroblasts demonstrates that citrullinemia patients have incorrectly spliced (and therefore nonfunctional) mRNA for argininosuccinate synthetase.

fragment 13.0 kilobases long, whereas the DNA of people with normal hemoglobin produces a β-globin gene fragment 7.6 kilobases long (Figure 17-5). This linkage allowed the prenatal diagnosis of sickle-cell anemia in fetuses of U.S. black couples at risk. Subsequently the sickle mutation in U.S. blacks was found to be linked to two other polymorphic restriction sites for the enzyme *Hin*dIII in neighboring genes. By analyzing the pattern of the three restriction sites, accurate prenatal diagnosis could be offered to over 80 percent of the families at risk. These indirect tests for the sickle mutation have, of course, been su-

perseded by the direct assay for the mutation using *Mst*II. Nonetheless, the experience with sickle-cell disease shows that even when a mutation cannot itself be detected directly with a restriction enzyme or an oligonucleotide probe, analysis of genetically linked restriction sites often forms the basis of a diagnostic test.

Incidentally, during the work that led to the recognition of specific linkage between the sickle mutation and the *Hpa*I and *Hin*dIII sites in the globin gene cluster in U.S. blacks, the genetic background of the sickle mutation in people of other ethnic groups was also analyzed and found to be different from that of U.S. blacks. The U.S. black population originates from West Africa, where the frequency of the sickle β-globin gene in some regions exceeds 10 percent (Figure 17-6). In Ghana and Nigeria, the sickle globin gene is linked to the mutated *Hpa*I site, so when DNA from afflicted people in these countries is digested with *Hpa*I, the result is a 13.0-kilobase fragment containing the globin gene. On the other hand, in Gabon, Kenya, South Africa, Saudi Arabia, and India, the sickle globin gene is not associated with this mutated *Hpa*I site. When DNA from afflicted people in these regions is digested with *Hpa*I, a 7.6-kilobase fragment containing either the sickle or the normal globin gene is liberated. This implies that in Ghana and Nigeria the sickle-cell mutation arose in a population in which the mutation of the *Hpa*I site had already occurred, whereas in other parts of the world the sickle mutation arose independently in populations with a normal *Hpa*I site. Interestingly, in Ivory Coast and Sierra Leone the sickle-cell population is mixed, some people having the linked *Hpa*I mutation and others not.

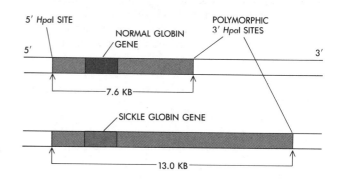

Figure 17-5
*Hpa*I-site polymorphism is diagnostic for the sickle β-globin gene in humans.

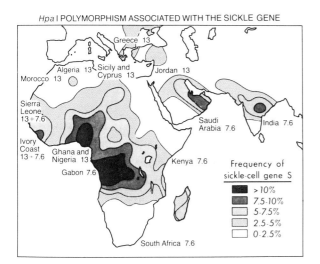

Figure 17-6
The sickle-cell mutation arose independently in different populations in Africa, in the Mediterranean region, and in western Asia (Kan, 1982).

The slave ships from Ghana and Nigeria called in at these places before crossing the Atlantic, and the sickle-cell mutation linked to the *Hpa* I mutation was introduced into a population with the sickle-cell mutation associated with a normal *Hpa* I site. Apparently, therefore, during the course of human history the GLU → VAL mutation at position 6 in the β-globin chain has occurred independently in different populations, and has spread through them because of the partial resistance the sickle-cell trait confers against malaria.

The Search for Mutated Genes

As soon as the normal genes for the proteins that are defective in other recessive genetic diseases have been cloned and sequenced, we will hopefully be able to devise either direct or indirect tests, at the DNA level, for the responsible mutations. For example, attempts are now being made, using all the relevant techniques of DNA manipulation and somatic-cell genetics, to pinpoint the gene defects that cause cystic fibrosis and muscular dystrophy. These are two of the most common and most devastating genetic diseases, and their biochemical bases are still a mystery.

Recently, the gene for human hypoxanthine–guanine phosphoribosyl transferase (HGPRT) has

been cloned, and this may lead eventually to new methods for diagnosing Lesch-Nyhan syndrome, an X-linked disorder caused by the lack of HGPRT and characterized by such symptoms as mental retardation and self-mutilation. HGPRT is an enzyme that is used to reutilize or "salvage" free guanine and hypoxanthine from the breakdown of nucleic acids. HGPRT-deficient individuals produce excessive levels of purines and excrete high levels of uric acid. It is, however, unclear why this leads to the clinical symptoms described above. A large fragment of the human HGPRT gene was obtained by screening a human genomic library with a cDNA for mouse HGPRT. Preliminary analysis using the human HGPRT gene fragments as hybridization probes to DNA from normal and Lesch-Nyhan individuals has thus far not revealed any consistent differences in restriction enzyme patterns between the two groups. Analysis of HGPRT mRNA from normal and mutant fibroblast lines, however, shows clear-cut differences between the two groups. Further analysis of the HGPRT gene in normal and Lesch-Nyhan individuals will hopefully reveal diagnostic differences at the DNA or RNA level.

Mapping Human Chromosomes

Both for the diagnosis of many genetic diseases and for the development of potential cures by gene transfer therapy, high-resolution genetic maps of each human chromosome will be essential. The first human gene ever to be mapped was one that causes color blindness; as long ago as 1911, it was proven by studies of the inheritance of color blindness in male and female members of afflicted families that the genetic trait causing this disability is sex-linked (resides on the X chromosome). Between 1911 and 1970, several other genes were mapped to the X chromosome; such genes are relatively easy to map by conventional Mendelian genetics because of the linkage to sex. It was, however, not until 1970 that a human gene was mapped to an autosomal (nonsex) chromosome by a purely Mendelian approach. Today some 500 human genes have been mapped, about a third of them to the X chromosome, and on the average the mapping of five more human genes is being reported each month. The reasons for this sudden increase in the rate of mapping human genes are

the development of somatic-cell genetics and, more recently, the advent of gene cloning.

Somatic-Cell Genetics

Mendelian genetics relies on analyzing the segregation of genetic traits among the progeny of sexual reproduction. With any species that breeds slowly, such breeding analysis is a protracted business, which is why experimental geneticists have turned to *E. coli,* yeast, *Drosophila,* or the mouse. Before somatic-cell genetics was developed, however, human geneticists had no option but to bide their time and study family pedigrees for the segregation of genetic markers (often genetic diseases). Today they are no longer restricted to this approach. Current methods allow

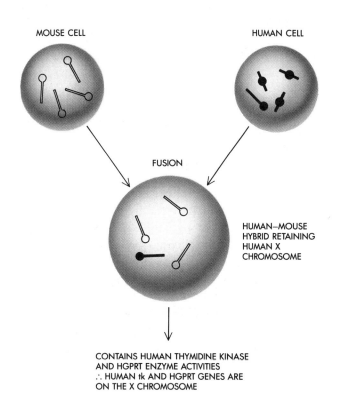

CONTAINS HUMAN THYMIDINE KINASE
AND HGPRT ENZYME ACTIVITIES
∴ HUMAN tk AND HGPRT GENES ARE
ON THE X CHROMOSOME

Figure 17-7
Assigning genes to chromosomes by somatic-cell hybridization. Human and mouse cells are fused (by using either Sendai virus or polyethyleneglycol), and the hybrid cells progressively shed chromosomes until they retain only one human chromosome. In cell hybrids retaining the human X chromosome, human thymidine kinase and HGPRT enzyme activities are found, indicating that these genes are on the X chromosome.

human cells in culture to be fused with cultured cells of other species. Moreover, segments of DNA ranging from whole chromosomes to cloned individual genes can be transferred into cultured human and animal cells. By exploiting these artificial modes of gene exchange, it is possible to correlate the presence of a human gene with the presence of a particular human chromosome or chromosomal fragment.

Human–Mouse Hybrid Cells

When human and mouse cells in culture are fused, the resultant hybrid cells survive and divide but progressively shed the human chromosomes. By correlating the presence of human protein—enzymes, structural proteins, or surface antigens, for example—in the hybrid cells with the presence of particular human chromosomes, genes can sometimes be assigned to individual chromosomes (Figure 17-7). Clearly, once a gene has been definitively located on a chromosome in this way, any other gene known from pedigree studies to be genetically linked to it can also be assigned to that chromosome. This procedure, which has led to the mapping of well over 100 human genes, depends on the *expression* in the hybrid cell of the gene being mapped. That is a severe restriction, despite the success of the method.

However, once a gene or segment of DNA has been cloned, it can be used as a hybridization probe for the same sequences on a chromosome in a hybrid cell; dependence on gene expression is obviated. That this approach works was first shown by using cloned human α-globin and β-globin cDNAs to map the α-globin gene to chromosome 16 and the β-globin gene to chromosome 11 (Figure 17-8). Through such procedures, 12 putative oncogenes already have been mapped to specific chromosomes (Figure 17-9).

Subchromosomal Localization of Genes

Once genes have been mapped to a particular chromosome, the next step is to establish their order along the chromosome. In classical genetics this is done by measuring the frequency of recombination during meiosis between pairs or sets of genes. The closer two genes are, the less frequently recombination occurs between them. The

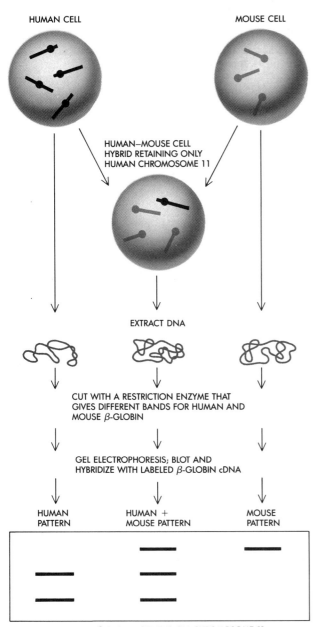

Figure 17-8
Human–mouse cell hybrids are made as described in Figure 17-7. If the gene under study is not expressed in the hybrid cells, its presence can be detected by Southern blot hybridization with a suitable probe. Human–mouse cell hybrids retaining human chromosome 11 are found to have the human β-globin gene.

somatic-cell geneticist has to resort to alternative phenomena that break the linkage between genes on the same chromosome, including spontaneous deletions and reciprocal translocation as well as experimentally induced chromosome fragmentation.

Spontaneous reciprocal translocations in which segments of two different chromosomes have been exchanged are relatively frequent in humans, where they manifest themselves by causing congenital abnormalities or high rates of spontaneous abortion. Over 300 human reciprocal translocations have been identified and cells from the affected individuals have been stored. The positions of reciprocal translocations can be identified because they alter the diagnostic pattern of bands that are revealed along each chromosome under the light microscope following appropriate staining. Once a gene has been mapped to a particular chromosome, its position can be determined by studying the set of reciprocal translocations that together span the chromosome. If the gene is absent from chromosomes with a particular translocation, it must normally reside in the region spanned by the translocated segment. Over 100 human genes have been assigned to particular subchromosomal regions in this way (Figure 17-10, page 222).

An alternative approach to the problem of establishing gene linkage in somatic cells is to induce random chromosome fragmentation by irradiating cells before fusing them with a recipient cell, and then following in the hybrid the retention of two

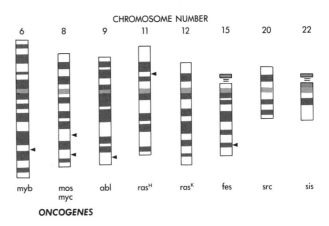

Figure 17-9
Mapping human oncogenes. The chromosomal location of several human oncogenes has been determined by somatic-cell hybridization.

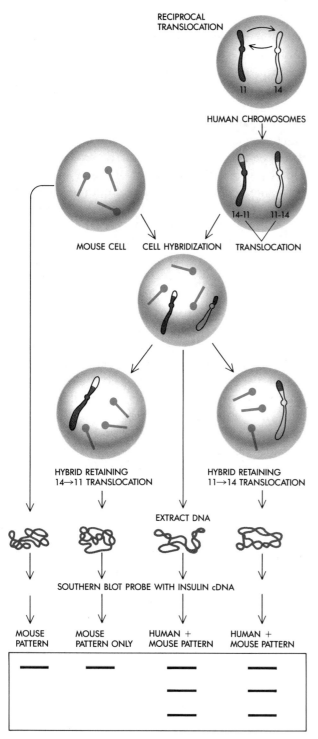

RECIPROCAL
TRANSLOCATION

11   14

HUMAN CHROMOSOMES

MOUSE CELL   CELL HYBRIDIZATION   TRANSLOCATION

14-11   11-14

HYBRID RETAINING   HYBRID RETAINING
14→11 TRANSLOCATION   11→14 TRANSLOCATION

EXTRACT DNA

SOUTHERN BLOT PROBE WITH INSULIN cDNA

MOUSE   MOUSE   HUMAN +   HUMAN +
PATTERN   PATTERN ONLY   MOUSE PATTERN   MOUSE PATTERN

∴ THE HUMAN INSULIN GENE IS LOCATED AT THE TOP OF
THE SHORT ARM OF CHROMOSOME 11

Figure 17-10
Subchromosomal localization of genes. The position of a gene along a chromosome can be determined by using naturally occurring chromosomal translocations. In this example, a human cell line that contained a reciprocal translocation exchanging the distal portions of the short arms of chromosomes 11 and 14 was obtained. The cell line was fused to mouse cells, and hybrids retaining either the 11 → 14 translocation or the 14 → 11 translocation were isolated. Southern blot hybridization was used to determine that the human insulin gene (which had been mapped to chromosome 11) was found in the cell hybrid that retained the distal portion of the short arm of chromosome 11.

genes mapped to the same chromosome. The closer the two genes are, the more frequently will both be retained together in the hybrids, because they will both be on the same chromosome fragment. With this approach, the genes for thymidine kinase and galactose kinase have been shown to be about 1.2 centimorgans (1,200,000 base pairs) apart on the human X chromosome (Figure 17-11).

Chromosomal Translocations and Cancer

Cytogenetic studies have indicated that many different leukemias and lymphomas are correlated with specific chromosomal translocations. It has long been thought that these translocations are somehow responsible for the malignant phenotype. With the discovery that the human genome contains genes that, if expressed inappropriately, can cause cancer (the oncogenes, Chapter 14), it seemed reasonable to speculate that the chromosomal translocations in these lymphomas and leukemias transpose silent oncogenes to new locations on the chromosomes (perhaps near active promoters), where the genes are turned on. Recently, recombinant DNA technology has been used to show that this is the case in several mouse leukemias and in at least one human disease, Burkitt's lymphoma. The endogenous mouse gene (called c-myc) analogous to the avian MC29 tumor virus oncogene was cloned and used as a probe to study the location of this gene in normal cells and in lymphoma cells. In several mouse lymphomas, the c-myc gene has been found to be translocated from its normal location on chromosome 15 into the antibody heavy-chain locus on chromosome

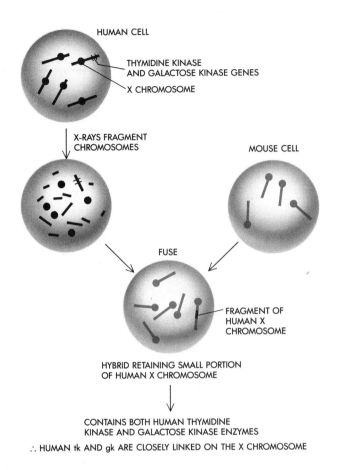

HUMAN CELL

THYMIDINE KINASE
AND GALACTOSE KINASE GENES

X CHROMOSOME

X-RAYS FRAGMENT
CHROMOSOMES

MOUSE CELL

FUSE

FRAGMENT OF
HUMAN X
CHROMOSOME

HYBRID RETAINING SMALL PORTION
OF HUMAN X CHROMOSOME

CONTAINS BOTH HUMAN THYMIDINE
KINASE AND GALACTOSE KINASE ENZYMES

∴ HUMAN tk AND gk ARE CLOSELY LINKED ON THE X CHROMOSOME

Figure 17-11
X-Rays can be used to fragment the human
chromosomes into small pieces before fusion
with mouse cells. In cell hybrids retaining only a
small fragment of the human X chromosome,
both human thymidine kinase and galactose
kinase enzyme activities are found, indicating
that these genes are closely linked on the X
chromosome.

12. (This 15 → 12 translocation had been shown
by cytogenetic evidence to occur in many mouse
plasmacytomas.) In the human disease Burkitt's
lymphoma, the c-myc gene is moved from its normal location on chromosome 8, and, in at least one
cell line, it is found associated with the immunoglobulin heavy-chain region on chromosome 14
(Figure 17-12, page 224).

It is perhaps not surprising that the immunoglobulin regions are associated with the chromosomal aberrations. In these B-cell-derived tumors,
the heavy-chain region is a "hot spot" for recombination (Chapter 9). In this region, a low frequency
of aberrant recombination in which an active

heavy-chain promoter is mistakenly placed next to
a cellular oncogene could lead to malignant transformation.

In Situ Hybridization

As we discussed above, cloned DNA can be used
in Southern blot hybridization experiments as a
probe for human genes in the total DNA from
mouse–human hybrid cells containing particular
human chromosomes. Such experiments can reveal on which chromosome a gene resides, but
they tell us nothing about the position of the gene
along the chromosome. A cloned DNA can, however, be used in a much more informative way,
namely for *in situ* hybridization that reveals not
only which chromosome carries a particular gene
but also where the gene is located along that chromosome.

In essence, *in situ* hybridization involves radio-
labeling the cloned DNA probe to a very high
specific activity, and hybridizing it to cells arrested
in mitosis in such a way that each chromosome is
distinguishable under the light microscope. Autoradiography is then used in conjunction with
chromosome staining to reveal directly to which
band on which chromosome the probe has hybridized.

This procedure is used routinely to map genes
on *Drosophila* salivary gland chromosomes. These
are particularly amenable to *in situ* hybridization
because each gene is amplified some 1000-fold:
Each salivary gland chromosome consists of about
1000 copies of a chromosome held together in
lateral register. To be useful for mapping human
genes on single copies of human chromosomes,
the DNA probe for the *in situ* hybridization must
be made extremely radioactive to give a detectable
signal. Recently DNA probes labeled with the iso-
tope ^{125}I have been introduced for this purpose,
and new methods of labeling DNA—through biotin derivatives of cytosine or other bases that can
be identified by antibiotin antibodies labeled
fluorescently, isotopically, or enzymatically—hold
great promise. They may well solve the problem
of amplifying the signal from one cloned fragment
of DNA, perhaps 20,000 base pairs long, hybridized to one complementary sequence in the 10^9
base pairs of an average-sized human chromosome.

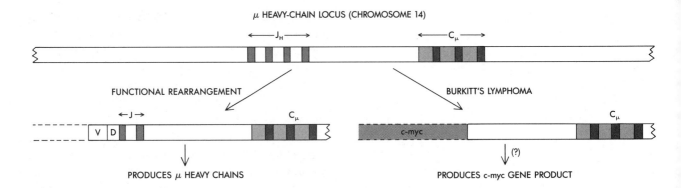

Figure 17-12
Gene rearrangements in Burkitt's lymphoma. In cell lines obtained from a Burkitt's lymphoma patient, the c-myc gene was found to be translocated from its normal position on chromosome 8 into the μ heavy-chain locus on chromosome 14. This may lead to overexpression of the c-myc gene.

Alternatively, some investigators have obtained convincing data on the localization of specific genes by using relatively *low*-specific-activity gene probes labeled with ³H. In this technique, only a few autoradiographic grains are present in any one spread (set of metaphase chromosomes). However if 50 to 100 spreads are analyzed and the number of grains on each chromosome is totaled up, a very clear picture of the location of the specific gene often emerges (Figure 17-13).

Already several genes have been mapped by *in situ* hybridization, including a V-region gene of the human immunoglobulin light chain located close to the centromere of the short arm of chromosome 2, and the human insulin gene on the distal protein of the short arm of chromosome 11.

Cloning Individual Chromosomes

Because it is now possible to sort human chromosomes physically into discrete size classes by a procedure known as fluorescence-activated cell sorting (FACS), it is possible to clone the DNA of individual chromosomes. For example, beginning with a human-cell line with an abnormal karyotype (four X chromosomes), it was possible to sort enough human X chromosomes free of autosomes to make a λ phage library of the X chromosome DNA. Together the recombinant phages in the library have most, if not all, of the X chromosome DNA. Therefore, all the techniques described earlier can be used to analyze the X chromosome

DNA, map restriction sites, identify the repetitive Alu sequences, and search for the genes known to be carried on the X chromosome. One task that can be undertaken now is to search for restriction site polymorphisms to find out if any are linked tightly to any of the genetic diseases mapped to the X chromosome. If, for example, it could be shown that a particular pattern of restriction enzyme sites in X chromosome DNA is tightly linked to Duchenne muscular dystrophy (Table 17-1) and is not present in normal individuals, it might well become possible to diagnose this common and lethal genetic disease even before we know which mutated gene is responsible for it.

Bridging the Gap Between Cytogenetics and Molecular Genetics

An average-sized human chromosome—the X chromosome, for example—contains on the order of 10^8 base pairs. By studying the banding patterns of stained chromosomes in the light microscope, markers about 5×10^6 base pairs apart can be resolved. Mapping genes using spontaneous translocations or by inducing chromosome fragmentation gives a resolution on the order of 10^6 base pairs. The largest DNA fragment that can be cloned (in cosmid vectors) is about 50,000 base pairs, and by studying differences in the restriction enzyme patterns of homologous pairs of chromosomes in an individual, differences as small as single base-pair changes can be detected. In short, in the next few years, the gap between traditional

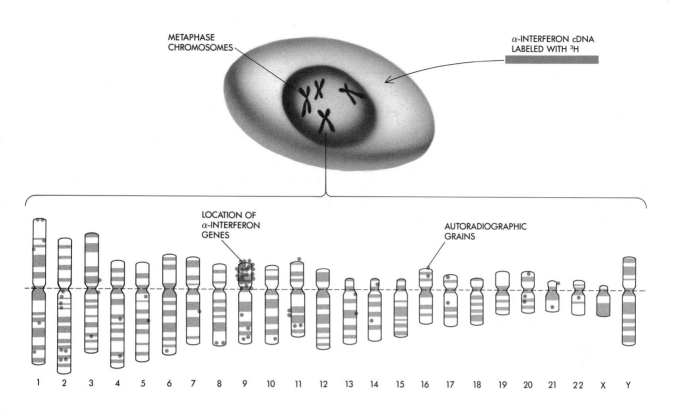

Figure 17-13
Mapping the human α-interferon gene by *in situ* hybridization. Metaphase spreads of human cells were prepared and hybridized to a cloned human α-interferon cDNA labeled with ^{3}H. Autoradiographic grains from 50 such spreads were totaled and clearly showed that the α-interferon genes are on chromosome 9.

human genetics and cytogenetics (with a resolution measured in millions of base pairs) and molecular genetics (with a resolution of single base pairs) can be bridged by exploiting the opportunities provided by somatic-cell genetics and gene cloning. We can confidently look forward to medium-resolution maps of entire human chromosomes, and high-resolution maps of interesting and medically important segments.

Prospects for Gene Therapy

Much has been written about the possibility that genetic diseases will eventually be cured by gene transplantation. The one attempt so far—to cure two terminally sick patients suffering from β-thalassemia by removing some of their bone marrow cells, transfecting them with normal human β-globin genes, and reimplanting the marrow—received worldwide publicity in 1981. Unfortunately, that initial foray into gene replacement therapy was premature. It was not the culmination of exhaustive experiments with cells in culture and laboratory animals; it was, rather, a long shot, and it failed. In fact, gene replacement therapy is still a very remote prospect. For there to be any real chance of its succeeding, the following tools and skills will be needed:

1. An abundant supply of the normal gene and an understanding of how its expression is regulated.
2. An efficient method of introducing genes into the chromosomes of cells in such a way that once they are integrated, they are expressed in a normally regulated way.
3. Vectors that will transport the replacement genes to the specific differentiated cells in the particular tissues and organs in which that gene normally functions.
4. The ability to introduce the replacement genes into a significant proportion of the appropriate cells. This is essential for there to be any significant medical benefit.

To help the victims of many genetic diseases, all this has to be done either *in utero* or during the first months after birth, before irreparable damage has occurred.

At present, none of these things can be done in a controlled, reproducible way. Techniques for transfecting cells are still primitive. Vectors that will carry genes efficiently to a particular cell type within an animal are lacking, although there are hopes that viruses may be modified and then exploited as cell- or tissue-specific vectors of DNA. Likewise, we will need to provide the genes we wish to introduce into cells with regulatory sequences that will ensure their appropriately controlled expression. These are formidable requirements, and they are presently beyond our capabilities. How long they will remain so is hard to predict, but the pace of discovery is fast.

Induction of Fetal Hemoglobin in Thalassemias

Normal adult hemoglobin consists of a pair of α-globin chains and a pair of β-globin chains (hemoglobin A is $\alpha_2\beta_2$). β-Thalassemia results from the production of an insufficient number of β-globin subunits. The clinical symptoms of β-thalassemia result in large part from the accumulation of excess free α-globin chains.

During fetal life, the β-globin chain is not made even in normal individuals; instead, fetuses make a γ-globin chain that associates with the α-chain to give fetal hemoglobin ($\alpha_2\gamma_2$). After about 30 weeks of gestation, synthesis of γ-globin begins to decrease and that of β-globin to increase. By a few weeks after birth, fetal hemoglobin synthesis has been completely switched off in normal people, and in thalassemia sufferers it is at this point that disease symptoms begin. However, we know,

from studies of individuals who have a rare disease called "hereditary persistence of fetal hemoglobin," that people can live essentially normal lives without β-globin if after birth they continue to make γ-globin. These people retain fetal $\alpha_2\gamma_2$ hemoglobin all their lives.

The switching off of the γ-globin gene in normal individuals is associated with an increase in the methylation of γ-globin DNA. Experiments with baboons showed that the drug 5-azacytidine, which inhibits DNA methylation, can induce a transient but striking increase in the amount of fetal hemoglobin. Two clinical research groups have given 5-azacytidine to a small number of patients with severe β-thalassemia; the idea was to alleviate the thalassemia by enhancing γ-globin synthesis and therefore fetal hemoglobin production. After seven days of the 5-azacytidine treatment, the γ-globin synthesis had increased more than five-fold, and there was a temporary increase both in the number of red blood cells and in their hemoglobin content. These changes were correlated with a decrease in methylation of the DNA in the region of the γ-globin gene.

Of course, many more studies now have to be done to determine the efficacy and risks of this experimental therapy, and in particular the long-term toxicity of 5-azacytidine, which may be switching on *many* genes—including perhaps oncogenes—along with γ-globin. Whether or not 5-azacytidine proves to be of therapeutic value remains to be seen, but whatever the outcome, new treatments will continue to be developed as the molecular physiology of genes becomes better and better understood. As an editorial in the *New England Journal of Medicine* put it: "The era when the physician will require a working knowledge of this discipline [molecular genetics] and an ability to apply its power is no longer approaching; it is here."

READING LIST

Books

Caskey, C. T., and R. L. White, eds. *Recombinant DNA Applications to Human Disease,* Banbury Report 14. Cold Spring Harbor Laboratory, Cold Spring Harbor, N.Y., 1983 (in press).

Dickerson, R. E., and I. Geis. *Hemoglobin: Structure, Function, Evolution, and Pathology.* Benjamin-Cummings, Menlo Park, Cal., 1983.

Ellis, R., ed. *Inborn Errors of Metabolism.* Croom Helm, London, 1980.

Galjaard, H. *Genetic Metabolic Diseases.* Elsevier/North-Holland, New York, 1980.

Original Research Papers (Reviews)

EARLY DIAGNOSIS OF GENETIC DISEASE

Friedmann, T. "Prenatal diagnosis of genetic disease." *Sci. Am.,* 225(5): 34–42 (1971).

Fuchs, F. "Genetic amniocentesis." *Sci. Am.,* 242(6): 47–53 (1980).

Orkin, S. H. "Genetic diagnosis of the fetus." *Nature,* 296: 202–203 (1982).

THALASSEMIAS

Chang, J. C., and Y. W. Kan. "β^0-Thalassemia, a nonsense mutation in man." *Proc. Natl. Acad. Sci. USA,* 76: 2886–2889 (1979).

Lauer, J., C. K. J. Shen, and T. Maniatis. "The chromosomal arrangement of human α-like globin genes: Sequence homology and α-globin gene deletions." *Cell,* 20: 119–130 (1980).

Busslinger, M., N. Moschonas, and R. A. Flavell. "β^+-thalassemia: Aberrant splicing results from a single point mutation in an intron." *Cell,* 27: 289–298 (1981).

Orkin, S. H., and S. C. Goff. "Nonsense and frameshift mutations in β^0-thalassemia detected in cloned β-globin genes." *J. Biol. Chem.,* 256: 9782–9784 (1981).

Orkin, S. H., S. C. Goff, and R. L. Hechtman. "An intervening sequence splice junction mutation in man." *Proc. Natl. Acad. Sci. USA,* 78: 5041–5045 (1981).

Spritz, R. A., P. Jagadeeswaran, P. V. Choudary, P. A. Biro, J. T. Elder, J. K. deRiel, J. L. Manley, M. L. Gefter, B. G. Forget, and S. M. Weissman. "Base substitution in an intervening sequence of a β^+-thalassemic human globin gene." *Proc. Natl. Acad. Sci. USA,* 78: 2455–2459 (1981).

Felber, B. K., S. H. Orkin, and D. H. Hamer. "Abnormal RNA splicing causes one form of α-thalassemia." *Cell,* 29: 895–902 (1982).

Fukumaki, Y., P. K. Ghosh, E. J. Benz Jr., V. B. Reddy, P. Lebowitz, B. G. Forget, and S. M. Weissman. "Abnormally spliced messenger RNA in erythroid cells from patients with β^+-thalassemia and monkey cells expressing a cloned β^+-thalassemic gene." *Cell,* 28: 585–593 (1982).

Goossens, M., K. Y. Lee, S. A. Liebhaber, and Y. W. Kan. "Globin structural mutant $\alpha^{125\text{Leu}\rightarrow\text{Pro}}$ is a novel cause of α-thalassemia." *Nature,* 296: 864–865 (1982).

Grosveld, G. C., E. deBoer, C. K. Shewmaker, and R. A. Flavell. "DNA sequences necessary for transcription of the rabbit β-globin gene *in vivo.*" *Nature,* 295: 120–126 (1982).

Kinniburgh, A. J., L. E. Maquat, T. Schedl, E. Rachmilewitz, and J. Ross. "mRNA-deficient β^0-thalassemia results from a single nucleotide deletion." *Nuc. Acids Res.,* 10: 5421–5427 (1982).

Orkin, S. H., H. H. Kazazian Jr., S. E. Antonarakis, H. Ostrer, S. C. Goff, and J. P. Sexton. "Abnormal RNA processing due to the exon mutation of β^ϵ-globin gene." *Nature,* 300: 768–769 (1982).

Treisman, R. A., N. J. Proudfoot, M. Shander, and T. Maniatis. "A single-base change at a splice site in a β^0-thalassemic gene causes abnormal RNA splicing." *Cell,* 29: 903–911 (1982).

Weatherall, D. J., and J. B. Clegg. "Thalassemia revisited." *Cell,* 29: 7–9 (1982). (Review.)

Spritz, R. A., and B. G. Forget. "The thalassemias: Molecular mechanisms of human genetic disease." *Am. J. Human Genet.* (1983) (in press).

SICKLE-CELL ANEMIA

Kan, Y. W., and A. M. Dozy. "Antenatal diagnosis of sickle-cell anaemia by DNA analysis of amniotic-fluid cells." *Lancet,* 2: 910–912 (1978).

Kan, Y. W., and A. M. Dozy. "Evolution of the hemoglobin S and C genes in world populations." *Science,* 209: 388–391 (1980).

Phillips III, J. A., S. R. Panny, H. H. Kazazian Jr., C. D. Boehm, A. F. Scott, and K. D. Smith. "Prenatal diagnosis of sickle cell anemia by restriction endonuclease analysis: *Hin*dIII polymorphisms in γ-globin genes extend test applicability." *Proc. Natl. Acad. Sci. USA,* 77: 2853–2856 (1980).

Chang, J. C., and Y. W. Kan. "Antenatal diagnosis of sickle-cell anaemia by direct analysis of the sickle mutation." *Lancet,* 2: 1127–1129 (1981).

Geever, R. F., L. B. Wilson, F. S. Nallaseth, P. F. Milner, M. Bittner, and J. T. Wilson. "Direct identifica-

tion of sickle cell anemia by blot hybridization." *Proc. Natl. Acad. Sci. USA,* 78: 5081–5085 (1981).

Kan, Y. W. "Hemoglobin abnormalities: Molecular and evolutionary studies." *Harvey Lec.,* 76: 75–93 (1982).

α_1-ANTITRYPSIN

Keuppers, F. "Inherited differences in α_1-antitrypsin." In S. Litwim, ed., *Lung Biology and Disease,* vol. 11: *Genetic Determinants of Pulmonary Disease.* Mercel Dekker, New York, 1978, pp. 23–74.

Kurachi, K., T. Chandra, S. J. F. Degen, T. T. White, T. L. Machioro, S. L. C. Woo, and E. W. Davie. "Cloning and sequence cDNA coding for α_1-antitrypsin." *Proc. Natl. Acad. Sci. USA,* 78: 6826–6830 (1981).

Leicht, M., G. L. Long, T. Chandra, K. Kurachi, V. J. Kidd, M. Mace, E. W. Davie, and S. L. C. Woo. "Sequence homology and structural comparison between the chromosomal human α_1-antitrypsin and chicken ovalbumin genes." *Nature,* 297: 655–659 (1982).

DIAGNOSIS OF POINT MUTATIONS USING SYNTHETIC OLIGONUCLEOTIDES

Conner, B. J., A. A. Reyes, C. Morin, K. Itakura, R. L. Teplitz, and R. B. Wallace. "Detection of sickle cell β^S-globin allele by hybridization with synthetic oligonucleotides." *Proc. Natl. Acad. Sci. USA,* 80: 278–282 (1983).

THE LESCH-NYHAN SYNDROME

Lesch, M., and W. L. Nyhan. "A familial disorder of uric acid metabolism and central nervous system function." *Am. J. Med.,* 36: 561–570 (1964).

Seegmiller, J. E., F. M. Rosenbloom, and W. N. Kelley. "Enzyme defect associated with a sex-linked human neurological disorder and excessive purine synthesis." *Science,* 155: 1682–1683 (1967).

Caskey, C. T., and G. D. Kruh. "The HPRT locus." *Cell,* 16: 1–9 (1979). (Review.)

Kruh, G. D., R. G. Fenwick Jr., and C. T. Caskey. "Structural analysis of mutant and revertant forms of Chinese hamster hypoxanthine-guanine phosporibosyl-transferase." *J. Biol. Chem.,* 256: 2878–2886 (1981).

Brennand, J., A. C. Chinault, D. S. Konecki, D. W. Melton, and C. T. Caskey. "Cloned cDNA sequences of the hypoxanthine/guanine phosphoribosyl-transferase gene from a mouse neuroblastoma cell line found to have amplified genomic sequences." *Proc. Natl. Acad. Sci. USA,* 79: 1950–1954 (1982).

Konecki, D. S., J. Brennand, J. C. Fuscoe, C. T. Caskey, and A. C. Chinault. "Hypoxanthine-guanine phosphoribosyltransferase genes of mouse and Chinese hamster: Construction and sequence analysis of cDNA recombinants." *Nuc. Acids Res.,* 10: 6763–6775 (1982).

Melton, D. W., J. Brennand, D. H. Ledbetter, D. S. Konecki, A. C. Chinault, and C. T. Caskey. "Phenotypic reversion at the HPRT locus as a consequence of gene amplification." In R. T. Schimke, ed., *Gene Amplification.* Cold Spring Harbor Laboratory, Cold Spring Harbor, N.Y., 1982, pp. 59–65.

CITRULLINEMIA

Su, T.-S., A. L. Beaudet, and W. E. O'Brien. "Increased translatable messenger ribonucleic acid for argininosuccinate synthetase in canavanine-resistant human cells." *Biochemistry,* 20: 2956–2960 (1981).

Su, T.-S., H.-G. O. Bock, W. E. O'Brien, and A. L. Beaudet. "Cloning of cDNA for argininosuccinate synthetase mRNA and study of enzyme overproduction in a human cell line." *J. Biol. Chem.,* 256: 11826–11831 (1981).

Ruddle, F. H. "Dispersion of argininosuccinate synthetase-like human genes to multiple autosomes and the X chromosome." *Cell,* 30: 287–293 (1982).

Su, T.-S., H.-G. O. Bock, A. L. Beaudet, and W. E. O'Brien. "Molecular analysis of argininosuccinate synthetase deficiency in human fibroblasts." *J. Clin. Invest.,* 70: 1334–1339 (1982).

HUMAN DNA POLYMORPHISMS

Kan, Y. W., and A. M. Dozy. "Polymorphism of DNA sequence adjacent to human β-globin structural gene: Relationship to sickle mutation." *Proc. Natl. Acad. Sci. USA,* 75: 5631–5635 (1978).

Bishop, D. T., and M. H. Skolnick. "Numerical considerations for linkage studies using polymorphic DNA markers in humans." In J. Cairns, J. L. Lyon, and M. Skolnick, eds., *Cancer Incidence in Defined Populations,* Banbury Report 4. Cold Spring Harbor Laboratory, Cold Spring Harbor, N.Y., 1980, pp. 421–433.

Botstein, D., R. White, M. Skolnick, and R.W. Davis. "Construction of a genetic linkage map in man using restriction fragment length polymorphisms." *Am. J. Hum. Genet.,* 32: 314–331 (1980).

Little, P. F. R., G. Annison, S. Darling, R. Williamson, L. Camba, and B. Modell. "Model for antenatal diagnosis of β-thalassaemia and other monogenic disorders by molecular analysis of linked DNA polymorphisms." *Nature,* 285: 144–151 (1980).

White, R. "In search of DNA polymorphisms in hu-

mans." In J. Cairns, J. L. Lyon, and M. Skolnick, eds., *Cancer Incidence in Defined Populations,* Banbury Report 4. Cold Spring Harbor Laboratory, Cold Spring Harbor, N.Y., 1980, pp. 409–420.

Wyman, A. R., and R. White. "A highly polymorphic locus in human DNA." *Proc. Natl. Acad. Sci. USA,* 77: 6754–6758 (1980).

Orkin, S. H., H. H. Kazazian Jr., S. E. Antonarakis, S. C. Goff, C. D. Boehm, J. P. Sexton, P. G. Waber, and P. J. V. Giardina. "Linkage of β-thalassaemia mutations and β-globin gene polymorphisms with DNA polymorphisms in human β-globin gene cluster." *Proc. Natl. Acad. Sci. USA,* 296: 627–631 (1982).

White, R., and M. Skolnick. "DNA sequence polymorphism and the genetics of epilepsy." In V. E. Anderson, W. A. Hauser, J. K. Penry, and C. F. Sing, eds., *Genetic Basis of the Epilepsies,* Raven, New York, 1982, pp. 311–316.

White, R., M. Schafer, D. Barker, A. Wyman, and M. Skolnick. "DNA sequence polymorphism at arbitrary loci." In B. Bonne-Tamir, T. Cohen, and R. N. Goodman, eds., *Human Genetics: The Unfolding Genome.* Alan R. Liss, New York, 1982, pp. 67–77.

White, R. "DNA polymorphism: New approaches to the genetics of cancer." *Cancer Surveys,* 1: 175–186 (1982).

GENE MAPPING

Owerbach, D., G. I. Bell, W. J. Rutter, and T. B. Shows. "The insulin gene is located on chromosome 11 in humans." *Nature,* 286: 82–84 (1980).

Owerbach, D., W. J. Rutter, J. A. Martial, J. D. Baxter, and T. B. Shows. "Genes for growth hormone, chorionic somatomammotropin, and growth hormone-like genes on chromosome 17 in humans." *Science,* 209: 289–292 (1980).

Harper, M. E., and G. F. Saunders. "Localization of single copy DNA sequences on G-banded human chromosomes by in situ hybridization." *Chromosoma,* 83: 431–439 (1981).

Harper, M. E., A. Ullrich, and G. F. Saunders. "Localization of the human insulin gene to the distal end of the short arm of chromosome 11." *Proc. Natl. Acad. Sci. USA,* 78: 4458–4460 (1981).

Owerbach, D., W. J. Rutter, N. E. Cooke, J. A. Martial, and T. B. Shows. "The prolactin gene is located on chromosome 6 in humans." *Science,* 212: 815–816 (1981).

Owerbach, D., W. J. Rutter, J. L. Roberts, P. Whitfeld, J. Shine, P. H. Seeburg, and T. B. Shows. "The proopiocortin (adrenocorticotropin/β-lipotropin) gene is located on chromosome 2 in humans." *Somat. Cell Genet.,* 7: 359–369 (1981).

Owerbach, D., W. J. Rutter, T. B. Shows, P. Gray, D. V. Goeddel, and R. M. Lawn. "Leukocyte and fibroblast interferon genes are located on human chromosome 9." *Proc. Natl. Acad. Sci. USA,* 78: 3123–3127 (1981).

Ruddle, F. H. "A new era in mammalian gene mapping. Somatic cell genetics and recombinant DNA methodologies." *Nature,* 294: 115–120 (1981).

Williamson, B. "The cloning revolution meets human genetics." *Nature,* 293: 10–11 (1981).

Dalla-Favera, R., M. Bregni, J. Erikson, D. Patterson, R. C. Gallo, and C. M. Croce. "Human c-*myc onc* gene is located on the region of chromosome 8 that is translocated in Burkitt lymphoma cells." *Proc. Natl. Acad. Sci. USA,* 79: 7824–7827 (1982).

DeKlein, A., A. G. van Kessel, G. Grosveld, C. R. Bartram, A. Hagemeijer, D. Bootsma, N. K. Spurr, N. Heisterkamp, J. Groofen, and J. R. Stephenson. "A cellular oncogene is translocated to the Philadelphia chromosome in chronic myelocytic leukemia." *Nature,* 300: 765–767 (1982).

Gusella, J. F., C. Jones, F.-T. Kao, D. Housman, and T. T. Puck. "Genetic fine-structure mapping in human chromosome 11 by use of repetitive DNA sequences." *Proc. Natl. Acad. Sci. USA,* 79: 7804–7808 (1982).

Langlois, R. G., L.-C. Yu, J. W. Gray, and A. V. Carrano. "Quantitative karyotyping of human chromosomes by dual beam flow cytometry." *Proc. Natl. Acad. Sci. USA,* 79: 7876–7880 (1982).

McBride, O. W., D. C. Swan, E. Santos, M. Barbacid, S. R. Tronick, and S. A. Aaronson. "Localization of the normal allele of T24 bladder carcinoma oncogene to chromosome 11." *Nature,* 300: 773–774 (1982).

Neel, B. G., S. C. Jhanwar, R. S. K. Chaganti, and W. S. Hayward. "Two human c-*onc* genes are located on the long arm of chromosome 8." *Proc. Natl. Acad. Sci. USA,* 79: 7842–7846 (1982).

Sakaguchi, A. Y., S. L. Naylor, R. A. Weinberg, and T. B. Shows. "Organization of human proto-oncogenes." *Am. J. Hum. Genet.,* 34: 175A (1982).

Shows, T. B., A. Y. Sakaguchi, and S. L. Naylor. "Mapping the human genome, cloned genes, DNA polymorphisms, and inherited disease." *Adv. Hum. Gen.,* 12: 341–452 (1982). (Review.)

Taub, T., I. Kirsch, C. Morton, G. Lenoir, D. Swan, S. Tronick, S. Aaronson, and P. Leder. "Translocation of the c-*myc* gene into the immunoglobulin heavy chain locus in human Burkitt lymphoma and murine plasmacytoma cells." *Proc. Natl. Acad. Sci. USA,* 79: 7837–7841 (1982).

Trent, J. M., S. Olson, and R. M. Lawn. "Chromosomal localization of human leukocyte, fibroblast, and

immune interferon genes by means of *in situ* hybridization." *Proc. Natl. Acad. Sci. USA,* 79: 7809–7813 (1982).

GENE THERAPY

Cline, M. J., H. Stang, K. Mercola, L. Morse, R. Ruprecht, J. Browne, and W. Salser. "Gene transfer in intact animals." *Nature,* 284: 422–425 (1980).

Wade, N. "UCLA gene therapy racked by friendly fire." *Science,* 210: 509–511 (1980). (News report.)

Wade, N. "Gene therapy caught in more entanglements." *Science,* 212: 24–25 (1981). (News report.)

Williamson, B. "Gene therapy." *Nature,* 298: 416–418 (1982).

MODIFICATION OF GENE EXPRESSION

Ley, T. J., J. DeSimone, P. Anagnou, G. H. Keller, R. K. Humphries, P. H. Turner, N. S. Young, P. Heller, and A. W. Nienhuis. "5-Azacytidine selectively increases γ-globin synthesis in a patient with β^+-thalassemia." *N. Eng. J. Med.,* 307: 1469–1475 (1982).

Benz Jr., E. J. "Clinical management of gene expression." *N. Eng. J. Med.,* 307: 1515–1516 (1982). (Editorial.)

18

The Science Used in the
Recombinant DNA Industry

The molecular biologists who developed gene cloning were the first to realize that the new techniques had great commercial as well as scientific potential. With the advent of recombinant DNA, the prospect of converting bacteria or yeast into fermenters for the production of valuable proteins immediately opened up. Throughout the "recombinant DNA debate" and the politicking—first to impose special safety regulations on recombinant DNA research and then to dismantle the regulations—the potential practical applications of the methods were stressed time and again. To reap the full commercial benefits of recombinant DNA will require much hard work and persistence, not to mention large capital investments. The long-term rewards, however, are certain to be very large for those with the staying power.

Commercial Potential of Recombinant DNA

The aims of the recombinant DNA industry are to harness for commercial ends our ability to manipulate, change, and transfer genes. The hope is that existing products—primarily proteins—can be made more efficiently, more cheaply, and more cleanly by gene cloning. One such example is making human insulin by putting an insulin gene into an appropriate microorganism. Another prospect is that of making available in commercial amounts proteins that would be of pharmaceutical value if enough of them could be produced. Examples are the antiviral agent interferon; plasminogen activator, a potential destroyer of unwanted blood clots; and the blood-clotting factor VIII, which hemo-

philiacs lack. Yet another prospect is to harness microorganisms to make viral proteins for use as safer vaccines. Particularly valuable should be vaccines against viruses that cannot be grown in cultured animal cells, for example, hepatitis B virus, which causes serum hepatitis. Outside the pharmaceutical industry, there are hopes that bacteria and other microorganisms can be genetically engineered to carry out processes of importance to the chemical, petroleum, and mining industries. The potential of gene cloning for plant breeding and agriculture has already been mentioned (Chapter 13).

Before these ambitions can be realized, several developments have to occur. The first, which we discuss in this chapter, is essentially scientific. The gene for the desired protein product has to be isolated and introduced into a microorganism in such a way that it will be expressed and its protein made. The subsequent requirements, which need not concern us here, include transferring the recombinant DNA into a microorganism suitable for growth on a commercial scale, which may mean hundreds of thousands of liters; purifying the product to meet required standards; testing its efficacy; and so on.

Methods of Commercial Gene Cloning

For the most part, the methods used by gene cloning companies to obtain recombinant DNA consisting of the desired gene in an expression vector are the same as those used in basic research laboratories (the methods we have described throughout this book). Since many of the commer-

cially valuable proteins are present only in very small amounts in animal cells and tissues, and since expression of the cloned gene is essential, much of the commercial work has centered upon the cDNA cloning of mRNAs present in very small amounts in cells. Another approach, used to obtain human insulin, has been the chemical synthesis of a gene.

Human Insulin from Bacteria

Human insulin is the only animal protein to have been made in bacteria in such a way that its structure is absolutely identical to that of the natural molecule. Considering the complexity of insulin synthesis in cells of the pancreas, this is a remarkable achievement. We owe our success with genetically engineered insulin to the unusual amino acid composition of the protein. Unfortunately, the method used cannot be generalized for the genetic engineering of larger proteins with more complicated amino acid sequences, although it could be used for other small proteins and peptides that do not contain methionine or tryptophan.

The Structure of Human Insulin

Insulin, which is secreted from the cells of the pancreas into the bloodstream, is one of the proteins that has a signal sequence used to lead the polypeptide chain across intracellular membranes. During this transport, the signal peptide is cleaved from the remainder of the polypeptide, which is stored within membrane-bound vesicles in the pancreatic cells. This storage form of insulin is known as proinsulin, to distinguish it on the one hand from preproinsulin, the complete polypeptide chain with the "pre" sequence still attached, and on the other hand from the physiologically active mature insulin. Proinsulin and insulin differ quite dramatically (Figure 18-1). Proinsulin is a single polypeptide chain bent into a complicated loop, with the two stems held together by three disulfide cross-bridges. Mature insulin consists of two separate chains, one of 21 amino acids (the A chain) and one of 30 amino acids (the B chain), held together by the same disulfide cross-bridges. Proinsulin is converted to insulin inside vesicles in the pancreatic cells by the enzymatic cleavage of

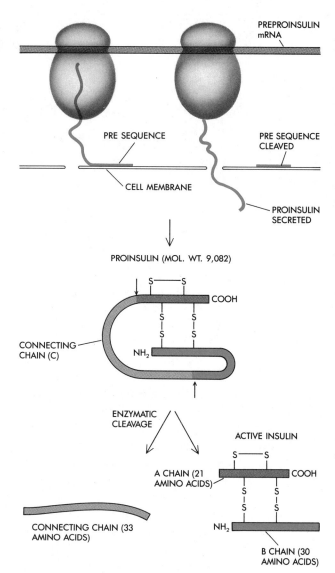

Figure 18-1
Insulin is synthesized on the ribosomes as preproinsulin; during secretion, the "pre" sequence is cleaved off, leaving proinsulin. The 33-amino-acid C chain of proinsulin is then removed enzymatically, liberating biologically active insulin consisting of A and B chains connected by disulfide bridges.

the connecting loop of polypeptide, 35 amino acids long, known as the C peptide. The insulin is then stored in the cells as a zinc salt crystal ready for export. In other words, insulin has three forms: preproinsulin, the primary product of translation, with the signal sequence and the A, B, and C segments all intact; proinsulin, which lacks the leader

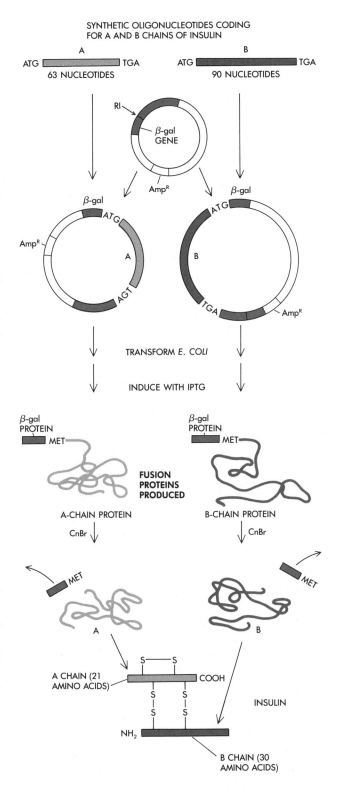

SYNTHETIC OLIGONUCLEOTIDES CODING
FOR A AND B CHAINS OF INSULIN

Figure 18-2
Production of insulin in bacteria by using synthetic insulin genes. Oligonucleotides coding for the A and B chains of insulin were synthesized chemically, and the trinucleotide ATG was added to the 5′ ends. These oligo- nucleotides were cloned downstream from the β-galactosidase promoter and part of the β-gal protein-coding sequence in pBR322. When the clones were introduced into *E. coli*, the bacteria produced hybrid proteins consisting of the N-terminal portion of the β-gal protein, followed by a methionine residue (from the ATG codon) fused to either the A or the B chain of insulin. The β-gal protein portion was removed by treatment with CnBr, which cleaves polypeptides at methionine, liberating free A or B insulin chains. The two chains were then combined to produce biologically active insulin.

sequence; and insulin, consisting of the separate A and B chains held together by three disulfide cross- bridges. Insulin does not carry any sugar side chains.

Synthetic Insulin "Genes"

How do we go about making insulin in bacteria? It would be very inefficient to begin with the chromosomal gene, since not only would that specify preproinsulin, but it also contains a non- coding intervening sequence.

To overcome these difficulties, DNA chains with the correct nucleotide sequences to specify the A and B polypeptide chains of insulin were chemically synthesized. This was feasible first of all because the amino acid compositions of both chains were known, so the required DNA se- quence could be worked out from the genetic code; and secondly because the chains are short. Sixty-three nucleotides were required for the A chain and 90 for the B chain, plus a codon was needed at the end of each chain to signal termina- tion of protein synthesis. A methionine codon was placed at the beginning of each chain to allow removal of the insulin polypeptide from prokar- yotic amino acids (as discussed below).

The synthetic A- and B-chain "genes" were then separately inserted into the gene for a bacte- rial enzyme, β-galactosidase, which was carried in a plasmid cloning vector (Figure 18-2). Care was taken to ensure that the codons of the synthetic genes were in phase with those of the β-galactosid-

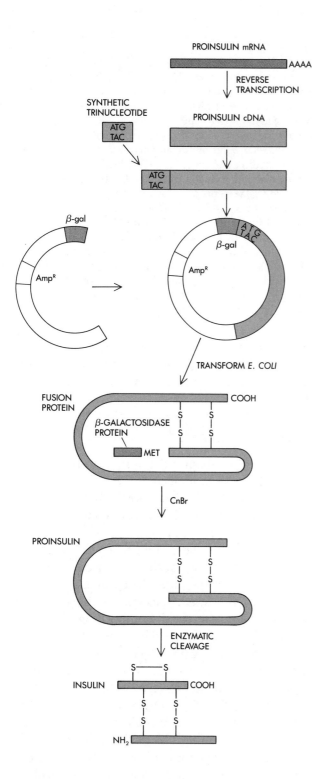

Figure 18-3
Production of proinsulin in bacteria. A full-length proinsulin cDNA was synthesized from proinsulin mRNA. An ATG triplet codon was added to the 5′ end chemically, and this construct was cloned downstream from the β-gal promoter and protein-coding sequence in pBR322. The hybrid plasmid, when introduced into *E. coli,* produced a β-gal–proinsulin fusion protein. The β-gal amino acids were removed by CnBr treatment, and the resulting proinsulin was enzymatically treated to give biologically active insulin.

ase. The recombinant plasmids were introduced into *E. coli.* In the bacterial cells they replicated, and under the control of the signal sequences of the β-galactosidase gene (its promoter and ribosome-binding site), mRNA was made and read into protein, which consisted of part of the β-galactosidase fused by the additional methionine residue to either the A or B chain of insulin. At this point the unusual amino acid composition of insulin and the methionine linker residue were exploited. Neither methionine nor another amino acid, tryptophan, occurs in either the A or the B chains. By means of a chemical reagent (cyanogen bromide) that destroys methionine and, to a lesser extent, tryptophan, the A and B chains were released from the β-glactosidase fragment. After purification, the two chains were mixed and reconnected together in a reaction that formed the disulfide cross-bridges. The result was pure human insulin. The only snag is that this trick, which uses methionine as a link that can be specifically broken between an animal polypeptide chain and a bacterial enzyme carrier, can be used only for proteins and peptides that contain neither methionine nor tryptophan. This severely limits its usefulness.

Proinsulin cDNA

In a second approach to the synthesis of human insulin, mRNA was copied into cDNA, and a methionine codon (ATG) was chemically synthesized and attached to the 5′ end of the proinsulin cDNA (Figure 18-3). This was hitched to a bacterial gene in a plasmid vector and grown in *E. coli.* The proinsulin fused to the bacterial enzyme could be released by destroying the methionine linker residue. The proinsulin chain folded into its natu-

ral three-dimensional structure, disulfide cross-bridges were formed, and finally the C peptide was cleaved away with enzymes, again to yield pure human insulin.

Bacteria That Secrete Proinsulin

As soon as bacteria that make a desired protein or polypeptide chain have been obtained, various genetic manipulations are tried to see if yields can be improved. For purposes of commercial exploitation, the yield per cell of the desired protein is crucial. The ideal situation would be to have large amounts of the foreign protein efficiently secreted into the medium by the engineered bacteria. Continuous-flow fermentation could then be used, and the product would be recovered from the culture medium without the necessity of harvesting the cells and destroying them in order to get at the desired protein.

One approach to this problem is to insert a cDNA copy of rat insulin mRNA into a bacterial β-lactamase gene carried in a plasmid vector (Figure 18-4). β-Lactamase, an enzyme that inactivates penicillin, is naturally secreted by bacteria into their culture medium. When this recombinant plasmid was introduced into *E. coli* and grown, the bacterial cells secreted into their medium a protein consisting of the rat proinsulin polypeptide chain and part of the β-lactamase molecule. The β-lactamase acted as a carrier for the secretion of the proinsulin. This result suggests that eventually it may be possible to produce clones of bacteria that not only synthesize insulin but that also secrete it into the medium.

Cloning Human Growth Hormone

The production of human growth hormone in *E. coli* draws attention to some of the vagaries of genetic engineering. Human growth hormone (HGH) is a single polypeptide chain that has 191 amino acids and that is produced in the human pituitary gland. Like insulin, it is not glycosylated. Growth hormone controls the growth of our bodies; the small stature of dwarfs is due to a deficiency of growth hormone.

Making the Growth Hormone "Gene"

Through the use of a combination of chemical synthesis of DNA and enzymatic synthesis of cDNA, a sequence that codes for this protein has been produced (Figure 18-5, page 236). A DNA

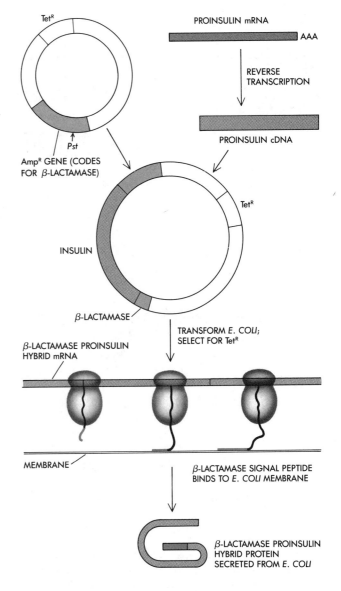

Figure 18-4
Prokaryotic signal sequences allow fusion proteins to be secreted. A proinsulin cDNA cloned into the β-lactamase gene results in a β-lactamase–proinsulin fusion protein that is secreted from *E. coli*, because of the presence of the signal peptide on the N-terminal portion of the β-lactamase protein.

fragment coding for amino acids 1 through 24 was synthesized chemically. Immediately in front of the first codon, a triplet of bases (ATG) specifying the amino acid methionine was added. Once the beginning of the gene had been chemically synthesized to ensure a correct start to the protein, a DNA sequence coding for the remainder of the

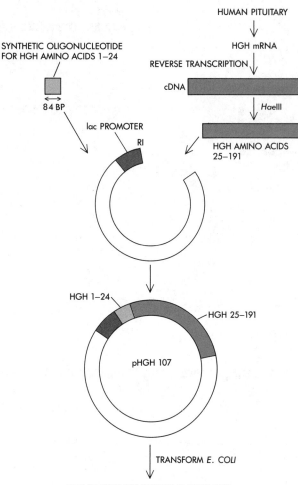

SYNTHETIC OLIGONUCLEOTIDE
FOR HGH AMINO ACIDS 1–24

84 BP

lac PROMOTER

RI

HUMAN PITUITARY

HGH mRNA

REVERSE TRANSCRIPTION

cDNA

HaeIII

HGH AMINO ACIDS
25–191

HGH 1–24

HGH 25–191

pHGH 107

TRANSFORM E. COLI

HUMAN GROWTH HORMONE PRODUCED

Figure 18-5
Production of human growth hormone in *E. coli.*
An oligonucleotide coding for the first 24 amino
acids of human growth hormone (HGH) was
synthesized chemically. HGH cDNA was then
prepared and cut with *Hae* III, which cleaves the
nucleotide sequence exactly in front of the
codon for amino acid 25 of HGH. These two
HGH fragments were then cloned downstream
from the lac promoter in pBR322. This plasmid
produces large amounts of HGH when
introduced into *E. coli.*

polypeptide chain, amino acid residues 25 through
191, was obtained by making cDNA copies of
mRNA preparations from human pituitary cells.
These two DNA fragments were then separately
cloned. The DNA fragments were repurified and
linked together to yield the complete DNA sequence for human growth hormone, beginning
with a methionine initiator codon, followed by the

sequence for the 191 amino acids in the mature protein, and ending with a signal to stop protein synthesis. The "gene" was then inserted in an expression
vector and introduced into *E. coli,* where it directed the synthesis of human growth hormone.

The Methionine Problem

The only drawback to the growth hormone made
in bacteria is that the initiator methionine residue
is not removed from the rest of the growth hormone polypeptide in the bacterial cells. *E. coli* has
enzymes that can do this, but, presumably, the
precise chemical composition of the growth hormone chain and the structure it folds into prevent
the bacterial enzymes from working. So bacteria
provide human growth hormone that is identical
in all respects, including in biological activity, to
the hormone made in human pituitary glands, except that the hormone from bacteria starts with an
additional methionine residue. This is an example
of our having to settle for a very good second-best.

Different Types of Interferons

Cloning the genes for insulin and growth hormone
was a straightforward procedure in comparison to
cloning interferon genes, because a great deal was
already known about the former two proteins, including their amino acid sequences. The situation
with the interferons was very different. Prior to
the advent of gene cloning, three different types of
interferon had been identified: leukocyte interferon (α), fibroblast interferon (β), and so-called
immune interferon (γ). But since nothing was
known about their amino acid sequences or the
structure of their genes, any attempt to chemically
synthesize DNA that could specify an interferon
was out of the question. The antiviral activity of
interferons was the only property of the molecules
that we could measure—the only handle we had
on them.

The key to the first successful isolation of α-
interferon cDNA was the prior discovery that the
very large oocytes of the toad *Xenopus* synthesize
measurable quantities of interferon when they are
injected with mRNA from interferon-making
cells. The approximate size of α-interferon mRNA
could thus be determined by fractionating poly-A
RNA (mRNA) from interferon-producing leuko-

cytes (white blood cells), injecting samples of the various fractions into *Xenopus* oocytes, and measuring the interferon (antiviral) activity excreted by the oocytes. Such experiments revealed the 12S mRNA fraction to have the highest activity, and it was used to make cDNA for later insertion into a plasmid pBR322 library.

On the assumption that the percentage of interferon mRNA in total leukocyte mRNA was between 0.1 and 0.01 percent, 10,000 clones of *E. coli* transformed by that cDNA library were picked individually and tested to see if they contained α-interferon-specific DNA sequences (Figure 18-6). Such sequences were looked for by observing whether the individual cDNA clones could hybridize to α-interferon-specific mRNA. If such interferon mRNA were bound to cDNA immobilized on a filter, the cDNA could be subsequently eluted and then injected into oocytes to see whether it might induce detectable interferon synthesis. Because of the expected low yield of interferon-containing clones, large pools of different cDNA clones were first tested. When a positive pool was found, it was subdivided into smaller groups until finally a single DNA clone that specifically hybridized to α-interferon mRNA was isolated. To show definitely that this clone was specific to α-interferon and not to a related molecule that lacked antiviral activity, the coding region was subsequently placed in an expression vector that enabled it to synthesize biologically active α-interferon, allowing the unambiguous identification of a functional interferon cDNA clone. Following this initial success, a large number of different interferon cDNAs were isolated in several different industrially supported laboratories.

From the nucleotide sequences of the cDNAs it was possible, using the genetic code, to deduce the amino acid sequences of several interferons. It turned out that fibroblast and leukocyte interferons have 166 amino acids, but they are synthesized with leader polypeptides of 21 and 23 amino acids, respectively, that effect membrane transport. Further experiments revealed that there is a family of at least 13 genes for different subtypes of leukocyte α-interferon as well as a number of inactive pseudogenes. In contrast, so far only one fibroblast β-interferon gene has been detected and cloned. All the human α-interferon genes and pseudogenes and the single β-interferon gene are located

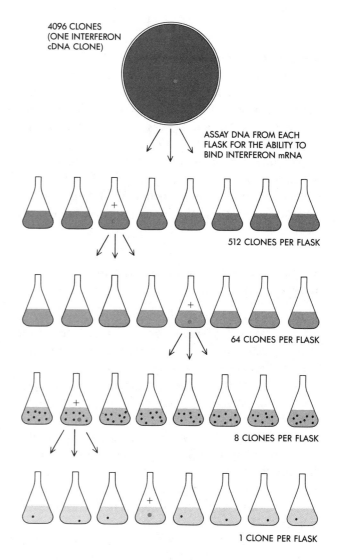

Figure 18-6
Cloning of human interferon cDNA. A cDNA library was prepared from human leukocytes. Interferon represents less than 0.1 percent of the total mRNA in these cells. The cDNA library was broken down into eight groups, and total plasmid DNA was prepared from each group. The presence of interferon cDNA in each group was determined by using the hybrid-selection procedure (Figure 6-6). When an interferon⁺ group of clones was identified, it was subdivided into groups containing fewer clones, and the process was repeated. Four such subdivisions were necessary to obtain a pure cDNA clone for human interferon. (The number of clones shown per flask is only approximate, of course.)

on chromosome 9. Several of these genes and pseudogenes are located within a few kilobases of each other; others are less closely linked.

Interestingly, none of the α- and β-interferon genes have introns. Fibroblast (β-) interferon and some but not all of the leukocyte (α-) interferons are glycosylated, and it was a stroke of luck that interferon made in *E. coli* and therefore lacking sugar side chains retains antiviral activity. Had glycosylation been essential for this function, the bacteria carrying interferon cDNA and making the unglycosylated protein would never have been detected.

High-Level Expression of Human α-Interferon in *E. coli*

The first successful expression of α-interferon in *E. coli* was at the level of about 50 molecules per cell. Higher levels of expression were subsequently obtained by precisely positioning the interferon cDNA sequences adjacent to appropriate *E. coli* promoter sequences. One of the most effective constructions fused a segment containing the lactose promoter and extending to the initiation codon with the human sequences coding for mature interferon (interferon lacking the signal sequences). This led to an *E. coli* strain making more than 1000 times more interferon per cell. By now there exist even more efficient interferon-making *E. coli* factories—some that yield over a mg of interferon per liter of culture. Such *E. coli*-made interferon is relatively easily purified, and it has been crystallized and already shown to be nontoxic at many times the dosage required to prevent infections in monkeys by the poxvirus vaccinia. Such synthetic interferon is currently being tested in humans, with the preliminary evidence suggesting that it could be effective in preventing common colds.

Cloning of γ- (Immune) Interferon

γ-Interferon (immune interferon) has proved more difficult to work with, but recently it has been cloned. Messenger RNA preparations from human spleen cells induced to make γ-interferon by exposure to a bacterial toxin were injected into frogs' oocytes, and the mRNAs were translated. The translation products were then biologically assayed for γ-interferon activity. The mRNA frac-

tions (18S) that specified γ-interferon in the oocytes were then copied into cDNA and cloned in a vector with both SV40 and plasmid sequences. The recombinants were propagated in *E. coli*, and then isolated and transferred to COS cells in which the SV40 sequences caused the vector to be amplified and the cDNA to be expressed (Figure 15-4, page 193). Production of γ-immune interferon could be assayed and the clones carrying the interferon cDNA were isolated and subsequently used to select the chromosomal gene from a library of human DNA in lambda vectors; the chromosomal γ-interferon gene was then sequenced. Unlike the several α- and β-interferon genes, the γ-interferon gene has three introns; it appears to be present in the human genome as a single copy.

Why the α- and β-interferon genes lack introns and the γ-interferon has three is a matter for speculation. Perhaps since the α- and β-interferons are made in cells infected by viruses, they have evolved to avoid splicing, which could be impaired in virally infected cells. The γ-interferon that is produced following mutagenic and allogenic induction appears to be more like an orthodox cellular gene.

Now that the interferon genes have been cloned, the preoccupation of the cloning companies is to devise host vector systems in which large amounts of interferon are synthesized, so that the protein can be obtained in commercially valuable amounts and tested for its therapeutic properties in the treatment of viral infections and cancer.

Viral Proteins for Vaccines

Viral vaccines consist either of virulent virus particles that have been inactivated, or of live particles of virus strains that have been weakened or attenuated so that they no longer cause disease but do immunize against the virulent strains. With inactivated virus vaccines there is always a very small chance that one or more virus particles have survived inactivation; vaccination may therefore lead to isolated cases of the disease. Such accidents have occurred more than once, for example in cowherds vaccinated against foot- (hoof-) and-mouth-disease virus. Moreover, because the viruses for both classes of vaccines are grown in animal cells, the vaccines are sometimes contaminated with cellular material that can cause ad-

verse, occasionally fatal reactions in a very small population of the recipients.

Since the surface proteins of the virus particles are the major antigens that induce immunity, it should be possible to use the proteins rather than the virus particles as vaccines, and thus avoid some of the risk connected with the vaccines that contain the whole virus.

Cloning Foot-and-Mouth-Disease Virus

Producing viral proteins from recombinant DNAs in yeast and *E. coli* should in principle be less expensive than propagating the virus in animal cells, and might lead to safer and cheaper vaccines. One particularly attractive virus for such investigations is foot-and-mouth-disease virus (FMDV) of cattle. The large vaccine market for this disease is presently met with conventional vaccines of inactivated particles. Complementary DNA copies of the single-stranded genomic RNA (MW 2.6×10^8) of FMDV have been cloned in expression vectors in *E. coli,* and synthesis of the major antigenic viral protein (VPI) has been achieved in quantities of 1 to 5 million molecules per bacterial cell. However, both VPI produced in this way and VPI isolated from virus particles grown in animal cells are poor antigens compared to virus particles, and the level of immunization they afford is low. This draws attention to a problem with purified protein vaccines: disaggregated pure protein may be less immunogenic than macromolecular assemblies of the same protein; the way the protein antigen is presented to the immune system may be crucial.

Synthetic Peptide Vaccines

Once cDNA of FMDV had been cloned and sequenced, the amino acid sequences of the viral proteins were deduced. While this work was going on, other experiments with hepatitis B virus and influenza virus had shown that short peptides corresponding to segments of the viral coat proteins were antigenic. Furthermore, the antibodies induced by some of the short peptides reacted with the corresponding viral proteins, although they only partly neutralized the infectivity of the intact virus.

Two laboratories have now produced, by chemical synthesis, peptides ranging from 8 to 41 amino acids in length, and corresponding to seg-

ments of the amino acid sequence of VPI of FMDV. The segments chosen were either known to be involved in the antigenicity of VPI, or predicted to be from structural considerations. The synthetic peptides were then chemically linked to protein carriers to increase their immunogenicity. When they are injected into cattle, guinea pigs, or rabbits, some of these peptides induce neutralizing antibody that is specific to the serotypes of FMDV being investigated. A single injection of one of the synthetic peptides linked to limpet hemocyanin, one of the most immunogenic proteins known, protects guinea pigs from virulent FMDV.

These results are very encouraging, much more so than the results obtained with whole viral proteins made by gene cloning methods. The peptides used in these first experiments were synthesized chemically, but they could easily be produced in large amounts by introducing into *E. coli* recombinant DNAs carrying the appropriate DNA sequence. The prospects for a totally synthetic peptide-carrier vaccine for foot-and-mouth disease are promising, but to be a commercial success it must compete with existing vaccines in terms of purity, safety, and efficiency.

Vaccines for Human Hepatitis B Virus

For studying and possibly developing vaccines against viruses that cannot be grown in cultured cells, gene cloning techniques are of the utmost importance. For example, hepatitis B virus causes serum hepatitis, which is a very severe and widespread disease against which there is no generally available vaccine because the virus does not multiply in any cultured animal cells. The only way to study the virus before the advent of recombinant DNA was to isolate particles from the blood of infected individuals, and the only "vaccine" was the antibodies from the serum of people who were carriers of the virus; these antibodies could be used to passively vaccinate acutely infected patients.

Hepatitis B Antigen by Gene Cloning

In 1979 the complete hepatitis B viral genome, a partially single-stranded DNA molecule of about 3200 base pairs, was cloned and sequenced. This allowed deduction of the amino acid sequence of the two known viral proteins: the core protein with which the DNA is associated in the virus, and

the major surface antigenic protein, which is part of the membranelike envelope of the virus.

Once the genome had been cloned and the viral protein genes had been identified, attempts were made to synthesize the viral protein in useful amounts by linking the genes to expression vectors. Viral core protein is now being made this way in *E. coli* in commercial quantities, and is being sold in diagnostic labs in some countries. The core antigen from *E. coli* proves to be at least as sensitive as that from human liver for the diagnosis of anticore antibodies in the blood of infected people.

Although improved diagnosis of serum hepatitis is a significant achievement, the real goal is a vaccine, and there is some reason for hope that eventually one will become available. Attempts to obtain high expression of the major surface antigen in *E. coli* have been disappointing, even when the viral gene has been inserted behind powerful bacterial promoters. However, recently the gene was linked to a yeast promoter in an autonomously replicating yeast plasmid and introduced into yeast cells. A particle isolated from the transformed yeast closely resembled in structure and antigenicity the so-called "22-nm particle" found in the blood of human carriers of the virus. These particles are in effect empty virions, consisting of the surface antigen protein and phospholipid but no core protein of DNA. The particles from yeast contain no glycosylated form of the hepatitis surface antigen, but this lack of glycosylation apparently does not alter their structure or immunogenicity. It should be possible to obtain surface antigen particles in large amounts from yeast, and their high immunogenicity raises hopes that they may form the basis of a vaccine.

READING LIST

Original Research Papers (Reviews)

EXPRESSION OF EUKARYOTIC GENES IN BACTERIA

Villa-Komaroff, L., A. Efstratiadis, S. Broome, P. Lomedico, R. Tizard, S. P. Nabet, W. L. Chick, and W. Gilbert. "A bacterial clone synthesizing proinsulin." *Proc. Natl. Acad. Sci. USA,* 75:3727–3731 (1978).

Goeddel, D. V., D. G. Kleid, F. Bolivard, H. Heyneker, D. Yansura, R. Crea, T. Hirose, A. Kraszewski, K. Itakura, and A. Riggs. "Expression in *Escherichia coli* of chemically synthesized genes for human insulin." *Proc. Natl. Acad. Sci. USA,* 76: 106–110 (1979).

Chan, S. J., B. Noyes, K. Agarwal, and D. Steiner. "Construction and selection of recombinant plasmids containing full-length complementary DNAs corresponding to rat insulin I and II." *Proc. Natl. Acad. Sci. USA,* 76: 5036–5040 (1979).

Goeddel, D., H. Heyneker, T. Hozumi, R. Arentzen, K. Itakura, D. Yansura, M. Ross, G. Miozzari, R. Crea, and P. Seeburg. "Direct expression in *Escherichia coli* of a DNA sequence coding for human growth hormone." *Nature,* 281: 544–548 (1979).

Murray, K. "Genetic engineering: Possibilities and prospects for its application in industrial microbiology." *Phil. Trans. R. Soc. London B,* 290: 369–386 (1980).

THE CLONING OF INTERFERON GENES

Nagata, S., H. Taira, A. Hall, L. Johnsrud, M. Streuli, J. Escodi, W. Boll, K. Cantell, and C. Weissmann. "Synthesis in *E. coli* of a polypeptide with human leukocyte interferon activity." *Nature,* 284: 316–320 (1980).

Derynck, R., E. Remaut, E. Saman, P. Stanssens, E. De Clercq, J. Content, and W. Fiers. "Expression of human fibroblast interferon gene in *Escherichia coli.*" *Nature,* 287: 193–197 (1980).

Goeddel, D., E. Yelverton, A. Ullrich, H. Heyneker, G. Miozzari, W. Holmes, P. Seeburg, T. Dull, L. May, N. Stebbing, R. Crea, S. Maeda, R. McCandliss, A. Sloma, J. Tabor, M. Gross, P. Familletti, and S. Pestka. "Human leukocyte interferon produced by *E. coli* is biologically active." *Nature,* 287: 411–416 (1980).

Gray, P. W., D. W. Leung, D. Pennica, E. Yelverton, R. Najarian, C. C. Simonsen, R. Derynck, P. J. Sherwood, D. M. Wallace, S. L. Berger, A. D. Levinson, and D. V. Goeddel. "Expression of human immune interferon cDNA in *E. coli* and monkey cells." *Nature,* 295: 503–508 (1982).

Devos, R., H. Cheroutre, Y. Taya, W. Degrave, H. Van Heuverswyn, and W. Fiers. "Molecular cloning of human immune interferon cDNA and its expression in eukaryotic cells." *Nuc. Acids Res.,* 10: 2487–2501 (1982).

VIRAL ANTIGEN CLONING

Valenzuela, P., A. Medina, W. J. Rutter, G. Ammerer, and B. D. Hall. "Synthesis and assembly of hepatitis B virus surface antigen particles in yeast." *Nature,* 298: 347–350 (1982).

Kupper, H., W. Keller, C. Kurz, S. Forss, H. Schaller, R. Franze, K. Strohmaier, O. Marquardt, V. G. Zaslavsky, and P. H. Hofschneider. "Cloning of a cDNA of major antigen of foot and mouth disease virus and expression in *E. coli.*" *Nature,* 289: 555–559 (1981).

Stahl, S., P. MacKay, M. Magazin, S. A. Bruce, and K. Murray. "Hepatitis B virus core antigen: Synthesis in *Escherichia coli* and application in diagnosis." *Proc. Natl. Acad. Sci. USA,* 79: 1606–1610 (1982).

Recombinant DNA Dateline

1871 Discovery of DNA in the sperm of trout from the Rhine River.

1943 DNA proves to be a genetic molecule capable of altering the heredity of bacteria.

1953 Postulation of a complementary double-helical structure for DNA.

1956 Genetic experiments support the hypothesis that the genetic messages of DNA are conveyed by its sequence of base pairs.

1958 Proof that DNA replication involves separation of the complementary strands of the double helix.

1958 Isolation of DNA polymerase I, the first enzyme that makes DNA in a test tube.

1960 Discovery of RNA polymerase, an enzyme that makes RNA chains on the surface of single-stranded DNA.

ca. 1930s OSWALD AVERY, shown in his lab where DNA was found to be a genetic molecule. *(Photo by Walther Goebel, courtesy the Rockefeller University Archives)*

1947 EDWARD TATUM (left) and JOSHUA LEDERBERG (center), who first sexually crossed mutant bacteria, at the 1947 Cold Spring Harbor Symposium on Nucleic Acids, together with SOL SPIEGELMAN (right), who later pioneered nucleic acid hybridization methods. *(Photo courtesy the Cold Spring Harbor Laboratory Library Archives)*

ca. 1950s FRANK STAHL (left), who with Matthew Meselson proved that the two strands of a DNA molecule separate when it replicates, photographed with MAX DELBRÜCK during a meeting at Cold Spring Harbor. *(Photo courtesy the Cold Spring Harbor Laboratory Library Archives)*

1960 Discovery of messenger RNA, and demonstration that it carries the information that orders amino acids in proteins.

1961 Use of a synthetic messenger RNA molecule (poly-U) to work out the first letters of the genetic code.

1965 Realization that genes conveying antibiotic resistance in bacteria are often carried on small supernumerary chromosomes called plasmids.

1966 Establishment of the complete genetic code.

1967 Isolation of the enzyme DNA ligase, which can join DNA chains together.

1970 Isolation of the first restriction enzyme, an enzyme that cuts DNA molecules at specific sites.

1972 The joining enzyme DNA ligase is used to link together DNA fragments created by restriction enzymes. The first recombinant DNA molecules are generated at Stanford University.

1973 Foreign DNA fragments are inserted into plasmid DNA to create chimeric plasmids. It is found that they can be functionally reinserted into the bacterium *E. coli.* The potential now exists for the cloning in bacteria of any gene.

1953 SALVADOR LURIA (standing) and MAX DELBRÜCK (seated), in front of the Cold Spring Harbor Laboratory where they had done many of their early experiments on the genetics of bacteria and bacterial viruses. *(Photo courtesy the Cold Spring Harbor Laboratory Library Archives)*

1955 AL HERSHEY (foreground), who showed that DNA is the genetic component of phages, sailing with JIM WATSON on Cold Spring Harbor. *(Photo by Jill Hershey, courtesy the Cold Spring Harbor Laboratory Library Archives)*

1958 MATTHEW MESELSON next to the analytical ultracentrifuge used in his classic experiments with Frank Stahl. *(Photo courtesy the Archives, California Institute of Technology)*

1973 First public concern that recombinant DNA procedures might generate potentially dangerous, novel microorganisms.

1974 Call for a worldwide moratorium on certain classes of recombinant DNA experiments.

1975 A government report in the United Kingdom calls for special laboratory precautions for recombinant DNA research.

1975 An international meeting at Asilomar, California, urges adoption of guidelines regulating recombinant DNA experimentation. Call for the development of safe bacteria and plasmids that cannot escape from the laboratory.

1976 Release of the first guidelines by the National Institutes of Health; restriction of many categories of recombinant DNA experimentation. Rising public concern that the guidelines might not be effective. A *New York Times Magazine* article urges prohibiting the awarding of the Nobel Prize for recombinant DNA research.

ca. 1960s ARTHUR KORNBERG, the master pioneer of the enzymology of DNA replication, photographed soon after he moved his lab to Stanford. *(Photo courtesy Stanford University)*

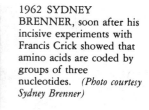

1962 SYDNEY BRENNER, soon after his incisive experiments with Francis Crick showed that amino acids are coded by groups of three nucleotides. *(Photo courtesy Sydney Brenner)*

1966 GOBIND KHORANA (left), the first to chemically synthesize a gene, photographed at the 1966 Cold Spring Harbor Symposium on the Genetic Code, together with FRANCIS CRICK (center) and MARIANNE GRUNBERG-MANAGO (right), who was the first to isolate an enzyme capable of making ribonucleic acid molecules. *(Photo courtesy the Cold Spring Harbor Laboratory Library Archives)*

1977 Formation of the first genetic engineering company (Genentech), specifically founded to use recombinant DNA methods to make medically important drugs.

1977 Creation of the first recombinant DNA molecules containing mammalian DNA, and the discovery of split genes.

1977 Development of procedures for the rapid sequencing of long sections of DNA molecules.

1978 The Nobel Prize in Medicine is awarded for the discovery and use of restriction enzymes.

1978 Somatostatin becomes the first human hormone produced by using recombinant DNA.

1979 General relaxation of the NIH guidelines allows viral DNAs to be studied by using recombinant DNA procedures.

1979 DNA from malignant cells is used to transform a strain of cultured mouse cells so that the cancer genes can be assayed in the malignant cells.

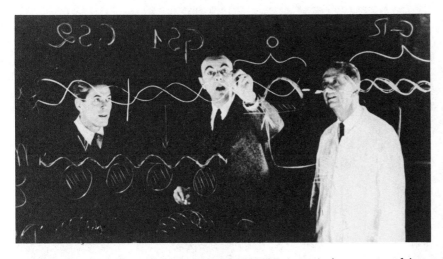

1966 FRANÇOIS JACOB (left) and JACQUES MONOD (center), the proposers of the operon model of bacterial genes, photographed at the Institut Pasteur with ANDRÉ LWOFF (right) soon after the announcement of their Nobel Prize. *(Photo from Paris-Match)*

1975 HERBERT BOYER, who together with Stanley Cohen and their collaborators developed the first practical methods for systematically cloning specific DNA fragments, photographed at the international conference on recombinant DNA held at Asilomar, California. *(Photo by Andrew A. Stern for the National Academy of Sciences)*

1975 STANLEY COHEN at the Asilomar conference. *(Photo by Andrew A. Stern for the National Academy of Sciences)*

1980 Construction work begins on the first industrial plant designed to make insulin by using recombinant DNA procedures.

1980 The Nobel Prize in Chemistry is awarded dually for the creation of the first recombinant DNA molecules and the development of powerful methods for sequencing DNA.

1981 Offer to the general public of stock in the first recombinant DNA company (Genentech). Valuation by Wall Street in excess of $200 million.

1981 Gene cloning experiments using laboratory strains of *E. coli* and yeast as host for propagation of recombinant DNA molecules are exempted from the NIH guidelines

1981 Sickle-cell anemia becomes the first genetic disease to be diagnosed antenatally directly at the gene level, by restriction enzyme analysis of the DNA.

1978 HAMILTON O. SMITH (left) and DANIEL NATHANS (right) featured on the *Baltimore Sun* Sunday magazine cover a few weeks after the announcement that they, along with Werner Arber, were the recipients of the Nobel Prize in Medicine, "for the discovery of restriction enzymes and their application to the problems of molecular genetics." *(Photo reprinted with permission of the* Baltimore Sun*)*

1978 WERNER ARBER at the 1978 Cold Spring Harbor Symposium on DNA Replication and Recombination. *(Photo by Ross Meurer, courtesy the Cold Spring Harbor Laboratory Library Archives)*

1982 There are further wholesale relaxations of the NIH guidelines, but a move to make compliance voluntary rather than mandatory fails.

1982 "Supermice" that are twice the normal weight are created by injecting into fertilized mouse eggs the cloned gene for rat growth hormone linked to the control signals of an inducible mouse gene, and then reimplanting them in foster mothers.

1982 Human insulin produced by recombinant DNA methods goes on the market under the trade name Humulin.

1982 A foreign gene incorporated into the cells of a tobacco plant is shown to be transmitted in a simple Mendelian manner through the pollen and the egg cells.

1982 A cancer gene from human bladder cancer cells is isolated and cloned in *E. coli.* The base sequence of a human bladder cancer gene is found to differ from its normal counterpart by only a single base change that leads to a single amino acid change in the protein product.

1983 The complete sequence of the DNA of bacteriophage λ, a total of 48,502 base pairs, is published.

1980 PAUL BERG, who shared that year's Nobel Prize in Chemistry with Walter Gilbert and Frederick Sanger, attending a press conference at the Royal Academy of Sciences in Stockholm. Berg was cited for constructing the first recombinant DNA molecules, Gilbert and Sanger for independently developing DNA sequencing procedures. *(Photo from Wide World Photos)*

1980 WALTER GILBERT standing next to a large DNA molecule in his Harvard University lab. *(Photo by D. N. Leff/NSF-Mosaic)*

1980 FRED SANGER, who was the first person to determine the amino acid sequence of a protein and who later devised equally important rapid methods for analyzing the sequences of RNA and DNA, photographed after the announcement of his second Nobel Prize. *(Photo by Arthur Foster from The Observer Ltd.)*

APPENDIX A
Restriction Enzymes

A complete updated listing of restriction enzymes, along with their recognition sequences and the microorganisms from which they are isolated, appears annually in the first issue of *Nucleic Acids Research* [R. J. Roberts, "Restriction and modification enzymes and their recognition sequences," *Nuc. Acids Res.,* 11: r135–r167 (1983)]. Many restriction enzymes are now available commercially, primary suppliers being: Bethesda Research Laboratories, Gaithersburg, Maryland; Boehringer Mannheim Biochemicals, Indianapolis, Indiana; and New England Biolabs, Beverly, Maine.

Enzymes with multiple recognition sequences
(See table on facing page)

*Acc*I	GT↓AGAC		T↓GGCCG		GAGCT↓C
	GT↓ATAC		T↓GGCCA		GGGCC↓C
	GT↓CGAC				GGGCT↓C
	GT↓CTAC	*Gdi*II	C↓GGCCG		
			T↓GGCCG		
*Acy*I	GA↓CGCC			*Hin*dII	GTC↓AAC
	GA↓CGTC	*Hae*I	AGG↓CCA		GTC↓GAC
	GG↓CGCC		AGG↓CCT		GTT↓AAC
	GG↓CGTC		TGG↓CCA		GTT↓GAC
			TGG↓CCT		
*Afl*III	A↓CATGT			*Nsp*BII	CAG↓CTG
	A↓CGTGT	*Hae*II	AGCGC↓C		CCG↓CTG
	A↓CACGT		AGCGC↓T		CCG↓CGG
	A↓CGCGT		GGCGC↓C		CAG↓CGG
			GGCGC↓T		
*Aha*II	GG↓CGCC			*Nsp*CI	ACATG↓C
	GG↓CGTC	*Hgi*AI	GAGCA↓C		ACATG↓T
	GA↓CGCC		GAGCT↓C		GCATG↓C
	GA↓CGTC		GTGCA↓C		GCATG↓T
			GTGCT↓C		
*Ava*I	C↓CCGAG			*Xho*II	A↓GATCC
	C↓CCGGG	*Hgi*CI	G↓GCACC		A↓GATCT
	C↓TCGAG		G↓GCGCC		G↓GATCC
	C↓TCGGG		G↓GTACC		G↓GATCT
			G↓GTGCC		
*Cfr*I	C↓GGCCG				
	C↓GGCCA	*Hgi*JII	GAGCC↓C		

Adapted from the 1982–1983 New England Biolabs catalog.

PALINDROMIC TETRA- AND HEXA-NUCLEOTIDE RECOGNITION SEQUENCES AND CLEAVAGE SITES FOR RESTRICTION ENZYMES

	AATT	ACGT	AGCT	ATAT	CATG	CCGG	CGCG	CTAG	GATC	GCGC	GGCC	GTAC	TATA	TCGA	TGCA	TTAA
▼□□□□									**Mbol**							
▼□□□□						**Hpall**				**SciNI**				**Taql**		
□▼□□□			**Alul**				**FnuDII**		**Dpnl**		**Haelll**	**Rsal**				
□□▼□□										**Hhal**						
□□□▼□																
▼A□□□□T			**Hindlll**		*Aflll*		**Mlul** *Aflll*		**Bglll** *Xholl*							
A▼□□□□T														**Clal**		
A□▼□□□T											**Stul** *Hael*	**Scal**				
A□□▼□□T																
A□□□▼□T					*NspCl*					*Haell*						
▼C□□□□G					**Ncol**	**Xmal** *Aval*		**Avrll**			**Xmalll** *Gdill Cfrl*			**Xhol** *Aval*		**Aflll**
C▼□□□□G				**Ndel**												
C□▼□□□G			**Pvull** *NspBll*			**Smal**	*NspBll*									
C□□▼□□G							**Sacll**		**Pvul**							
C□□□▼□G															**Pstl**	
▼G□□□□C	**EcoRI**						**BsePl**		**BamHl** *Xholl*	*HgiCl*		*HgiCl*		**Sall**		
G▼□□□□C		*Ahall Acyl*								**Narl** *Ahall Acyl*		*Accl*	*Accl*			
G□▼□□□C				**EcoRV**		**Nael**								*Hindll*		**Hpal** *Hindll*
G□□▼□□C																
G□□□▼□C		**Aatll**	**Sacl** *HgiAl HgiJll*		*NspCl* **Sphl**					**Bdel** *Haell*	**Apal** *HgiJll*	**Kpnl**		*HgiAl*		
▼T□□□□A								**Xbal**	**Bcll**		*Cfrl*					
T▼□□□□A														**Asull**		
T□▼□□□A						**Nrul**				**Mstl**	**Ball** *Hael*					**Ahalll**
T□□▼□□A																
T□□□▼□A																

Note: The sequences at the top of each column, by convention, are written in the $5' \rightarrow 3'$ direction. The open squares at the left of each row are place holders for the nucleotides represented in the top sequences. Enzymes appearing in bold type recognize only one sequence; enzymes in light type recognize multiple sequences.

PALINDROMIC PENTANUCLEOTIDE RECOGNITION SEQUENCES AND CLEAVAGE SITES FOR RESTRICTION ENZYMES

	AA□TT	AC□GT	AG□CT	AT□AT	CA□TG	CC□GG	CG□CG	CT□AG	GA□TC	GC□GC	GG□CC	GT□AC	TA□TA	TC□GA	TG□CA	TT□AA
▼ □□N□□																
▼ □□N□□								Ddel	Hinfl		Asul					
▼ □□N□□						ScrFl				Fnu4HI						
▼ □□N□□																
▼ □□N□□																
▼ □□N□□																
▼ A □□T□□						EcoRII										
▼ A □□T□□											Avall					
▼A □□T□□						BstNI										
A▼ □□T□□																
A ▼ □□T□□																
A ▼ □□T□□																
▼ G □□C□□																
▼ G □□C□□																
▼G □□C□□						NcII										
G▼ □□C□□																
G ▼ □□C□□																
G ▼ □□C□□																

Enzymes of unknown cleavage location*

Ava III	ATGCAT
Eco DXI	ATCA(N)₇ATTC
Hgi EII	ACCNNNNNNGGT
Hin fIII	CGAAT(25)
Sdu I	G(A/G)GC(A/C)C (with G,C over A,T)
Sna I	GTATAC

Enzymes that recognize octanucleotide sequences

Not I	GC↓GGCCGC
Sfi I	GGCCNNNN↓NGGCC

Enzymes that recognize interrupted palindromes

Bgl I	GCCNNNN↓NGGC
Bst EII	G↓GTNACC
Mst II	CC↓TNAGG
Sau I	CC↓TNAGG
Tth 111I	GACN↓NNGTC
Xmn I	GAANN↓NNTTC

Enzymes that recognize nonpalindromic sequences†

Bbv I	GCAGC(N)8
	CGTCG(N)12
Bin I	GGATC(N)4
	CCTAG(N)5
Fok I	GGATG(N)9
	CCTAC(N)13

*The number (25) following the recognition sequence for *Hin* fIII indicates that cleavage occurs approximately 25 bases downstream of the recognition sequence.

†Both strands of DNA are shown with the top strand written in the $5' \rightarrow 3'$ direction. The cleavage site is immediately to the right of the last N (any nucleotide) in each strand.

*Hga*I	GACGC(N)5		*Mnl*I	CCTC(N)7
	CTGCG(N)10			GGAG(N)7
*Hph*I	GGTGA(N)8		*Sfa*NI	GCATC(N)5
	CCACT(N)7			CGTAG(N)9
*Mbo*II	GAAGA(N)8		*Tth*111II	CAAACA(N)11
	CTTCT(N)7			GTTTGT(N)9

Isoschizomers

Isoschizomers are enzymes that recognize the same recognition sequence. Prototypic restriction enzymes are represented in bold type. Subsequently discovered isoschizomers are listed below in light type. Known cleavage sites are indicated by arrows.

AcyI   G(A/G)↓CG(C/T)C
- *Aha*II   . . ↓ . . . .
- *Aos*II   . . ↓ . . . .
- *Ast*WI   . . ↓ . . . .
- *Asu*III   . . ↓ . . . .
- *Bbi*II   . . . .
- *Hgi*DI   . . ↓ . . . .
- *Hgi*GI   . . ↓ . . . .
- *Hgi*HII   . . ↓ . . . .

AluI   AG↓CT
- *Oxa*I   . . . . .

AsuI   G↓GNCC
- *Nsp*(7524)IV   ↓ . . . . .
- *Psp*I   . . . . . .
- *Sau*96I   . ↓ . . . .
- *Sdy*I   . . . . . .

AsuII   TT↓CGAA
- *Mla*I   . . ↓ . . .
- *Nsp*(7524)V   . . . . . .
- *Nsp*BI   . . . . . .

AvaI   C↓(C/T)CG(A/G)G
- *Agu*I   . . . . . .
- *Avr*I   . . . . . .
- *Nsp*(7524)III   . ↓ . . . .

AvaII   G↓G(A/T)CC
- *Afl*I   . ↓ . . . .
- *Bam*N$_x$I   . ↓ . . . .

- *Cau*I   . . . . .
- *Clm*II   . . . . .
- *Fdi*I   ↓ . . . .
- *Hgi*BI   ↓ . . . .
- *Hgi*CII   ↓ . . . .
- *Hgi*EI   ↓ . . . .
- *Hgi*HIII   ↓ . . . .
- *Sin*I   . . . . .
- *Nsp*HII   . . . . .

BamHI   G↓GATCC
- *Aac*I   . . . . . .
- *Aae*I   . . . . . .
- *Bam*FI   . . . . . .
- *Bam*KI   . . . . . .
- *Bam*NI   . . . . . .
- *Bst*I   . ↓ . . . .
- *Dds*I   . . . . . .
- *Gdo*I   . . . . . .
- *Gox*I   . . . . . .
- *Rhs*I   . . . . . .

BclI   T↓GATCA
- *Atu*CI   . . . . . .
- *Bst*GI   . . . . . .
- *Cpe*I   . . . . . .
- *Sst*IV   . . . . . .

BsePI   GCGCGC
- *Bss*HII   . ↓ . . . .

BstEII   G↓GTNACC
- *Asp*AI   . ↓ . . . . .

- *Bst*PI   . ↓ . . .
- *Eca*I   . ↓ . . .
- *Fsp*aI   . ↓ . . .

CauII   CC↓(C/G)GG
- *Aha*I   . . . . .
- *Bcn*I   . . ↓ . . .
- *Nci*I   . . ↓ . . .

CfrI   (C/T)↓GGCC(A/G)
- *Eae*I   . ↓ . . . .

ClaI   AT↓CGAT
- *Ban*III   . . . . . .

EcoRI   G↓AATTC
- *Rsr*I   . . . . . .

EcoRII   ↓CC(A/T)GG
- *Aor*I   . . ↓ . . .
- *Apy*I   . . ↓ . . .
- *Atu*BI   . . . . . .
- *Atu*II   . . . . . .
- *Bin*SI   . . . . . .
- *Bst*GII   . . . . . .
- *Bst*NI   . . ↓ . . .
- *Eca*II   . . . . . .
- *Ecl*II   . . . . . .
- *Mph*I   . . . . . .
- *Taq*XI   . . ↓ . . .

Fnu DII   CG↓CG
  Acc II   . . . .
  Bce RI   . . . .
  Hin 1056I   . . . .
  Tha I   . .↓. .

Fok I   GGATG
  Hin GUII   . . . . .

Hae II ^A_GGCGC↓^C_T
  Hin HI   . . . . . .
  Ngo I   . . . . . .

Hae III   GG↓CC
  Blu II   . . . .
  Bse I   . . . .
  Bsp RI   . .↓. .
  Bss CI   . . . .
  Bst CI   . . . .
  Bsu 1076I   . . . .
  Bsu 1114I   . . . .
  Bsu RI   . .↓. .
  Clm I   . . . .
  Clt I   . .↓. .
  Fnu DI   . .↓. .
  Hhg I   . . . .
  Mni I   . . . .
  Mnn II   . . . .
  Ngo II   . . . .
  Pal I   . . . .
  Sfa I   . .↓. .
  Ttn I   . . . .
  Vha I   . . . .

Hgi CI   G↓G$^{CA}_{TG}$CC
  Ban I   .↓. . . .
  Hgi HI   .↓. . . .

Hgi JII   G^A_GGC^T_C↓C
  Ban II   . . . . .↓

Hba I   GCG↓C
  Cfo I   . . . .
  Fnu DIII   . . .↓.
  Hin GUI   . . . .
  Hin P₁I   .↓. . .
  Hin S₁   . . . .
  Hin S₂   . . . .
  Mnn IV   . . . .
  Sci NI   .↓. . .

Hin dII   GT^C_T↓^A_GAC
  Chu II   . . . . . .
  Hin cII   . . .↓. . .
  Hin JCI   . . .↓. . .
  Mnn I   . . . . . .

Hin dIII   A↓AGCTT
  Bbr I   . . . . . .
  Bpe I   . . . . . .
  Chu I   . . . . . .
  Hin 173I   . . . . . .
  Hin bIII   . . . . . .
  Hin fII   . . . . . .
  Hin JCII   . . . . . .
  Hsu I   .↓. . . . .
  Mki I   . . . . . .

Hin fI   G↓ANTC
  Fnu AI   .↓. . .
  Hha II   . . . . .
  Nca I   . . . . .
  Nov II   . . . . .

Hpa I   GTT↓AAC
  Bse II   . . . . . .

Hpa II   C↓CGG
  Hap II   .↓. .
  Mni II   . . . .
  Mno I   .↓. .
  Msp I   .↓. .
  Sfa GUI   . . . .

Kpn I   GGTAC↓C
  Nmi I   . . . . .

Mbo I   ↓GATC
  Bsa PI   . . . .
  Bss GII   . . . .
  Bst EIII   . . . .
  Bst XII   . . . .
  Cpa I   . . . .
  Dpn I   . .↓. .
  Dpn II   . . . .
  Fnu AII   . . . .
  Fnu CI   ↓. . . .
  Fnu EI   ↓. . . .
  Mno III   . . . .
  Mos I   . . . .
  Mth I   . . . .

  Nde II   . . . .
  Nfl I   . . . .
  Pfa I   . . . .
  Sau 3AI   ↓. . . .

Mst I   TGC↓GCA
  Aos I   . . .↓. .
  Fdi II   . . .↓. .

Mst II   CC↓TNAGG
  Cvn I   . .↓. . . .
  Sau I   . .↓. . . .

Nae I   GCC↓GGC
  Nba I   . . . . . .
  Nbr I   . . . . . .
  Nmu I   . . . . . .
  Rlu I   . . . . . .

Nar I   GGCGC↓C
  Bbe AI   . . . . .↓
  Bde I   . . . . .↓
  Bin SII
  Nam I
  Nda I   . .↓. . .
  Nun II   . .↓. . .

Nsp CI ^A_GCATG↓^C_T
  Nsp HI   . . . . .↓.

Pst I   CTGCA↓G
  Bbi I   . . . . . .
  Bce 170I
  Bsu 1247I
  Noc I   . . . . . .
  Pma I   . . . . . .
  Sal PI   . . . . .↓.
  Sfl I   . . . . .↓.
  Xma II   . . . . . .
  Xor I   . . . . . .

Pvu I   CGAT↓CG
  Nbl I   . . . .↓.
  Rsh I   . . . .↓.
  Rsp I   . . . . . .
  Xni I   . . . . . .
  Xor II   . . . . .↓.

Sac I   GAGCT↓C
  Sst I   . . . . .↓

*Sac*II CCGC↓GG
  *Bac*I   . . . . . .
  *Csc*I   . . . ↓ . .
  *Ecc*I   . . . . . .
  *Mra*I   . . . . . .
  *Ngi*III   . . . . . .
  *Sbo*I   . . . . . .
  *Sfr*I   . . . . . .
  *Shy*I   . . . . . .
  *Sst*II   . . . . ↓ . .
  *Tgl*I   . . . . . .

*Sal*I G↓TCGAC
  *Hgi*CIII   . ↓ . . . .
  *Hgi*DII   . ↓ . . . .
  *Nop*I   . ↓ . . . .
  *Rhe*I   . . . . . .
  *Rhp*I   . . . . . .
  *Rrh*I   . . . . . .
  *Rro*I   . . . . . .
  *Xam*I   . . . . . .

*Sdu*I
$$\begin{matrix} & G & & C & \\ G&A&G&C&A&C \\ & T & & T & \end{matrix}$$
  Nsp(7524)II   . . . . . ↓.
  *Bsp*1286   . . . . . ↓.

*Sma*I CCC↓GGG
  *Xma*I   . ↓ . . . . .

*Stu*I AGG↓CCT
  *Aat*I   . . . .
  *Gdi*I   . . . ↓ . . .

*Taq*I T↓CGA
  *Tfl*I   . . . .
  *Tth*HB8I   . . . .

*Tth*111I GACN↓NNGTC
  *Tte*I   . . . .
  *Ttr*I   . . . .

*Xho*I C↓TCGAG
  *Blu*I   . ↓ . . . .
  *Bbi*III   . . . . . .
  *Bss*HI   . . . . . .
  *Bst*HI   . . . . . .
  *Bth*I   . . . . . .
  *Ccr*II   . . . . . .
  *Dde*II   . . . . . .
  *Msi*I   . . . . . .
  *Pae*R7I   . ↓ . . . .
  *Scu*I   . . . . . .
  *Sex*I   . . . . . .
  *Sga*I   . . . . . .
  *Sgo*I   . . . . . .
  *Sla*I   . ↓ . . . .
  *Slu*I   . . . . . .
  *Spa*I   . . . . . .
  *Xpa*I   . ↓ . . . .

APPENDIX B
Other Enzymes Used in Recombinant DNA Research

Reverse Transcriptase (Avian myeloblastosis virus)*: Synthesizes DNA from an RNA template; used to synthesize complementary DNA (cDNA) from messenger RNA.

DNase I (Bovine pancreas): Degrades DNA by nicking; used at low concentrations to introduce one or a few nicks in a cloned DNA molecule for subsequent mutagenesis.

E. coli DNA Polymerase I: Synthesizes DNA using DNA as template (requires a primer). Also contains both 5' to 3' and 3' to 5' exonucleolytic activities; used to label cloned DNA by "nick translation."

Klenow or Large Fragment of E. coli DNA Polymerase I: Prepared by subtilisin treatment of DNA pol I; does not contain the 5' to 3' exonuclease activity; used to "fill out" 5' or 3' overhangs at the ends of DNA molecules produced by restriction nucleases.

M. luteus Polymerase: A DNA polymerase which contains a 5' to 3' exonuclease activity; used to create small single-stranded regions in cloned DNA molecules starting from a nick.

T4 DNA Polymerase: Activities similar to *M. luteus* polymerase.

T4 DNA Ligase: Catalyzes the formation of a phosphodiester bond between a 5' phosphate and a neighboring 3' hydroxyl in DNA. Used to add "linkers" to linear DNA molecules, to attach two DNA molecules to each other, if these molecules contain "sticky ends" formed by a restriction enzyme, and to circularize linear DNA molecules.

Terminal Transferase (Calf thymus): Adds nucleotides to the 3' ends of DNA molecules; used to "tail" DNA for subsequent cloning.

S1 Nuclease (Aspergillus oryzae): A single-strand specific nuclease; used to clip the hairpin formed during the synthesis of double-stranded cDNA, to analyze mRNA-DNA hybrids, and, in concert with exo III (see below), to make deletions in DNA molecules.

Exonuclease III (E. coli): A 3' to 5' exonuclease which starts from each 3' end of duplex DNA and chews away that strand; used in concert with S1 to create deletions in cloned DNA molecules.

Bal 31 Exonuclease (Brevibacterium albidum): An exonuclease which chews away both 5' and 3' strands of linear DNA from each end; used to make deletions in cloned DNA molecules.

*The source of each enzyme, if not included in its name, appears in parentheses.

Index

Ac (Ds) elements, 144–145, 173
Actin, 4
Adapter hypothesis, 34
Addition mutations, 31
Adenine, 12
Adenovirus, 127, 176
 mRNA splicing, 91
Agarose gel electrophoresis, 26, 60
Agrobacterium tumefaciens, 168–174
Alcohol dehydrogenase, 144, 172
Alpha$_{2u}$ (α_{2u}), 84, 183
"Alu" family DNA, 100–101, 180, 184
Amber mutations, 181
Amniocentesis, 213
Antibiotic-resistance genes, 73, 181
Antibodies. *See* Immunoglobulin genes. *See also* Immunological
  screening
Anticodon, 36
Antigens
 HLA, 123–124
 H2 transplantation, 123–124
 in trypanosomes, 147–149
 See also Immunoglobulin genes
α_1-Antitrypsin, 216–217
APRT, 182
Argininosuccinate synthetase, 217–219
ARS. *See* Autonomously replicating
  sequence
Asilomar Conference, 69
Astbury, W., 17
Attenuation, 48
Autonomously replicating sequence, 154–155
Autosome, 8
Auxins, 169
Avery, O., 13
5-Azacytidine, 226

Bacterial transformation, 25
Bacteriophage, 14–15
 host range, 32
 lambda, 23–24
 M13, 81–83, 112–114
 ϕX174, 130
 T4, 32, 109–110
 T7, 22
 transducing phage, 24

 vectors, 72
Bal 31 exonuclease, 106–107
Baltimore, D., 54
Beadle, G., 9
Benzer, S., 32
Biotin labeling, 223
Bisulfite mutagenesis, 109–110
Blunt-ended DNA, 66
Bovine papilloma virus (BPV), 195–196
Brenner, S., 36
Bromodeoxyuridine (BrUdr), 176

Calcitonin, 97–98
Callus culture, 164
Calmodulin, 4
cAMP, regulation by, 40
Cancer, 135–136, 184–186, 222. *See also* Oncogenes
CAP, 48
Capping of mRNA, 52
Cassette model for mating-type switch, 146–147
cDNA cloning, 74–84
 of immunoglobulin genes, 121–123
 and immunological screening, 87–99
Cell sorting, 224
Cellulose, 165
Centromere DNA, 155
Cesium chloride gradients, 51
Chargaff, E., 16–19
Chase, M., 14–15
Chromatin, 53
 in yeast, 161
Chromosomal abnormalities
 in maize, 144
 See also Translocations
Chromosomes, 2
 autosomes, 8
 diploid number, 6
 in *E. coli,* 23, 32
 haploid number, 6
 linkage, 8
 mapping, 219–221
 in SV40, 53
 "walking," 80
 X chromosome, 8
 in yeast, 158–159
Cirrhosis, 216

Citrullinemia, 217–219
Class I genes (transplantation antigens), 123–124.
Class II (immune-response) genes, 123–124
Class III (complement) genes, 123–124
Cloning of animal cells, 206–208
Codons, 34
Cohesive ends of restriction fragments, 64
Colinearity, 32–33
Collagen, 4
Complement genes, 123–124
Complementation in yeast, 153
Consensus sequences for splicing, 93–94,
    214
Copia elements, 142–143
Corticotropin (ACTH), 101–102
COS cells, 193
Cosmids, 79–80, 172
Cotransformation, 178–179
Crick, F., 18, 34, 36
cro repressor, 48
Crossover, 8–9
    within genes, 32
    and transposition, 141
Crown gall plasmids, 164–173
Cruciform structures, 21
Cyanogen bromide, 233–234
Cystic fibrosis, 211–212
Cystonuria, 212
Cytokinin, 169
Cytoplasm, 2
Cytosine, 12
    5-azacytidine, 226
    deamination, 110
    5-methylcytosine, 22, 179–180
    N-4-hydroxycytosine, 110, 112

D (diversity) sequences of immunoglobulin genes,
    120–121
Deamination of cytosine, 110
Degeneracy of genetic code, 38
Delbrück, M., 15, 32
Deletion mutations, 31, 106
Delta sequences, 142
Dideoxy sequencing, 61–63, 82
Dintzis, H., 35
Diplococcus pneumoniae, 12
Diploid number of chromosomes, 6
Direct terminal repeats, 145
DNA
    "B" form, 25, 27
    backbone, 17
    in chromosomes, 11, 22
    complementarity of, 18
    crystalline structure of, 17
    denaturation, 21
    double helix, 17
    G–C content of, 21
    ligase, 65
    mitochondrial, 39
    polymerases, 35, 65, 109–110
    renaturation, 21
    repetitive, 100
    –RNA hybrid, 21, 41, 61, 74, 92
    structure, 13, 15
    supercoiling, 25

x-ray diffraction, 17
Z-DNA, 27
DNA-binding proteins, 51, 66
DNA polymerase
    Klenow fragment, E. coli, 109–110
    M. luteus, 110–113
    T4, 109–110
dnaB protein, 66
DNA tumor viruses, 176. See also Polyoma; SV40
DNase I, 13
    in vitro mutagenesis, 109
Dominance. See Genes
Double helix
    of DNA, 17
    stereochemistry, 25–26
Dreyer, W., and Bennett, C., 118
Drosophila, 8
    heat shock, 183
    movable elements in, 140, 142–144
Ds elements, 144–145, 173

E. coli, 14
    cloning in, 72
    DNA polymerase, 65–66, 109
    expressing yeast genes in, 153
    genetics of, 32
    movable elements in, 140
    phage resistance in, 32
    protein levels, 45
    restriction modification in, 59
    trp mutations in, 32
Elastase, 216
Emphysema, 216
Endorphins, 101
Enhancers, 131, 194
ENV glycoprotein, 132
Episomes
    in animal cells, 195–196
    in bacteria, 24
Ethidium bromide, 113
Eukaryotic cells, 2
Evolution, 6
    of multigene families, 99
Exons
    rearrangements in heavy-chain genes, 119
    See also Splicing
Exonuclease III, 106
Expression vectors
    eukaryotic, 177–178, 189–196
    in plants, 172–173
    plasmid, 86–88
    in yeast, 153

Fertilization, 6
fes oncogene, 134. See also RNA tumor viruses
Foot-and-mouth disease, 239
Frame-shift mutations, 214
Franklin, R., 18
Fusion proteins, 86

GAG, 132
Galactosemia, 212
β-Galactosidase in E. coli, 46
Gametes, 7
Garrod, A., 9

Gaucher's disease, 212
Gel electrophoresis
    in agarose, 26, 60
    and blotting, 83
    in polyacrylamide, 62
Gene duplication, 148
Gene families, 99–100
Gene transfer in mammalian cells, 176–186
Genes
    cloning, 72–76
    clusters, 99
    colinearity of, 33
    dominant, 6
    library, 74, 85
    of low abundance, 84
    pseudogenes, 96, 100
    recessive, 6
Genetic code, 38–39
Genetic diseases, 209–226
Genotype, 6
G418, 178
Gilbert, W., 61
β-Globin, 92
Glucocorticoid hormones, 183–184
Glycoproteins, 192
Griffith, F., 12
Growth hormone, 203, 235–236
Guanine, 12
    7-methylguanosine, 52

Hairpin loops, 156
Hammerstein, E., 17
Haploid number of chromosomes, 6
Harvey sarcoma virus, 186
HAT medium, 177
Heat-shock genes, 183
Heavy-chain sequences
    of immunoglobulin genes, 117–118
    switching, 120–121
Helicase, DNA, 66
Hemagglutinin, 191–193
Hemoglobin, 4
    mRNA splicing of, 97
    in sickle-cell anemia, 9, 31
    in β-thalassemia, 214, 226
Hemophilia, 211–212, 231
Hepatitis B, 239–240
Heredity
    Mendelian, 6, 211
    of methylation patterns, 205
Herpes simplex virus, thymidine kinase gene, 107,
    176–177, 202
Hershey, A., 14–15
Heteroduplex analysis, 92
HGPRT, 177–178, 219
HindII. See Restriction enzymes
Histones
    in DNA, 11
    in nucleosomes, 43
HLA antigens, 123–124.
HML. See Mating-type allele
HMR. See Mating-type allele
Holley, R., 58
Hormones, 4, 183, 203
    alpha$_{2u}$ (α_{2u}), 84, 183

corticotropin (ACTH), 101–102
    in gene regulation, 51
    HCG, 85
    β-lipotropin (β-LPH), 101–102
    α-melanotropin (α-MSH), 101–102
    pituitary, 101–102
    in plants, 165, 169
H2 transplantation antigens, 123–124.
Human chorionic gonadotropin (HCG), 85, 196
Hybrid cells, 165–167, 220
Hybrid dysgenesis, 143
Hydrogen bonding, 14, 18
Hydroxyapatite, 85
N-4-Hydroxycytosine, 110, 114

Immune-response genes, 123–124
Immunoglobulin genes, 117–124
    translocations, 223
Immunological screening, 87–88
In situ hybridization, 201, 216
Inducers
    IPTG, 87
    of transcription, 45
Ingram, V., 31
Insertion mutations, 108, 140
    in plants, 173
    See also Transposons
Insulin, 3–4, 97, 222, 232–235
Integral membrane proteins, 96, 192
Interferon, 70, 231, 236–238
Introgression, 164
Introns, in immunoglobulin genes, 119. See also Splicing
Inverted terminal repeats, 145
IPTG, 87

J (joining) segments of immunoglobulin genes, 119–120
Jacob, F., 45

Khorana, H. G., 38
Kirsten sarcoma virus, 186
Klenow fragment, DNA polymerase I, 109–110
Kornberg, A., 66

β-Lactamase, 86
Lambda bacteriophage, 23–24
    cloning in, 78, 185
    repressor, 48, 86
Leader sequences of membrane proteins, 78, 96, 192
Lederberg, J., 32
Lesch-Nyhan syndrome, 214, 219
Ligase, DNA, 65
Light-chain sequences of immunoglobulin genes, 117–118
Linkage, 8
    and gene mapping, 80, 219
Linkers. See Oligonucleotides
β-Lipotropin (β-LPH), 101–102
LTRs, retroviral, 132, 145
Luria, S., 32
Lwoff, A., 23
Lysogenic phage, 23. See also Lambda bacteriophage

MacLeod, C., 13
Macromolecules, 2
Maize, 140, 144–145
Major histocompatibility complex (MHC), 123–124

MAT. *See* Mating-type allele
Mating-type allele, 109, 146–147
Matthaei, H., 37
Maxam–Gilbert DNA sequencing, 61–62
McCarty, M., 13
McClintock, B., 140, 144
MC29 virus, 135
Meiosis, 6–7
α-Melanotropin (α-MSH), 75, 101–102
Mendel, G., 6
Mendelian inheritance, 6, 211
Meselson, M., 20
Messenger RNA, 35
    and northern blots, 83
    as probes for genes, 74
    secondary structure, 49
    splicing, 91–98
Metabolic diseases, 9, 209–226
Metachromatic leukodystrophy, 212
Metallothionine, 183, 202
Methionine-enkephalin, 101
Methylation
    in DNA, 21, 179–180
    inheritance, 205
    inhibition by 5-azacytidine, 226
    and restriction enzymes, 59
β₂-Microglobulin, 123–124
Microinjection of DNA, 178, 200
Mitochondria, DNA, 39
Mitosis, 6–7
    in yeast, 155
Mizutani, S., 54
MMTV, 183–184
Mobile genetic elements, 25, 140–149. *See also* Transposons
Monod, J., 38, 45
Morgan, T. H., 8
mos oncogene, 134. *See also* RNA tumor viruses
M13 bacteriophage, 81–83
    and oligonucleotide mutagenesis, 112–114
MuLV, 204–206
Muscular dystrophy, 212
Mutagenesis
    bisulfite, 109–110
    in vitro, 106–114, 136
Mutants
    deletion, 106
    in maize, 144–145
    in mating-type allele, 109
    in metabolism, 8, 214
    nonsense, 39
    in T4, 32
    tRNA-suppressor, 39–40, 113
Mutation
    mechanisms, 31
    somatic, in immunoglobin genes, 122–123
myc oncogene, 135, 222
Mycophenolic acid, 178
Myelomas, 117

Neisseria gonorrhoeae, 149
Neomycin, 178
Neurospora, mutants in, 9
Nirenberg, M., 37
Nitrocellulose filters
    and colony hybridizations, 76–77

for screening mutants in M13, 112–114
    and Southern blotting, 83
Nonsense mutants, 39, 214
Nopaline, 168–171
Northern blotting, 83
Nuclear transplantation, 208
Nucleoli, 11
Nucleosomes
    and enhancers, 131
    in eukaryotic cells, 53
    in SV40, 195
    in yeast DNA, 154
Nucleotide analogs, 110, 114
    BrUdr, 176
Nucleotides, 12
Nucleus, 2

Octopine, 168–171
Oligonucleotides
    chemical synthesis of, 63
    as probes for screening libraries, 85
    and site-specific mutagenesis, 107–109, 111–114
    synthetic linkers, 76
Oncogenes, 184–186, 221
    in vitro mutagenesis, 136
    and tumor viruses, 128, 132, 135
Open reading frames, 96
Operator sequences in *E. coli,* 47
Operon, bacterial, 46
Opines, 168–171
Ovalbumin, 93

"P" elements, 142–144
Palindromes, 21
Pauling, L., 9, 15
pBR322, 73
Pentosuria, 212
Phenotype, 6
Phenylketonuria, 9, 212
Phosphorylation by oncogenes, 133
Pili in *Neisseria gonorrhoeae,* 149
Pituitary hormones, 101–102
Plasmid rescue, 181–183
Plasmids
    in bacteria, 24
    and cloning foreign genes, 57–68
    crown gall, 164–173
    drug resistance in, 72–73
    expression vectors, 86–88
    replication of, 24–25
    Ti, 168–174
    as vectors, 67, 72, 233
    in yeast, 153–154
Plasminogen activator, 231
Plus-minus sequencing, 61
POL, 132
Poly-A polymerase, 99
Polyadenylation, 52
    and cDNA cloning, 74
    signals, 99
Polycistronic mRNA, 52
Polymerases
    DNA, 35, 65, 109–110
    in nucleic acid synthesis, 35
    RNA, 35, 41, 53

Polyoma, 53, 127
Polysaccharides, 5
Prenatal diagnosis, 213–219
Primase, 66
Promoters, 41
    in *E. coli,* 46–47
    in plants, 169, 172–173
    in plasmid vectors, 86
    and retroviral LTRs, 133
    for RNA polymerase II, 98
    for RNA polymerase III, 110
    in yeast, 161, 240
Prophage lambda, 23
Protamines, 11
Protein synthesis. *See* Translation
Protoplast fusion, 165–167
Provirus and retrovirus replication, 132
Pseudogenes, 96, 100
Ptashne, M., 48
Purine bases in nucleic acids, 12
Pyrimidine bases in nucleic acids, 12

ras oncogene, 134, 186. *See also* RNA tumor viruses
Recessive. *See* Genes
Reciprocal translocations, 221
Recombinant DNA guidelines, 69
Recombination, 129
    of immunoglobulin genes, 118, 121
rep protein, 66
Repetitive DNA, 75, 100–101, 180
Replication
    in COS cells, 193
    of DNA, 20
    in *E. coli,* 66
    origins, 131, 154
    of plasmid DNA, 24–25
    semiconservative, 20
    in SV40, 129, 191
    and transposition, 141, 145
    in yeast, 154, 156–158
Replicon fusion, 25
Repressors
    Iq repressor, 66
    in lac operon, 45–47
    lambda, 48
    of transposase mRNA, 142
Resolution of transposition, 141–142
Restriction enzymes, 58–61
Restriction mapping, 60
Restriction-modification in *E. coli,* 59
Retriever vectors, 159–160
Retrovirus. *See* RNA tumor viruses
Reverse transcriptase
    and cDNA synthesis, 74–75
    and RNA tumor viruses, 54, 132
    and transposition, 145
Ribosomal proteins, 50
Ribosomal RNA, 11, 35, 42, 50–51
    splicing in *Tetrahymena,* 94–95, 156
    in *Xenopus,* 51–52
Ribosomes, 35
    Shine-Dalgarno sequences, 50, 86
RNA
    –DNA hybrids, 21, 41, 61, 74, 92
    messenger RNA, 35–42, 49

    in nucleoli, 11
    polymerases, 35, 41, 53
    ribosomal, 11, 35, 42, 50–52
    secondary structure, 51
    7S, 101
    snRNA, 93–94
    transcription, 32–34
    transfer RNA, 35–37
RNA polymerases, 33, 41
RNA tumor viruses, 54, 127
    gene organization, 131, 135
    MC29 (myc), 222
    as movable genetic elements, 145
    MuLV, 204–206
    vectors, 196–197
RNase
    P, 93
    T1, 58
Robertson mutator (Mu), 173
Rous sarcoma virus (RSV), 132. *See also* RNA tumor viruses

Saccharomyces cerevisiae. See Yeast
Sanger, F., 3, 61
sarc. *See* src
Segregation distortion, 155
Sequencing of nucleic acids, 58, 61–63
    dideoxy-, 61–63, 82
    of mammalian genes, 93
    Maxam–Gilbert, 61–62
Shine–Dalgarno sequences, 50
    in expression vectors, 86
Shuttle vectors, 159–160, 193–194
Sickle-cell anemia, 9, 31, 212, 215–219
Signal peptidase, 96
Signal sequences, 96
Signer, R., 17
SIR. *See* Mating-type allele
snRNA, 93–94
Somatic-cell hybrids, 165–167, 220
Somatic mutation theory, 122–123
S1 nuclease
    and cDNA cloning, 74
    mapping, 91
Southern, E. M., 83
Southern blotting, 83
Spacer DNA, 52, 99
Spheroplasts, 152–153
Splicing, 91–98
    mutations, 214
    T antigen, 128
Splicing of mRNA, 69–70
    of immunoglobulin genes, 121–122
src, 133. *See also* RNA tumor viruses
Stahl, F., 20
Stem-loop structures, 21
Strand separation in DNA replication, 20
Substitution mutations, 31, 113
supF. *See* Suppressor tRNA
Supercoiling of DNA, 25–26
Superhelicity, 26
    of Z-DNA, 27
Suppressor rescue, 181, 186
Suppressor tRNA, 39–40, 113, 181
SV40
    deletion mutants, 106

SV40 (*continued*)
  life cycle, 53
  minichromosome, 129
  in mouse embryos, 200
  restriction enzyme mapping of, 60, 127
  splicing, 91
  T antigen, 53, 92, 191
  transcription, 130
  vectors, 177–178, 189–196
Synthetic peptides, 239

T antigen, 53, 92, 191
  splicing, 128–129
T DNA, 168–174
"TATA" box, 98, 108, 130, 160, 214
Tatum, E., 9, 32
Tay-Sachs disease, 211–212
Telomeres, 156
Temin, H., 54
Terminal transferase, 75
Testosterone, 84
Tetrahymena, rRNA splicing in, 94–95
β-Thalassemia, 212, 214, 226
Thymidine kinase, of HSV-1, 107, 176–177, 202
Thymine, 12
Ti plasmids, 168–174
Tn3, 143–144
Todd, A., 16
Topoisomerases, 26
Totipotency, 165
Transcription
  control of, 45–49, 98, 108, 181, 195, 203
  in vitro, 108
  mutants, 214
  in SV40, 130
  in yeast, 161
Transducing phage, 24
Transfer RNA, 35
  suppressors, 39–40, 113, 181
Translation, 34–42
  control of, 45–52, 98–99
  in vitro, 42, 102
Translocations, 221–223
Transposase, 140, 142
Transposons, 25, 140–146
  and LTRs, 133
Tropomyosin, 4
  and immunological screening, 87–88
trp operon, 48
Trypanosomes, 147–149

Tryptophan synthetase, 33
T7 bacteriophage, 22
Tumor virus, 176. *See also* Polyoma; SV40; RNA tumor viruses
Ty1 elements, 142

U1 snRNA, 93
Uracil, 12
  bromodeoxyuridine (BrUdr), 176

Vaccines, 238–240
Van der Waals forces, 14
Variable regions of immunoglobulin
      genes, 117–118
V–C joining hypothesis, 118
Vectors
  in animal cells, 176–177, 193–194
  bacteriophage, 72, 78, 185
  expression plasmid, 86–88, 189–197
  in plants, 170–171
  plasmid, 67, 72–76, 233
  viral, 177–178, 189–197
  in yeast, 153, 159–160
Viruses
  chromosomes in, 53
  DNA of, 14
  vectors, 189–197
  See also RNA tumor viruses; SV40
VSG, 147

Watson, J., 18
Wilkins, M., 17
"Wobble" hypothesis, 38

X chromosome, 8. *See also* Linkage
Xanthine, 178
XGPRT, 177–178

Yanofsky, C., 32, 48
Yeast, 152–161
  mating-type locus, 109, 146–147
  Ty1 elements in, 142

Z-DNA, 27